Benchmark Papers in Geology

Series Editor: Rhodes W. Fairbridge
Columbia University

PUBLISHED VOLUMES

RIVER MORPHOLOGY / *Stanley A. Schumm*
SLOPE MORPHOLOGY / *Stanley A. Schumm and M. Paul Mosley*
SPITS AND BARS / *Maurice L. Schwartz*
BARRIER ISLANDS / *Maurice L. Schwartz*
ENVIRONMENTAL GEOMORPHOLOGY AND LANDSCAPE CONSERVATION, VOLUME I: Prior to 1900 / VOLUME II: Urban Areas / VOLUME III: Non-Urban Regions / *Donald R. Coates*
TEKTITES / *Virgil E. Barnes and Mildred A. Barnes*
GEOCHRONOLOGY: Radiometric Dating of Rocks and Minerals / *C. T. Harper*
MARINE EVAPORITES: Origins, Diagenesis, and Geochemistry / *Douglas W. Kirkland and Robert Evans*
GLACIAL ISOSTASY / *John T. Andrews*
GLACIAL DEPOSITS / *Richard P. Goldthwait*
PHILOSOPHY OF GEOHISTORY: 1785–1970 / *Claude C. Albritton, Jr.*
GEOCHEMISTRY OF GERMANIUM / *Jon N. Weber*
GEOCHEMISTRY AND THE ORIGIN OF LIFE / *Keith A. Kvenvolden*
GEOCHEMISTRY OF WATER / *Yasushi Kitano*
GEOCHEMISTRY OF IRON / *Henry Lepp*
GEOCHEMISTRY OF BORON / *C. T. Walker*
SEDIMENTARY ROCKS: Concepts and History / *Albert V. Carozzi*
METAMORPHISM AND PLATE TECTONIC REGIMES / *W. G. Ernst*
SUBDUCTION ZONE METAMORPHISM / *W. G. Ernst*
PLAYAS AND DRIED LAKES: Occurrence and Development / *James T. Neal*
PLANATION SURFACES: Peneplains, Pediplains, and Etchplains / *George Adams*
SUBMARINE CANYONS AND DEEP-SEA FANS: Modern and Ancient / *J. H. McD. Whitaker*
ENVIRONMENTAL GEOLOGY / *Frederick Betz, Jr.*
LOESS: Lithology and Genesis / *Ian J. Smalley*
PERIGLACIAL DEPOSITS / *Cuchlaine A. M. King*
LANDFORMS AND GEOMORPHOLOGY: Concepts and History / *Cuchlaine A. M. King*
METALLOGENY AND GLOBAL TECTONICS / *Wilfred Walker*
HOLOCENE TIDAL SEDIMENTATION / *George deVries Klein*
PALEOBIOGEOGRAPHY / *Charles A. Ross*
MECHANICS OF THRUST FAULTS AND DÉCOLLEMENT / *Barry Voight*
WEST INDIES ISLAND ARCS / *Peter H. Mattson*
AIR PHOTOGRAPHY AND COASTAL PROBLEMS / *Mohamed T. El-Ashry*

Benchmark Papers in Geology/38

A BENCHMARK ® Books Series

AIR PHOTOGRAPHY AND COASTAL PROBLEMS

Edited by

MOHAMED T. EL-ASHRY

Environmental Defense Fund
Denver, Colorado

Dowden, Hutchinson & Ross, Inc.

STROUDSBURG, PENNSYLVANIA

Benchmark Papers in Geology, Volume 38
Library of Congress Catalog Card Number: 77–1554
ISBN: 0-87933-252-2

79 78 77 1 2 3 4 5
Manufactured in the United States of America.

LIBRARY OF CONGRESS CATALOGING IN PUBLICATION DATA

Main entry under title:

Air photography and coastal problems.

(Benchmark papers in geology ; 38)

Includes bibliographical references and indexes.

CONTENTS: Air photography in coastal studies: Dietz, R. S. Aerial photographs in the geological study of shore features and processes. Clos-Arceduc. Aerial photography and the study of coastal deposits. [etc.]

1. Aerial photography in geology—Addresses, essays, lectures. 2. Coasts—Addresses, essays, lectures. I. El-Ashry, Mohamed T.

QE33.2.A3A36 551.4′5 77-1554
ISBN 0-87933-252-2

Exclusive Distributor: **Halsted Press**
A Division of John Wiley & Sons, Inc.
ISBN: 0-470-99224-7

To the Memory of
Professor Harold R. Wanless

SERIES EDITOR'S FOREWORD

The philosophy behind the "Benchmark Papers in Geology" is one of collection, sifting, and rediffusion. Scientific literature today is so vast, so dispersed, and, in the case of old papers, so inaccessible for readers not in the immediate neighborhood of major libraries that much valuable information has been ignored by default. It has become just so difficult, or so time consuming, to search out the key papers in any basic area of research that one can hardly blame a busy man for skimping on some of his "homework."

This series of volumes has been devised, therefore, to make a practical contribution to this critical problem. The geologist, perhaps even more than any other scientist, often suffers from twin difficulties—isolation from central library resources and immensely diffused sources of material. New colleges and industrial libraries simply cannot afford to purchase complete runs of all the world's earth science literature. Specialists simply cannot locate reprints or copies of all their principal reference materials. So it is that we are now making a concerted effort to gather into single volumes the critical material needed to reconstruct the background of any and every major topic of our discipline.

We are interpreting "geology" in its broadest sense: the fundamental science of the planet Earth, its materials, its history, and its dynamics. Because of training and experience in "earthy" materials, we also take in astrogeology, the corresponding aspect of the planetary sciences. Besides the classical core disciplines such as mineralogy, petrology, structure, geomorphology, paleontology, and stratigraphy, we embrace the newer fields of geophysics and geochemistry, applied also to oceanography, geochronology, and paleoecology. We recognize the work of the mining geologists, the petroleum geologists, the hydrologists, the engineering and environmental geologists. Each specialist needs his working library. We are endeavoring to make his task a little easier.

Each volume in the series contains an Introduction prepared by a specialist (the volume editor)—a "state of the art" opening or a summary of the object and content of the volume. The articles, usually some twenty to fifty reproduced either in their entirety or in significant extracts, are selected in an attempt to cover the field, from the key papers of the last century to fairly recent work. Where the original works are in foreign languages, we have endeavored to locate or commission translations. Geologists, because of their global subject, are often acutely aware of the oneness of our world. The selections cannot, therefore, be restricted to any one country, and whenever possible an attempt is made to scan the world literature.

To each article, or group of kindred articles, some sort of "highlight commentary" is usually supplied by the volume editor. This commentary should serve to bring that article into historical perspective and to emphasize its particular role in the growth of the field. References, or citations, wherever possible, will be reproduced in their entirety—for by this means the observant reader can assess the background material available to that particular author, or, if he wishes, he,too, can double check the earlier sources.

A "benchmark," in surveyor's terminology, is an established point on the ground, recorded on our maps. It is usually anything that is a vantage point, from a modest hill to a mountain peak. From the historical viewpoint, these benchmarks are the bricks of our scientific edifice.

RHODES W. FAIRBRIDGE

PREFACE

The availability of air photographic coverage of coastlines, particularly after World War II, provided a wealth of information regarding the coastal zone not readily available by other means of investigation. Air photographs proved valuable in showing the larger features and circulation patterns which do not normally show from ground levels. A large number of papers and reports dealing with the use of air photography in the study and mapping of shoreline features and shoreline changes as well as in coastal engineering and coastal zone management exists today. However, technological developments in remote sensing are rapidly expanding at the present time, while developments in interpretation techniques and full utilization of remotely sensed data are still lagging. It is the intent of this volume to bring together typical examples of the use and applications of conventional air photography and imagery from other sensors in studies of coastal development and coastal problems for the purpose of reference, education, and stimulation of further research and study. The book contains twenty-four papers arranged in four distinct sections representing the areas that are of prime interest at the present time.

I would like to express my gratitude and appreciation to Prof. Rhodes Fairbridge (Series Editor) for his valuable suggestions and comments on the manuscript; to Profs. J. A. Steers and H. G. Gierloff-Emden for reading the introduction section of the manuscript and for supplying the addresses for information on air photographic coverage in England, Italy, and West Germany; to Prof. A. J. Brandenberger of Laval University (Canada) for most of the addresses of government agencies outside the United States and Canada; to Prof. K. Löfström of the University of Helsinki for the information on Finland; to Mr. William Lynn of the U.S. Geological Survey Technical Information Office and Mr. Don Hopkins of the U.S. Geological Survey Cartographic Service for information on sources of air photography; to the staff of the Library of the U.S. Army Coastal Engineering Research Center for making available their resources in the initail phase of research; to the authors and publishers who kindly agreed to the reproduction of their articles in this volume; and to Ms. Bernice Wisniewski, Production Editor at Dowden, Hutchinson & Ross, for her patience and cooperation in the production of the manuscript.

MOHAMED T. EL-ASHRY

CONTENTS

CONTENTS BY AUTHOR

AIR PHOTOGRAPHY AND COASTAL PROBLEMS

ERTS-1 photograph (1972) of the Nile Delta and the Mediterranean coast of Egypt. (NASA photo)

INTRODUCTION

A Chinese proverb states: "A picture is worth a thousand words." When the value of air* photography is considered, the same proverb could be rephrased to state: A photograph taken from the air is worth a thousand pictures taken on the ground.

The value of air photographs as a tool in intelligence gathering during military operations has been recognized since those taken during the Franco-Prussian War and the American Civil War using balloon-carried cameras. By World War I, air photography and photointerpretation proved their military significance through the development of improved cameras and airplanes for military use. In the last thirty years, peace-time applications of air photography have been developed at an increasing rate in many fields ranging from topographic mapping, natural resources surveys, and city planning to pollution detection. Air photographs have also proved to be an indispensable tool in the study and mapping of shoreline features and shoreline changes as well as in coastal engineering and coastal zone management.

* In this volume, for the sake of simplification, the word "air" instead of "aerial" will be used in describing photography, photographs, mosaics, surveys, etc., following the recommendation of Lahee (1951). The two words are essentially synonymous. However, "air" has been used in Canada and England for some time, but not usually in the United States, where one normally speaks of airport, airplane, aircraft, etc.

BACKGROUND INFORMATION

History of Air Photography and Photointerpretation

It is generally agreed that photography was originated by the Frenchman Louis Daguerre in 1839. He is credited to have invented a positive-image process for making portraits. Two years later, in 1841, an Englishman, William Henry Fox Talbot, developed the negative-positive process which with some refinement is still being used today. Talbot's book, *Pencil of Nature*, published in 1844, is considered the first to be illustrated with photographs (Amer. Soc. Photogramm., 1960).

The history of photography, however, goes back many years before the days of Daguerre. Aristotle described the principle of the "camera obscura" (a pinhole camera) more than 2300 years ago. Accounts of observations of eclipses utilizing pinhole cameras can also be found in the writings of the Arab scholar Ibn Al-Haitam from the eleventh century, and in the manuscripts of Leonardo da Vinci, who lived during the second half of the fifteenth and early part of the sixteenth century.

It is interesting to note that air photographs were used in mapping as well as in military reconnaissance before airplanes were invented. In the early 1850s a French Army engineer, Aimé Laussedat, took a number of air photographs from kites and captive balloons (Avery, 1968). In 1858, the photographer Gaspard Felix Tournachon, who was known later as "Nadar," took photographs of the land surface near Paris from a captive balloon at a height of several hundred meters with the purpose of producing a topographic map of that region. In the United States, the first air photographs were made in 1860 of Boston Harbor, Massachusetts, by Samuel A. King and J. W. Black from a captive balloon at a height of 360 m above sea level. In 1862, the Union Army under General McClellan employed balloon techniques to gather intelligence on Confederate positions in Virginia during the American Civil War. In Russia, pictures of the Kronstadt and Petersburg fortresses were taken in 1886 from a free flying balloon (Amer. Soc. Photogramm., 1960).

In the late 1880s a Major Elsdale of the British Engineer's Balloon Establishment introduced free unmanned balloons carrying self-releasing cameras which exposed several plates successively (Deuel, 1969). In 1891 another British Army officer, D. F. Close, who was serving in India, conceived the idea of producing an archeological map of the ruins near Agra by utilizing a similar photographic technique. The Scheimpflug-Kammerer eight lens camera mounted to the basket of a balloon was used in the German Army maneuvers of 1911 (Talley, 1938). This camera obtained eight exposures on that many plates which were transformed into a single composite by a universal transforming printer.

The first recorded photographs from an airplane were not taken until approximately six years after the first historic flight of the Wright brothers in 1903. They were obtained by Wilbur Wright on April 24, 1909, in the form of motion pictures of parts of Italy.

With the development of airplanes for military use, air photography and photointerpretation achieved great military significance during World War I as intelligence gathering tools. Following the war, improvements in planes, cameras, and film resulted in increasing peace-time photography. Most of the photography was used for illustrative purposes till 1922 when Lee pointed out the potential use of air photographs for data gathering with respect to both physical and cultural features on the earth's surface.

By about 1925, parts of the United States were covered by the U.S. Army Air corps for the purpose of topographic mapping, and by 1930 a program for air photographic coverage of the whole United States had begun.

It was, however, not until the Second World War that there was any notable advancement and refinement of the technology related to air photography and the techniques of photointerpretation. During the war, air photography was used primarily to locate enemy targets and to assess damage. The British pioneered the development of photo intelligence systems which were significant in the early detection of German V-2 rocket-lauching sites along the French coast. A good account of the history of British photo intelligence can be found in Babington-Smith (1957). Emphasis at that time was also given to the preparation and use of photointerpretation "keys" for the purpose of rapidly identifying items of military significance regardless of the interpreter's previous experience (Stone, 1974).

During World War II, also, radar instruments were developed to obtain images of almost photographic quality during day or night and through dense cloud covers. From these techniques, Side Looking Radar (SLR) emerged as an all-weather, all-time photographic technique to be used from airplanes. This method obtains a picture by means of a series of lines similar to those on a television screen, and the image can easily be recorded on photographic film. Radar is considered an "active" system whereby energy is sent out and the sensor records the portion of that energy reflected by the object (Scherz and Stevens, 1970). This is in contrast to conventional photographic methods, which are considered "passive" systems in that they record solar energy reflected or emitted directly from the object.

In the 1960s weather satellite systems, NIMBUS (started in 1958) and TIROS, as well as manned space flights, Gemini and Apollo, provided spectacular photography of the earth's surface from altitudes many times the flight height of jet airplanes. This was the beginning of the field of "remote sensing," which is considered an outgrowth of air photographic interpretation. The term remote sensing was coined by Evelyn Pruitt in 1960, and it was Evelyn Pruitt and Walter H. Baily who initiated efforts leading to the "First Symposium on Remote Sensing of the Environment" held at the University of Michigan in Ann Arbor in 1962 and held annually since then. The major advantage of space photographs is that the area of the surface shown in a single exposure is many times larger than that shown on a conventional air photograph. Typical coverage of a Gemini space photo, for example, is about 38,500 km^2, whereas typical coverage of a 9×9-inch vertical air photograph (at a scale of 1:20,000) is only about 20 km^2.

In the 1970s the supply of remotely sensed data about the surface of the earth, and the importance of remote sensing, greatly increased with the launching on July 23, 1972, of the unmanned ERTS-1 (Earth Resources Technology Satellite-1), now known as LANDSAT-1, which was placed in near polar orbit, 900 km above the earth. This was overlapped by the manned SKYLAB-A, in 1973, orbiting at 435 km above the surface of the earth.

Photographic remote sensing is only one type of data collected by these vehicles. Other remote sensing instruments were placed in these platforms and operate in parts of the electromagnetic spectrum well beyond the range of the human eye or that of special photography. The human eye has a spectral range of 0.4–0.7 microns, whereas the spectral response of photography is about three times as wide, ranging from 0.3–1.2 microns (Scherz and Stevens, 1970). It is beyond the scope of this volume to discuss techniques or applications of remote sensors that do not produce "imagery."

Air Photographs in Coastal Studies

A coastline represents a dynamic, fragile, physical and biological environment that is constantly changing in response to natural processes as well as human activities. It has held great attraction for man as it provided him with a variety of resources for food, construction, recreation, transportation, and aesthetics. It also provides a natural protective barrier for inland areas against violent storms as well as daily or seasonal attacks by wind and waves.

For many years, studies of coastal processes and landforms attempted to gain a better understanding of the complex interaction between land, air, and water. Nautical charts and topographic maps were used for gathering information regarding the coastal zone.

Comprehensive studies of air photographs of coral reefs were made in the East Indies by Umbgrove (1928 and 1929) and in the Great Barrier Reef of Queensland by Stephenson, Tandy, and Spender (1931). Both used air photographs as a check on ground observations and not as independent source of information.

The use of air photography in charting coastlines was initiated in Germany in 1928 and in the United States by the U.S. Coast and Geodetic Survey in the early 1930s. Air photographs were taken of Ten Thousand Islands region on the west coast of Florida, an area then quite inaccessible. These proved to be both efficient for obtaining data needed for charting purposes and also economical. During World War II, the landing at Tarawa revealed a major deficiency in American photo intelligence. The water over the fringing coral reefs was much shallower than was previously reported and many landing crafts ran aground at a distance from the shore resulting in the loss of many lives (Amer. Soc. Photogramm., 1960). Extensive air photo and field surveys were conducted by Fairbridge and Teichert to establish the characteristics of coral reefs in the southwest Pacific (Fairbridge and Teichert, 1948; Teichert and Fairbridge, 1950). The Naval Photo Interpretation Center also began intensive research on water depth determination and gradients of enemy held shorelines. Methods for measuring water depths from air photographs were devised (e.g., Williams, 1947, and Lundahl, 1948). The latter is reproduced in this volume, Part I. Such information was used later in the landing at Okinawa. Since World War II, air photographs have been increasingly used to a large extent in the study of coastlines and in coastal engineering.

Air photography has even been used for the exploration of undersea archeological sites. Much of ancient Baiae, imperial Rome's most luxurious resort, has been photographed beneath nine meters of water in the Bay of Naples (Solecki, 1960).

An air photograph provides an excellent record of the pattern of features on a coastline at the instant the photograph is made. The size and spacing of waves, direction of wave fronts, direction of littoral drift, distribution of bare sand areas resulting from recent erosion or deposition, and the distribution of shallow water sediment accumulations to depths of about seven meters may easily be determined. More details of configuration of coastal and nearshore features are discernible than are usually recorded in careful field surveys of a coastline.

The availability of comprehensive air photography of coastal areas, along with rephotographing of the same areas at intervals of time during the last forty years, has made it possible to observe changes in beaches, sand dunes, and inlets, as well as the effects of man-made structures, between dates of photography. Published topographic maps and coastal charts, as well as field sketches by coast surveys permitted observation and comparison of changes prior to the availability of air photography. From sequential air photography it is possible to measure changes resulting from sediment shift during the interval between dates of photography. The greatest difficulty encountered in interpreting the changes, however, results from the lack of continuous day-by-day photographic coverage of the coastline. If the photographs indicate that a new inlet has been opened, an old one closed, a new beach ridge formed, or a spit lengthened or cut away, all that can be said is that these changes have occurred between the dates of photography. The greatest advantage of sequential air photography, especially when supplemented by older surveys and charts, is in recording long-term trends in shoreline changes that help in planning shore protection programs.

Studies of hurricane effects and shoreline changes in selected areas along the Atlantic and Gulf Coasts of the United States by the volume editor in the last fifteen years have demonstrated that sequential air photography is an indispensable tool and have suggested various other applications as for example: river flood effects, changing outlines of deltas and their distributaries, shifting sand dunes, glacial fluctuations, landslides, earthquake effects, volcanic eruptions, and man-made changes.

In the last decade, weather satellites and manned as well as unmanned satellites and space crafts have been placed in orbit about the earth at altitudes many times the flight height of jet planes. Spectacular photographs were obtained from these flights, from which valuable information regarding the coastal zone has been obtained. Observations and measurements of ocean currents and sediment laden waters have been made from such photographs. The major advantage of space photographs over conventional air photographs is that the area of the land or water surface covered in a single exposure is many times larger, which helps in studying and mapping large regional features and their interrelationships. On the other hand, space photographs have the disadvantage of not showing the minute details that may appear on air photographs. Space photographs, when supplemented with air photographs, however, have become an excellent tool in studying coastal features and their changes. Sequential space photography similar to that obtained by ERTS-1 (LANDSAT-1) proved to be a valuable tool in studying rapidly changing shorelines and in applications similar to those suggested for air photography mentioned before (Nichols and Kelly, 1972, and El-Ashry, 1973).

Types of Air Photographs

Two basic types of air photographs are generally recognized based on the orientation of the camera axis relative to the surface of the earth. A *vertical* air photograph is obtained when the camera axis is pointing vertically down perpendicular to the surface of the earth. It should be noted that there is no such thing as a perfectly vertical air photograph as there is always a slight deviation from verticality known as tilt. Corrections (rectification) for tilt have to be made, or accounted for, when accurate measurements are needed from the photographs. An *oblique* air photograph is obtained when the camera axis is at considerable angle from the vertical. When the inclination is great enough, to the point that the horizon appears in the picture, it is called a *high oblique*. Those with less inclination of the camera axis that do not show the horizon are called *low obliques*.

Vertical photos are the ones commonly in use today and very little oblique photography is being done.

A *photo mosaic* is an assembly of a number of air photographs matched and pasted together systematically so as to form a composite view of the area covered by the individual photographs. There are basically two general classes of photo mosaics, controlled and uncontrolled. A *controlled mosaic* is a composite of rectified individual photographs of uniform scale and tone with sufficient ground control. Reliable measurements can be made directly from this type of mosaic. An *uncontrolled mosaic* is without ground control or uniform scale. Accordingly, no accurate measurements of distances, area, or directions can be made from them. A *photo index sheet* is similar to an uncontrolled mosaic but has the individual index number of the photographs indicated. These permit the location of individual photographs already on hand and provide easy reference for ordering additional ones.

The type of film used in air photography is one of the factors that determine the tone and sharpness of an object. According to

Colwell (1961), there are four primary factors that determine photographic tone. These are: (1) light reflectance from an object; (2) light sensitivity of the film used; (3) light transmission by the filter; and (4) light scattering by atmospheric haze, turbid water, or any other medium separating the object from the camera lens.

There are four principal types of film available for air photography. These are black-and-white, black-and-white infrared, conventional color, and color infrared or camouflage detection film (Avery, 1968).

The major type of film used for black-and-white air photography in the United States is panchromatic. Images on panchromatic photographs appear in different shades of gray, with each tone comparable to the color density of an object as it appears to the human eye. In coastal studies panchromatic film is considered superior because of its ability to show details of underwater features. The use of yellow filters on the camera lens enhances haze penetration and increases tonal contrast of the photographs (Way, 1973).

Black-and-white infrared film is mainly sensitive to blue-violet and infrared radiation. The gray tones in this type of photograph result from the amount of infrared radiation reflected by an object and do not reflect the object's true color as in the case of panchromatic film. Because of the fact that water bodies absorb infrared radiation to a great degree, they appear dark on this type of photographic film, except when they contain high sediment concentrations. The sharp demarcation between land and water, then, makes this type of film highly useful in mapping shorelines.

Conventional color film is sensitive to all visible radiation although with variations according to film types. Color film is superior in recording underwater sediment accumulations (Swanson, 1960). Shoals at the mouths of rivers and inlet channels as well as in lagoons and estuaries are better shown this way than on any other type of film. However, the cost of color photography (about three times that of black-and-white) limits its use to special purposes only.

Color infrared film is sensitive to green, red, and infrared radiation. Colors in this type of photography are false colors and do not reflect the true colors of the photographed objects. For example, an object that reflects ultraviolet, red, and infrared radiation will appear red, while another that reflects green will show blue, and one that reflects red will appear green. Color infrared film is quite useful in the identification of tree species or ground cover types as well as in water pollution studies.

Multispectral Scanners

Mechanical scanners are very sensitive scanning radiometers capable of scanning a scene from an airplane or a satellite and producing an image the same way a picture is shown on a television screen. A radiometer is a non-imaging device which measures the absolute energy emitted directly from an object or the amount of solar energy reflected from it. A radiometer may be sensitive to energy from 0.3 to 14 microns in wavelength (ultraviolet to far infrared) depending on the type of sensing element being used (Scherz and Stevens, 1970).

The value of a multispectral scanner stems from the fact that it allows an investigator to view imagery obtained from a number of different narrow band widths. Multispectral scanners utilize anywhere from three to eighteen or more channels and have the ability to look at all channels simultaneously and to pick those that are most useful for a desired discrimination (Scherz and Stevens, 1970). Images from a multispectral scanner can be treated in different ways. One way is to assign a different color to a number of selected channels. The resulting color images could be studied individually, or a composite color picture could be made by the superimposition of the different color images in which the intensity of individual colors may be varied as desired. Examples of some applications of this technique to different aspects of the marine environment include: (1) mapping of aquatic vegetation; (2) feasibility of bathymetric determination and detection of underwater features; (3) study of thermal effluents and associated water mass movements; and (4) detection and mapping of industrial discharges and large oil slicks (Amer. Soc. Photogramm., 1975).

The ERTS-1 Multispectral Scanner Subsystem (MSS) provided images of the earth's surface in four different bands of the electromagnetic spectrum (NASA, 1972). These are:

Band 4 (0.5–0.6 microns)—Green
Band 5 (0.6–0.7 microns)—Red
Band 6 (0.6–0.8 microns)—Near Infrared
Band 7 (0.8–1.1 microns)—Near Infrared

Band 4 has the capability of water penetration and thus is useful in studying underwater features; Band 5 defines the extent of turbid plumes and sediment laden waters best; Band 6 is useful in the differentiation between water and vegetation in marsh areas; and Band 7 is very useful in coastline mapping because it yields excellent demarcation between land and water.

Image "enhancement" involves the reworking of data after it has been received. Examples include electronic manipulation of the data, color enhancement by assigning brightly contrasting colors in place of the original subtle variations in tone, and combination of two or more bands to accentuate a particular color feature. Teleki, Rabchevsky, and White (1973), in their study of the near shore circulation of the Gulf of Carpentaria in Australia from ERTS-1 imagery combined bands 4 and 5 to accentuate the total content of water-borne sediments, and bands 5 and 6 to delineate high concentrations and dispersal of suspended materials at the shore. They found that these combinations permit the recognition of subaqueous topography in shallow water, charting the outflow from rivers, defining areas of upwelling, and mapping the distribution of salt marshes. In addition, some of the enhancements were very useful in the recognition of former shorelines.

ORGANIZATION OF THIS VOLUME

The twenty-four papers that constitute the main body of this volume are organized into four distinct sections, where the papers represented in each section are related in their general theme to one another. However, for a broad subject like the one covered by this volume, where the central theme is "Air Photography and Coastal Problems," it will be hard to claim that the sections, or even the papers contained in them, do not overlap or bear a general relationship to one another. It is only hoped that, in organizing the volume into four sections, we would offer a logical and convenient means for treating the subject and the areas that are of prime interest at the present time.

Part I: *Air Photography in Coastal Studies*. This section provides an overview of the utilization of different types of conventional air photographs as well as those from other sensors (radar, ERTS-1) in the basic aspects of coastal geomorphology and mapping. The seven papers in this section deal with the specific aspects of coastal features and landform analysis, coastal processes, coastal delineations and mapping, and water depth determination.

Part II: *Photointerpretation and Shoreline Changes*. Seven papers constitute the main body of this section. The aspects of shoreline development and shoreline changes are the common theme. Discussions on shoreline changes include those that are due to normal (fair weather) processes as well as those resulting from hurricanes and severe storms, earthquakes, and tsunamis.

Part III: *Coastal Engineering Problems*. In this section five papers are used to illustrate the importance of air photographs in planning coastal engineering structures (groins, breakwaters, jetties, etc.) as well as in monitoring and documenting the effects of these structures on beaches, inlets, estuaries, and harbors.

Part IV: *Coastal Zone Management*. This section deals with the use of air photography and other remote sensors in managing the coastal zone with emphasis on man's role in altering the fragile coastal environment. It includes five papers ranging from an outline of the problem, to specific examples of studies of barrier islands, wetlands, and land use planning.

In the majority of the papers used in this volume, air photographs constituted the primary tool that was used for the identification of coastal features, mapping of coastal areas, analysis of coastal processes, measurements of coastal changes, planning for coastal protection and other engineering structures, and documenting man-made changes. However, in a few of the papers, which are included because of their pertinence to the subject matter of this volume, air photographs were not the principal tool but were used as illustrative matter which in many cases was an integral part of the discussion.

References cited in the introduction and in the beginning of each section, in addition to other supplementary references, are listed at the end of the volume. This list of references could also serve as a selected bibliography on the subject of "Air Photography and Coastal Problems." It does not, however, include the twenty-four papers reproduced in this volume. Additionally, a list of sources of air photography is supplied in the Appendix. This list includes the addresses of those government or academic offices throughout the world from which air photographs or information pertaining thereto can be obtained.

Part I

AIR PHOTOGRAPHY IN COASTAL STUDIES

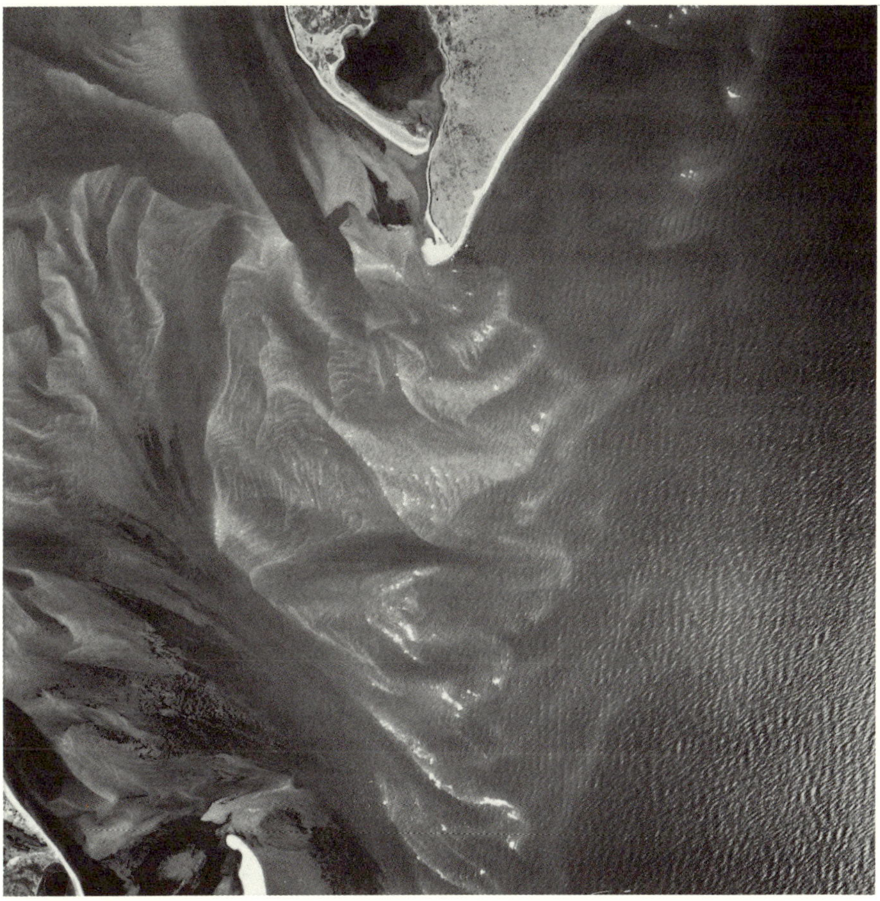

Air photograph (1960) showing underwater sedimentary features in the vicinity of Tuckernuck and Muskeget Islands, Massachusetts. (U.S.C. & G.S. photo)

Editor's Comments on Papers 1 Through 7

1 **DIETZ**
Aerial Photographs in the Geological Study of Shore Features and Processes

2 **CLOS-ARCEDUC**
Excerpts from *Aerial Photography and the Study of Coastal Deposits*

3 **MOFFITT**
History of Shore Growth from Aerial Photographs

4 **LUNDAHL**
Underwater Depth Determination by Aerial Photography

5 **LUEDER**
Excerpts from *A Method for Estimating Beach Trafficability from Aerial Photographs*

6 **MacDONALD, LEWIS, and WING**
Mapping and Landform Analysis of Coastal Regions with Radar

7 **MAGOON**
Use of Earth Resources Technology Satellite (ERTS-1) in Coastal Studies

Coastal zones have always held a special attraction for man. Approximately two-thirds of the world population live in coastal areas along the 448,000 km of coastline of the world. In the United States, about seventy-five percent of the population live in states bordering the oceans and the Great Lakes with a total of about 134,400 km of beaches and shorelines.

As the use of coastal areas was being intensified and the demand on ocean and beach resources was increasing, studies of coastal processes and landforms were initiated to gain better understanding of the complex interaction between land, air, and water. Field measurements and observations as well as laboratory modeling were, and remain, the major activities for collecting and interpreting data from coastal areas. Nautical charts and topographic maps were used for gathering information regarding the configuration of the coasts and the changes that take place along them. However, when air photographic coverage of coastlines became available, and specifically since World War II, air photographs and imagery from other remote sensors have provided a wealth of information in a relatively short time that was not available through other means of investigation. Air photographs, and other sensors, proved valuable in showing the larger features and circulation patterns which do not normally show on photographs taken from ground levels. Over shallow and clear waters, they serve to indicate shoals and channels which exist along many shorelines and to provide accurate means for revising and updating maps and nautical charts.

Nautical charts in the United States, are supposed to be resurveyed on a fifty-year cycle in eighty percent of the areas covered, and on a five-year, ten-year, or twenty-five-year cycle in the less stable areas that constitute the remaining twenty percent (Herbich and Hales, 1970). Nautical charting of the U.S. coastal waters has been conducted for more than one hundred years by the U.S. Coast and Geodetic Survey (now known as NOAA/National Ocean Survey). However, there are still many areas that were never surveyed, and others that are badly in need of revision. The U.S. Coast and Geodetic Survey has utilized air photography for over forty years for the following purposes: to position landmarks, aids in navigation, and controls for inshore hydrographic surveys; to map coastlines and coastal features; and to update the land information on nautical charts (Swanson, 1968).

In 1957, the U.S. Coast and Geodetic Survey pioneered the use of color air photography in the preparation and maintenance of nautical charts. It was realized that color photography permits the identification of objects that previously required field reconnaissance when only panchromatic film was used. Color photos show bottom features to depths of eighteen to twenty-one meters in some areas (Swanson, 1960).

For a number of years, air photographs of coastal areas were used mainly for illustrative purposes. One of the earliest studies to utilize air photographs in the interpretation of coastal features, including channels, shoals, and tidal deltas is that by Lucke (1934), which is partially reproduced in this volume (Paper 8). As greater numbers of air photographs became available, particularly after World War II, and as more and more people realized that valuable information can easily be obtained from them, an increased number of studies based on information extracted from air photography became available. Shepard, Emery, and LaFond (1941) identified rip currents from air photographs by gaps in the breaking waves and by areas of discolored water that is carried seaward. They also used air photographs to document the relation between rip currents and beach erosion in California. Munk and Traylor (1947) conducted a classic study on the relation between wave refraction and beach erosion using air photographs. They showed that variations in wave heights along shores are related to submarine topography.

The paper by Robert Dietz (Paper 1), was the first to present a comprehensive analysis of the importance of air photography in the study of shore features and shore processes. He pointed out the value of air photographs in studying underwater features, either directly or indirectly. Direct observation of such features is possible in shallow, calm, and clear waters. However, when the conditions do not permit direct observation of submerged features on air photographs, indirect methods, which include breaking waves, refraction of waves, decrease in wave length, or peaking of the waves, are utilized. The paper also pointed out the capability of air photographs to display a wealth of depositional features such as bars, beach ridges, tombolos, and spits, and erosional features such as wave-cut terraces, sea cliffs, and stacks. Dietz emphasized the usefulness of before and after photography in studying shoreline changes, and documented changes in the coastline that have occurred in nine years at Santa Monica, California, due to the construction of a breakwater.

The paper by Dietz was followed, in time, by several other papers emphasizing the value of air photography (conventional aircraft and satellite) as a tool in coastal studies. Examples include: McCurdy (1947), Coleman (1948), Shepard (1950a), Teichert and Fairbridge (1950), Zeigler and Ronne (1957), Sonu (1964), Clos-Arceduc (1964, Paper 2), El-Ashry and Wanless (Paper 12 and 1968), Nichols (1970), El-Ashry (1973), and Magoon (1973, Paper 7); as well as the books by Shepard and Wanless (1971) and Gierloff-Emden (1976).

Coastal features on air photographs are identified by their shape, tone, texture, and pattern. Beaches appear as greyish white bands. Changes in tone and texture of a beach can be indicative of local conditions. Answers to questions about the changes may be found in the parent material of the beach as present in rock outcrops, types of vegetation in the area, exposure and prevailing surf and wind conditions, slope, and capillarity of beach material, among others (Coleman, 1948). Beach features exposed above water normally appear lighter in tone than submerged ones. Dried sand and shingle reflect more light than wet ones.

Air photography is valuable, if not indispensable, in the geologic study of marshlands. Marshes appear as even-textured, nearly featureless expanses, whose tones range from light grey to black, according to vegetation and surficial water (Amer. Soc. of Photogramm., 1960). Cheniers are discernible as lines and bands of light tones, sometimes separated by narrow breaks. Areas of washover sediments in marsh areas appear even-textured and have light or mottled tone. Such deposits normally lack ponds or similar water bodies, and recently formed ones are devoid of vegetation.

The paper by Clos-Arceduc (Paper 2) represents an example from France of the systematic use of air photography in the study of coastal deposits. It portrays the integration of theoretical principles, observations from air photographs, and reproduction in laboratory models, in the identification and interpretation of coastal features, as well as in the prediction of occurrence of certain features in other coastal environments. The study also emphasizes the importance of the data and information obtained from air photographs in planning shore protection and other engineering measures.

A similar study on air photography and coastal geography of the German North Sea coast was published by Gierloff-Emden (1961), in which an analysis of the applications and problems of using air photographs in studying the coastal landforms was presented. A good example of a Russian study of coasts and submarine shoals with the aid of air photographs is the one by Guréva et al. (1968). In this study, the authors utilized air photography in studying the shoals in parts of the Caspian, Black, and Azov seas, and presented discussions on surveying submarine features and determination of water depth over the shoals, as well as compilation of sea bottom maps.

Ikramuddin and Laming (1966), utilized air photography at different tidal stages to demonstrate the instability of channels in the intertidal delta of the Alma River in the Bay of Fundy; and Tricart (1967) utilized air photointerpretation in studying lagoon deltas at the mouth of the Rio Santa Maria in Panama.

One of the major advantages of air photographs is their utilization in mapping projects in any geologic environment. However, in mapping coastlines or in revising older maps, there is always the problem of where to place the shoreline on air photographs. Should it be placed at the water's edge? This cannot be done because of the difference between the horizontal position of the high (tide) water line and the low (tide) water line, which can be considerable in the case of very gradually sloping coastlines. Should it be placed at the low water line? This cannot be done either, because air photography in future or past dates may have been taken at a high tidal stage where the low water line would be submerged. The only choice, then, is to place the shoreline at the high water line. Shorelines appearing on most nautical charts are the high water lines. The exception to this rule is where vegetation, such as mangrove, nipa palm, and prominent marsh grass, grows in the area between low and high tide (McCurdy, 1950). The outer limit of the vegetation growth becomes delineated, then, as the shoreline.

The exact position of the high water line is not always easy to determine on air photographs. The method that is commonly used is based on a combination of photographic tone, texture, and alignment patterns (McBeth, 1956; Stafford and Langfelder, 1971). The high water line is identified by a change in color tone. The seaward side of the line has a relatively darker tone than the landward side. This is related to the water content of the sand.

Use of the high water line in delineating shorelines is not without problems. Wind tides normally affect the location of the high water line. However, nothing can be done about this, and the effect is small and can be neglected. Also, under conditions of a gradually sloping coast with considerable tidal variation, it might be difficult to fix the exact position of the high water line. The use of a magnifiying glass could help in this case, where two lines of tonal variation can be seen along the beach. The landward line represents the wave runup on the beach slope, and the seaward line is the high water line.

The paper by Professor Francis Moffitt (Paper 3) presents two alternative systems to fix the position of a shoreline on air photographs of the same area but at different dates. A direct system is one in which the shoreline is related directly to a horizontal datum such as a state plane coordinate system. Correlation points are chosen on the photographs. The points must be of nearly equal elevation, otherwise corrections for relief displacement must be made. The second system is devised to be used when the absolute position of the shoreline is not required, as in the case in determining rates of erosion or deposition. In this system, direct scaling from correlated features or points to the shoreline can be made.

However, correction of the position of the shoreline to some standard such as mean sea level will have to be made. This can be done if the time of photography, tidal stage, and the beach profile are known. An easier way would be to locate the shoreline at the high water line, as described before, if possible.

An important area of application of air photography is in information gathering for military purposes. During World War II, air photographic interpretation of coasts and beaches was of great interest to the military. Coleman (1948) stated: "A great deal of World War II was fought along enemy-held coastlines . . . . Not one of the thousands of assault troops or landing craft personnel had seen those beaches before . . . . Aerial photography was the greatest, and in some cases the only source of information" regarding the coastlines.

One area of interest during the war was the determination of gradients on enemy-held beaches. This was important to determine the type of landing craft and equipment and the distance from the shore at which a craft would ground. The methods used to determine beach gradients, or underwater depth, from air photographs are mentioned in Williams (1947) and Lundahl (1948). There is a great similarity between the two papers, and Lundahl's paper, which is shorter in length, is reproduced in this volume (Paper 4).

One method for the determination of water depth utilizes measurements of horizontal parallax of sea surface and bottom details from the air photographs. This method was abandoned early for several reasons. Flying altitudes of seventy-five hundred meters did not allow heights that are correct to thirty or sixty centimeters to be obtained by the parallax equation. Also, the lack of beach texture and the large expanses of water surface on the photographs made it difficult to achieve control.

Good results were obtained by film density measurements. A good correlation usually exists between the density of a photographic image and water depth since more light is reflected through a thin layer of water than through a thick one. An exception to such correlation exists when the sea surface is rippled by waves and wind currents. Other methods of water depth determination, based mainly on wave and sea surface analysis, are also discussed by Williams and Lundahl.

Water depth determination is also of interest in the preparation and revision of nautical charts for navigation as well as engineering purposes. Accordingly, refinements in techniques, films, and filters, are continuously being made. Tewinkel (1963) discussed the factors that normally cause errors in depth determinations from air photographs. The main factor is the refraction of light at the air-water boundary. He derived an equation for a factor to be used to correct for refraction.

The availability of color air photography provided superior definition of underwater features to depths of eighteen to twenty-one meters (Swanson, 1960). It was also reported that water depths obtained from color photos that were corrected for refraction using a Kelsh Plotter were within plus or minus five percent of values obtained from hydrography in a test near the Virgin Islands. In a similar test, Geary (1968) reported that seventy-five percent of water depth values obtained from color photos of a study area on Wake Island were within plus or minus thirty centimeters of the measured depths, and ninety-five percent were within plus or minus sixty centimeters. He also reported on the use of black-and-white photography as well as on some satellite photography in revising nautical charts.

Advances in remote sensing techniques and multispectral scanning systems in recent years permitted accurate and automatic determinations of water depth. Studies by Polcyn and Sattinger (1969), Polycyn, Brown, and Sattinger (1970), Polycyn (1970), and Brown et al. (1971) are good examples. Basically two techniques are emphasized in these studies. The first is based on wave refraction patterns from shallow depths. The other is obtained from multispectral scanner imagery and is based on reflected radiation from the water surface and the sea bottom. The latter technique can be performed by a computer and computer-generated depth charts can be obtained.

The availability of high quality satellite imagery and specialized sensors has also permitted mapping of large areas to depths of over ten meters. A study by Polcyn and Lyzenga (1974) utilized ERTS-1 (LANDSAT-1) data in the updating of coastal and navigational charts. Depth charts for two geographical areas were constructed, representing different solar illumination and water transparency conditions. Absolute depth calculations for water depth to eight meters have been demonstrated by the authors for the Little Bahama Bank. Depth charts were also constructed for areas in Lake Michigan. They stated that the demonstration of shoreline depth mapping suggests the application of the techniques to problems involving changes in beaches and in erosion processes as well as storm induced changes in coastlines.

Another example of the military applications of air photointerpretation is one by Professor Lueder (Paper 5). The study is part of a series of reports produced in connection with a contract between the Office of Naval Research and Cornell University, under the direction of Professor D. J. Belcher. Technical report number 3 by Belcher et al. (1951) was concerned with the interpretation of ground conditions and characteristics of various types of beaches from air photographs. Other reports dealt with the determination of specific beach conditions. The applications of the material in these reports to civilian investigations of shores and beaches from air photographs are of great value. Unfortunately, the limited availability of the reports have restricted their use. It is hoped that the reproduction of the report on estimating beach trafficability in this volume will draw the attention of interested investigators to the amount of valuable information that is contained in these reports.

As mentioned before, remote sensing is considered an outgrowth of air photointerpretation. The term was introduced in 1960, when more information regarding the earth's surface became available from remote distances and with the use of other sensors in addition to conventional black-and-white photography. The study by Tuyahov and Holz (1973) is a good example, in which a combination of thermal-infrared, color, and color-infrared imagery proved to be more valuable for studying Padre Island, Texas, than imagery from a single source that operates at a specific wavelength. Thermal-infrared photographs provided a valuable tool in determining general trends in vegetation cover, water detection, and sediment movement. Color-infrared imagery provided the best means for moisture and vegetation detection and delineation. Conventional color photographs were most effective in water penetration and analysis. The study shows that shore and beach investigations are enhanced when remote multisensor imagery is used. Each sensor provides valuable information to the data collected.

Side Looking Radar (SLR) has been advanced as a remote sensing technique with great, yet few proven, possibilities. Although radar imagery gives the appearance of a black-and-white photograph with a lower degree of illumination, it is not a photograph, but an electronically constructed image of the various ways in which surface features reflect the radar energy (Cannon, 1974). Due to the strong contrast in the reflective ability of land and water, coastlines usually appear quite sharp on radar imagery. Water associated features such as harbors, ship channels, spoil banks, jetties, and the like appear also quite well on radar imagery.

The paper by MacDonald, Lewis, and Wing (Paper 6) is an example of a coastal study utilizing radar imagery. The value of radar as a tool to the coastal geomorphologist is pointed out. It has the ability to scan a broad band of terrain (65 km) with a single pass and of presenting the imaged area on a continuous strip of film.

Radar imaging systems have a near all-weather, twenty-four-hour imaging capability, which is an advantage of special importance for coastal studies of areas masked by darkness as in polar regions, or by clouds as in tropical regions, where conventional air photography can be completely useless.

Observable features on radar imagery include most of those observed on air photographs, for example, mud flats, beach ridges, beach features, and breaker zones. Sequential radar mapping of remote coastlines will provide a practical and rapid method of updating maps and charts and monitoring coastal processes. The spectral resolution of radar imagery, however, is less than that of air photography. Wave height and dune shape and orientation, for example, are not always easy to observe, and water depth information is not available. However, radar imagery provides a basis for measuring net changes in shape or dimension of coastal features and is an important source of information in cloud covered regions.

Paper 7 by Orville Magoon and a study by Nichols (1970), demonstrate the varied areas of application of space photography (Gemini, Apollo, and ERTS-1). Shoreline features, including barrier islands, beach ridges, dunes, salt marshes, tidal flats, tidal inlets, and deltas, are portrayed clearly on space photographs. Changes in shoreline configuration and identification of major areas of erosion or deposition can also be obtained from space photography. Considerable underwater penetration is possible and thus the utilization of space photographs in water depth determination. Information regarding littoral drift direction can be obtained from certain shoreline features or changes as well as from sediment plumes discharged from river mouths and bays. The possibility of widespread offshore movement of nearshore sediments was appreciated after space photography became available.

Watermass variability patterns, shown by color space photographs can provide valuable oceanographic data (Mairs, 1970). Space photography has also been demonstrated to be a valuable tool in studying and understanding circulation patterns and mixing on continental shelves.

An understanding of coastal circulation patterns is needed in studies of near-shore problems such as erosion, sedimentation, and pollution. Remote sensing techniques have been demonstrated to be valuable tools in providing the required data and coverage.

1

Reprinted from *Photogramm. Eng.* **13**(4):537–545 (1947)

AERIAL PHOTOGRAPHS IN THE GEOLOGICAL STUDY OF SHORE FEATURES AND PROCESSES

Robert S. Dietz, U. S. Navy Electronics Laboratory, San Diego 52, California

INTRODUCTION

IN RECENT years aerial photography has received ever increasing application to scientific work. Because aerial photos have many uses, they become an important and sometimes an indispensable geological tool. The aerial or bird's-eye view of terrain is an exceedingly useful one because from an aerial photo one can, at a glance, see patterns which would require long and tedious groundwork to unravel. Color, tone, shadow, and form on photos bring out much geomorphic and some bedrock data. Besides showing geological detail, the photos serve as useful base maps for plotting, and assist in ground travel as well. The aerial camera can be rapidly transported to even the most inaccessible areas. Also, aerial photographs permit taking the "field" into the laboratory, where specialists can make observations and measurements; then the photo may be filed as a permanent record. The writer's wartime flying experience with the 311th Photo Mapping Wing of the Army Air Forces has indicated to him that certain types of geological groundwork can be greatly simplified and expedited if high-altitude vertical photographs can be obtained, and if one can make low altitude visual reconnaissance flights during which the photographing of selected subjects is possible.

Aerial photos are especially useful in studying shore features and processes because of the presence of a wealth of physiographic features and because of the active processes which constantly change the shoreline. This is fortunate, as the geologist is keenly interested in this three-way boundary between the atmosphere, hydrosphere, and lithosphere. So much natural energy is expended in this zone that the shoreline is a birthplace of many physiographic features resulting from sedimentation or erosion. The geologist frequently refers to the present-day shoreline for help in unravelling the record of the ancient sequences of marine sedimentary rocks that now cover a large portion of the continents.

SHORE FEATURES

The most striking aspect of air photographs of shorelines is the camera's ability to see underwater features in shoal water. The depth to which submerged features may be photographed depends upon the intensity and the angle of incidence of light, the roughness of the sea surface, the nature of the bottom, and the transparency of the water. Photos taken away from the sun on clear days with quiet sea conditions show features at a considerable depth. Judging from the depth at which a white object, such as a Secchi disk, can be seen when lowered overside from a vessel, the depth at which the bottom is photographically discernible varies from a few feet in turbid water to 10 to 40 feet in phytoplankton-rich coastal water, and to 40 to 100 feet in oceanic water. In usually clear water, the bottom may be visible at even greater depth. When Navy ships first arrived at Bikini Atoll for the Atom Bomb tests, objects on the bottom could be seen at a depth of 180 feet. Oceanic water is most transparent to blue light with a wave length of 0.48μ, whereas coastal water is most transparent to green light of 0.53μ wave length or higher; thus film which is selective and is sensitive to such wave lengths is most satisfactory for photographing underwater features. Infrared photography, which is so useful in penetrating haze, is of no value in under-

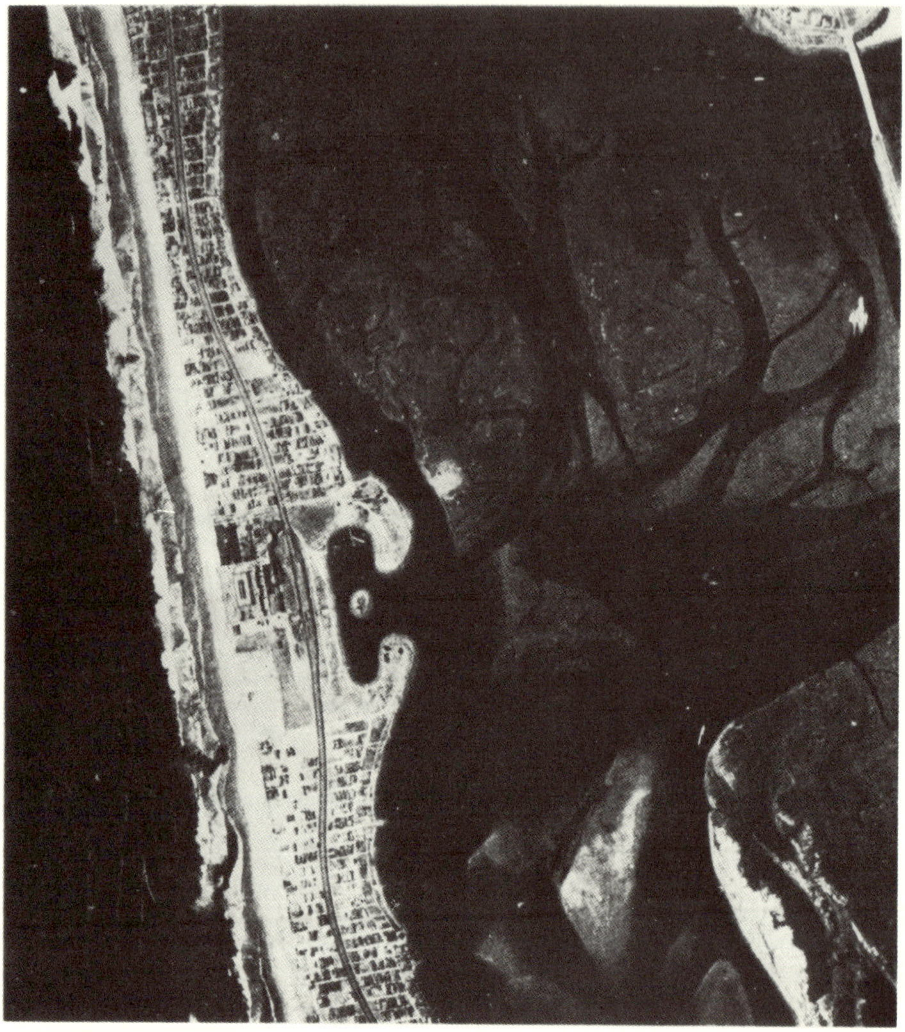

FIG. 1. A portion of Mission Bay, California, shows a tidal distributary system and a variety of other underwater features revealed by an air photo. Only the lightest areas are above water. U. S. Navy photo.

water work because such wave lengths are absorbed in the upper few centimeters of water.

Submerged features in shallow water bays and lagoons show up especially well from the air (Fig. 1). The coral reefs and the lagoon bottoms of Pacific atolls are also striking from the air. The Great Bahama Bank shoal often has the appearance of dry land, and the giant bottom ripples with the wave length and amplitude of sand dunes that it displays would excite the interest of any geomorphologist or sedimentologist. Along the coast of Florida on the Gulf side, bottom features can be seen many miles from land, and, near Miami, large saucer-shaped depressions are present which, above water, would have the appearance of the "hammocks" of the Everglades or of the "bays" of South Carolina. The aerial photographic coverage of such areas has great scientific value.

Even if aerial photos do not show the bottom, much information can be gleaned by indirect methods. Breaking waves usually indicate shallow reefs and banks; the deeper ones are indicated by refraction of the waves, by a decrease in wave length, or by a peaking of the waves. A reef may appear to be a source of circular waves. Some data regarding the depth of the bottom can be obtained by means of plunging breakers, the depth being equal approximately to 1.3 times the

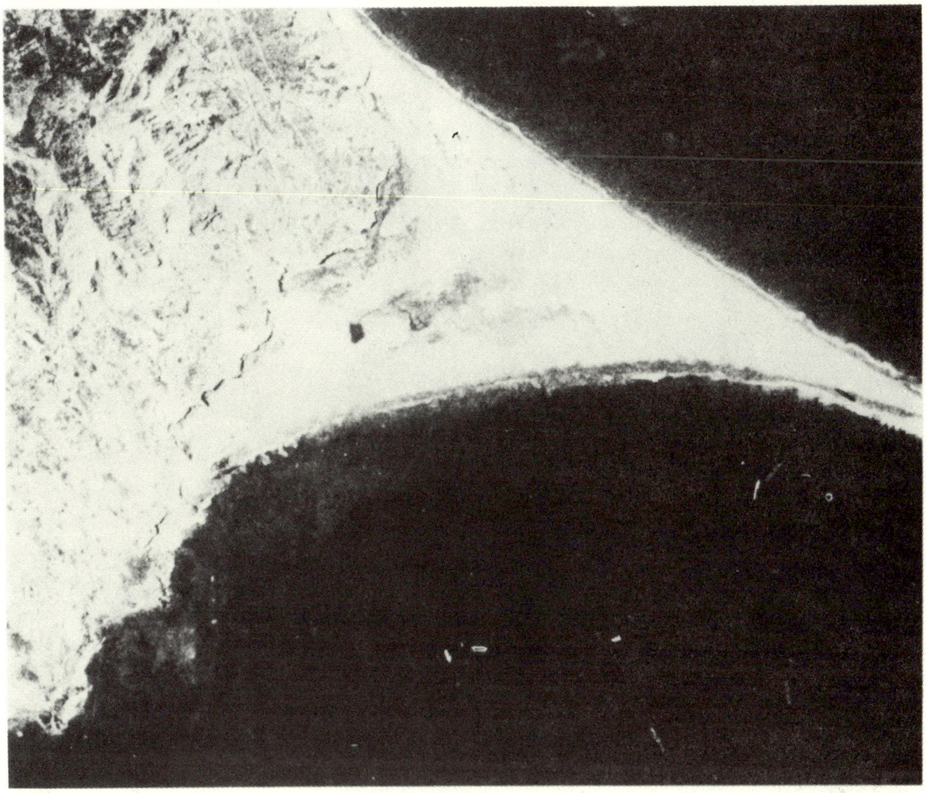

FIG. 2. This sand spit on San Nicolas Island off California is one type of feature formed by shore processes. Note the kelp bed (appearing as a black area) and the beach cusps on the concave beach. The cusps are ephemeral features produced by waves striking the beach at right angles and are destroyed by a change in wave direction. U. S. Navy photo.

height of the breaker. Information regarding the shore profile can be obtained by comparing aerial photographs taken at high tide with those taken at low tide. Along some coast lines depth of water and type of bottom is indicated by kelp; for instance, the large kelp Macrocystis usually grows on rock bottoms that vary in depth from about 25 to 100 feet (Fig. 2). Rea (1) states that kelp adhering to the off shore outcrops of Monterey shale delineates the Elwood anticline off the California coast at Santa Barbara and is clearly discernible on aerial photographs. For determining the suitability of beaches for landing operations during World War II, the Navy made extensive use of such indirect methods of obtaining bottom information from aerial photos.

Above water, shorelines display a wealth of both depositional features, such

FIG. 3. Shoreline in the vicinity of Morro Bay, California, photographed from 12,000 feet during a period of large waves. Note large silty swirls of a rip current carrying sediment offshore. U. S. Navy photo.

as bars, beach ridges, tombolos, and spits, and erosional features, such as wave-cut terraces, sea cliffs, and stacks (Figs. 2 and 3). Such features are frequently found elevated well above sea level due to local earth movements or because sea level is at present lower than during previous interglacial times. Aerial photos are useful in determining the geographic distribution of these physiographic features. In this manner, much information can be obtained regarding the migrations of the wandering shoreline or the degree to which a shoreline has achieved equilibrium at its present position. From an aerial photo alone, one can usually classify a shoreline according to the system suggested by Shepard (2).

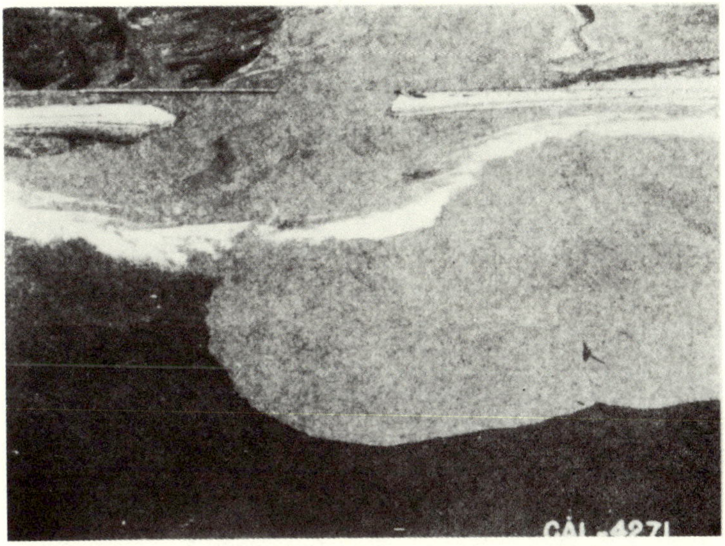

FIG. 4. Extent and direction of a movement of a water mass containing sediment transported in the ocean off the mouth of the Santa Clara River, California, is clearly outlined by an air photo. Photo by M. L. Natland.

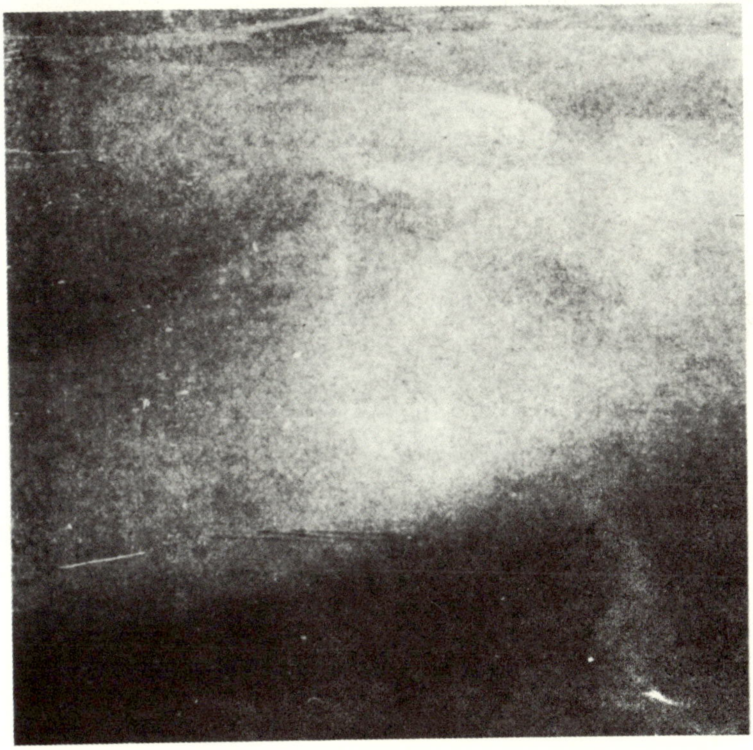

FIG. 5. Turbid water mass off the mouth of the Fraser River, British Columbia. Three boats traversing the muddy water leave a clear wake which indicates that this silty layer is floating on denser (more saline and/or colder) sea water. Photo by R. E. McFarland.

SHORE PROCESSES

The study of shore processes is another phase of shoreline science in which aerial photographs are of importance. Waves constantly cut back the headlands and straighten the shoreline. Sediment carried out from the continents is sorted and transported. Aerial photos can be used for studying the transportation of this sediment along the shore, as areas of discolored water can be readily mapped especially if color-sensitive film is used. Also, it is possible to map the extent to which suspended sediment is transported out to sea by rivers (Figs. 4 and 5). In times of flood, discolored water has been reported hundreds of miles from land.

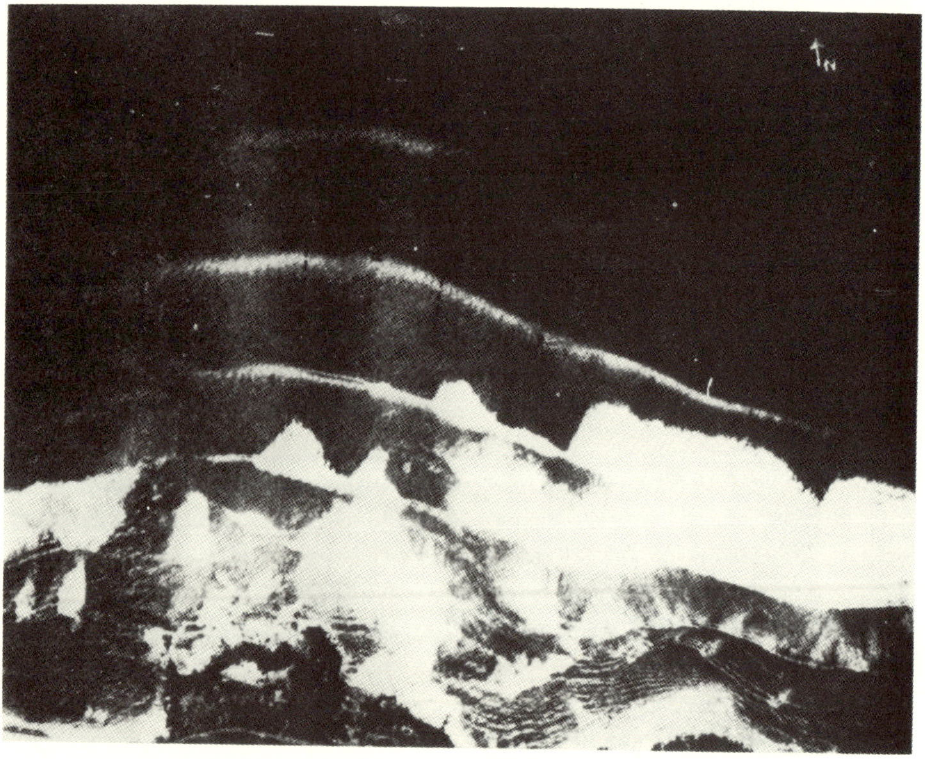

FIG. 6. Wave refraction and convergence caused by a submarine ridge extending out from a headland on the north side of Oahu, Hawaii. Air Forces photo.

Another interesting shore process that shows up from air is the transportation of sediment out through the breaker zone by rip currents (Fig 3).

Munk and Traylor (3) recently made an excellent study of the association of wave refraction and beach erosion by using air photos. Along the California coast, it was noted that offshore ridges, reefs, and submarine canyons refract waves which, in turn, caused both divergences and convergences. These writers showed that variation in wave height along the shore is controlled, in part, by sea-floor topography. Figure 6 shows that a submarine ridge extending out from a headland causes a convergence or a focusing of the waves, which results in rapid erosion of the headland. Figure 7 shows the effect of a submarine canyon on wave refraction.

FIG. 7. Waves forming a cross pattern north of La Jolla, California. A highly refracted group of waves deflected by a shoal are crossing unrefracted waves moving into shore in the deep water over a submarine canyon. Photo by W. H. Munk.

SHORE CHANGES

An especially fruitful field for aerial photos is in the study of shoreline changes. A photograph is a permanent and detailed record which can be readily compared with other photographs of the same area taken at a different time.

Air photos taken of the same area of shoreline many years apart can be used to determine the rate of prograding or retrograding of a shoreline, or to determine the sand fill behind a breakwater or in a harbor. Diastrophic uplifting or sinking of the land can also be detected. The advent of aerial photography is so recent that there has not been a sufficient lapse of time to make such observations in most places so that comparative studies have had to make use of maps or ground photographs. Yet the comparative photos of Figures 8 and 9 display a

FIG. 8. Beach at Santa Monica, California, in June 1925 showing a slightly cuspate form, apparently caused by the construction of a pier. Spence Air Photo.

FIG. 9. Beach at Santa Monica, California, in October 1934 showing an extensive building out of the beach caused by the construction of a breakwater. Spence Air Photo.

marked change in the shoreline that has occurred in nine years at Santa Monica, California, due to the construction of a breakwater. Aerial photographs taken of the present shoreline will obviously be valuable in the future if properly filed and made available to the general public.

Important changes in the shoreline take place in a period as short as one year. Along the California coast many beaches undergo a cyclic building up during the summer and a cutting down in the winter; other beaches appear to have a seasonal shift in position. There is also a monthly cycle of slight building up and destruction of beaches related, to the periods of spring tides. Aerial photos taken at proper intervals would yield much information on these cyclic sand movements.

FIG. 10. Low altitude air photo of damage and sand transport inland by the Hawaiian tsunami (tidal-wave) of 1 April 1946. The sea lies to the left. U. S. Navy photo.

Photographs taken at intervals as short as a single day show interesting changes on the beach. For example, the beach cusps shown by Figure 2 are ephemeral features, developed by waves striking normal to the coast, and are destroyed by a change in wave direction. Excellent use can be made of aerial photos in studying shoreline changes caused by related tsunamis (so-called tidal waves) and by hurricanes or other storms (Fig 10).

ACKNOWLEDGEMENTS

The writer wishes to express his appreciation to F. P. Shepard and W. H. Munk of the Scripps Institution of Oceanography and to E. C. LaFond and D. W. Pritchard of the Navy Electronics Laboratory for helpful suggestions for preparing this paper.

REFERENCES

1. Rea, H. C., Photogeology, Bull. Amer. Assoc. Petroleum Geologists, vol. 25, No. 9, pp. 1, 798, 1941.
2. Shepard, F. P., Revised Classification of Marine Shorelines, vol. 45, No. 6, 602–624, 1937.
3. Munk, W. H. and M. A. Traylor, Refraction of ocean waves: A process linking underwater topography to beach erosion, J. Geology, vol. 55, No. 1, pp. 1–26, 1947.

2

AERIAL PHOTOGRAPHY AND THE STUDY OF COASTAL DEPOSITS

Clos-Arceduc

These excerpts were translated expressly for this Benchmark volume by R. L. Rivera-Reyes from pp. 1–9, 38–50 of "La photographie aérienne et l'étude des dépots prélittoraux," Étud. Photo-Interprétation, No. 1:1–53 (1964), *with the permission of the publisher, Institut Géographique National, Paris.*

ABSTRACT

Coastal and near-shore deposits are related to the dynamic geomorphology of a region. Their study, however, is closely linked to matters of human geography. Along the river banks, human establishments are frequently situated upon the flood-plains and coastal plains bordering the sea. Tourism utilizes the beaches; roads make paths over the coastal dunes and the peduncles of the tombolos. The depositional structures, not very high, are fragile, sensitive to storms, and are covered during surges. Through the ages they have been the scene of catastrophes such as the submergence of the Zuyder Zee and so many other disasters.

Actually, where the coast is of depositional origin, that is to say formed of movable material, it poses problems to humans: protection of roads and towns menaced by erosion and choice of favorable sites for watering places. Their solution implies purely physical problems related to erosion and sedimentation.

In the study that will follow, abundant photographic documentation has been used from the Institut Géographique National to obtain useful principles regarding coastal deposits. The utilization of these principles for the estimation and explanation of a certain number of shapes or forms observed on the aerial photographs has served afterwards as their verification. It has been sought finally to apply these principles to matters of regulation and protection of coasts.

All of this is only an indication of what we can expect to obtain in these matters from the systematic study of aerial photographs.

This use is also subject to verification. The author establishes all the fundamental results by several means: theoretical reasoning, study of aerial photographs, and reproduction of natural aspects in the laboratory.

Moreover, the formulated theories permit the prediction of which forms will exist under certain conditions. The fact that on the aerial photographs the scalloped coastline with the triangular points (capes) in the predicted site have been found constitutes a verification of theory. In addition, this procedural control can be generalized in the geographic studies made by photointerpretation.

AERIAL PHOTOGRAPHY AND THE PROBLEM OF COASTAL DEPOSITS

With regard to coastal erosion and deposition, the problem to be resolved is the following: "Given a coast with known outline and nature, determine the actions upon it of the body of water that borders it, as well as the nature of and the position of the landforms to be built."

Its solution requires:

1. The establishment of relations between the coastal landforms, including beaches, and a certain number of parameters characteristic of the body of water, be it sea, lake, or shallow basin;
2. The knowledge of how to procure the parameters in question;
3. The ability to make the final calculation.

There are a great number of studies dealing with the subject of local phenomena. But these are frequently so complex that it is hard to deduce the general principles from them.

The study that follows is an attempt to establish the basic elements. The extent of aerial photographic coverage made these days for topographic mapping purposes allow us to chose a site for study of the function of the features selected for the end being pursued. This end, being the research of principles, in the sense given in physics to the term, requires locating on the photographs the most susceptible forms for a simple geometrical description.

The aerial views reveal many features little known or unknown until what is known as the age of complete aerial coverage. This, for France, started around 1962. It has been necessary in what follows to explain the forms observed on the aerial photographs by invoking some theorems well-known to hydrographers as well as the results of experimental verification made in laboratories or on site.

The vertical photographs make visible, at the same time, the nearshore formations, the trace of the beach, and the position of the crest of the swell. These have been utilized in an attempt to relate the depositional features to the oscillating state of the water.

I — THE STABLE FORMS OF SHORES IN THE PRESENCE OF A SWELL

It has been recognized that a shore is stable when it has appreciably conserved its outline during a period in which it could be observed with precision. As a reference for such observations in France, we have the charts of the General Staff and those which have followed; the shores of Africa or Madagascar were observed for a shorter period. Gulf shores, wherever they may be, are all stable.

The observations made this way over the whole of the French coasts and a good portion of the African coasts verify that wherever the long

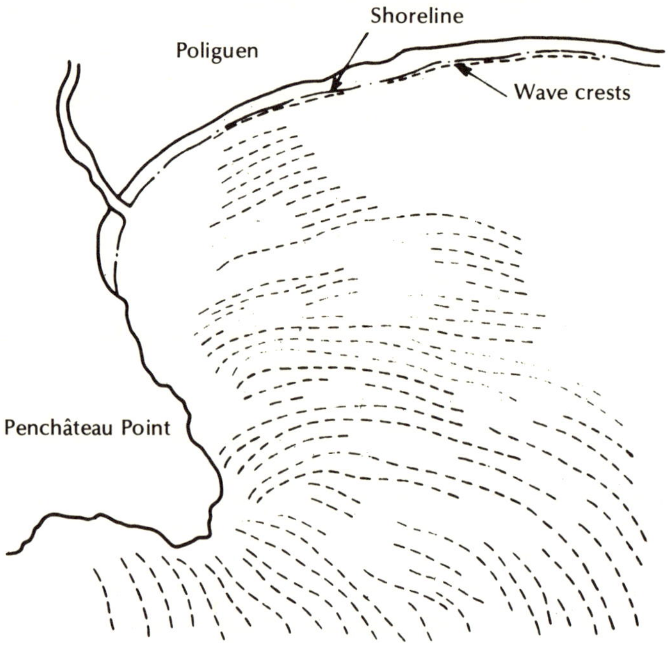

Fig. 1. Refraction of the swell over the high bottoms of the Penchâteau Point, resulting in the crests paralleling the shore at Pôliguen. (Tracing of Saint-Nazaire-Paimboeuf 1958, No. 120.)

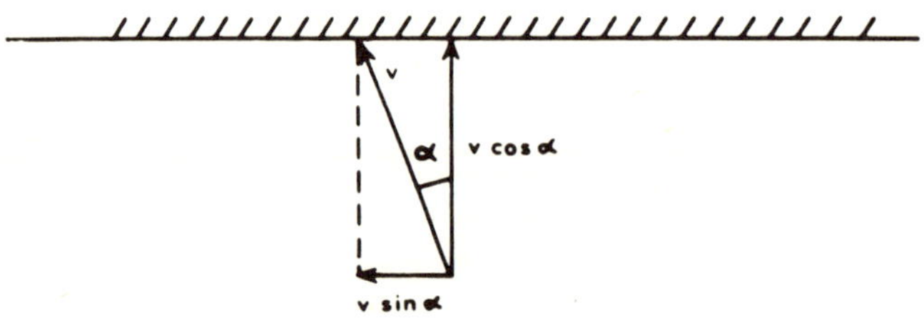

Fig. 2. Effects of erosion and transport by a swell attacking a beach at an angle α.
$v \cos \alpha$: erosion effect component.
$v \sin \alpha$: transport effect component.

oceanic swell is observed, its last crest in front of a shore is parallel to it. The refractions or sustained defractions have no more effect than to complete this parallelism. By means of photography or flying over in a plane at any given time, one can conclude that the established coincidence is permanent, which leads one to formulate the two experimental principles:

(a) An eroding or accreting beach is stable when it is parallel to the crest of the long swells bordering it.

(b) At the very least, in the oceans, the direction of the long swells originating at a great distance (old swells) is practically constant.

One can, to be scrupulously rigorous, add to this second statement "during good weather." The I.G.N. photography planes avoid, in principle, bad weather.

Figure 1, a tracing from the Saint-Nazaire Paimboeuf 1958, No. 120 photograph, shows both the refraction of the swell on the shoals and their final parallelism to the shore of Pouliguen.

Statement (a) above differs little from that of the Lewis law. It is more limited in the sense that it is applied to a particular instant. Its extension in time is justified by the verified stability of the long oceanic swells. A second experimental justification of this is that the short swells, having their origin in the local winds, are often in discordance with the outline of the shore and that, down to a slight depth, they are not refracted. It can be concluded that either their action is weak or that, due to their variability, their final consequence is negligible as compared to that of the long oceanic swell.

Statement (a) can be deduced from simple reasoning. In effect, let us say v is the mean speed of the mass of water thrown by the swell toward the beach. If the crest of the swell makes an angle α with the perpendicular to the shore, this speed v, in addition to its component $v \cos \alpha$, which produces the impact against the shore, also starts the movement of the materials with a component $v \sin \alpha$, which has the effect of transporting them laterally. Having a constant direction, it has no need to be strong to produce large displacements in the long run. If this component exists, the shore will be unstable and will not maintain itself unless it is being fed with replacement materials.

Stability demands then that $v \sin \alpha = 0$ (no transport), which implies $\alpha = 0°$, and the case of the Lewis law is as follows: $v \cos \alpha = 0$ (no erosion) in which $\alpha = 90°$. In this case, the coast is normal to the crests of the swell which pass in front of it (Fig. 2).

A shore oblique to the swell that has been protected from erosion by a groin rapidly realizes both cases side by side, replacing the oblique outline by a staircase outline (ph. Noirmoutier-Saint-Philbert de Granlieu 1958, No. 107–108).

The case where the crest of the swell is parallel to the shore is one only to be found on shores with some promontories. The action of a swell parallel to the coast cannot be seen sometimes in the photography, due to the reflection of the sun. This is frequent for the long swells with weak amplitude. It is often observed in the Mediterranean. It happens that this sea, due to the weakness of its tides, is the most likely to present simple coastal landforms.

II — OFFSHORE BARS

A certain number of photographs show submarine deposits in the form of bars, more or less regular, parallel to the line of the beach.

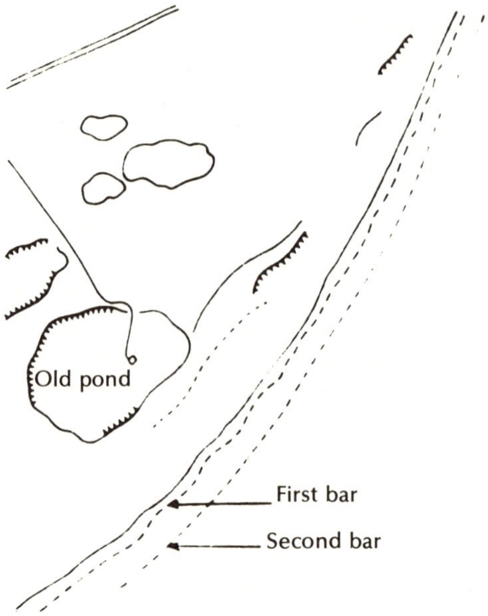

Fig. 3. Offshore bars east of Grau du Roi.

The first from the shore is separated from it by a hollow; the length of the sides is very regular, with a faint curvature turning its convexity toward the open sea. The bars can be multiple; one may frequently count up to three. Figure 3 represents a tracing taken from the I.G.N. photograph of a beach of Carmargue formed by sediments contributed by the Rhone. On the photograph, scale is approximately 1/25,000; the first offshore bar is approximately 1.8 mm from the shore, the second approximately 6 mm from the shore, in the part situated northeast of the old pond. The distance between the two bars is then 4.2 mm: that is to say, a little more than double that of the first one to the beach. This experimental finding appears to be constant for multiple offshore bars, and the law for their spacing can be stated in the following form:

D (shore to first bar) $= a$;
D (first to second bar) $= b$ where $b > 2a$;
D (second to third) $> D$ (first to second).

This law coincides with the one for the spacing in constant depth, starting from the beach, of the nodes of a clapotis. This is defined as the mode of stationary oscillation provoked by the interference of the approaching swell with the reflected swell. In constant depth, the spacing starting from the beach is $\lambda/4$, $\lambda/2$, $\lambda/2 \ldots \lambda$ (wave length or spacing of the crests of the swell) being itself a constant (see Figure 4).

Now for lesser depths, this spacing of crests is proportional to the square root of the depth (approximately the formula of Lagrange: $V = 3.13\sqrt{h}$). The spacing between crests is, in effect, the distance covered

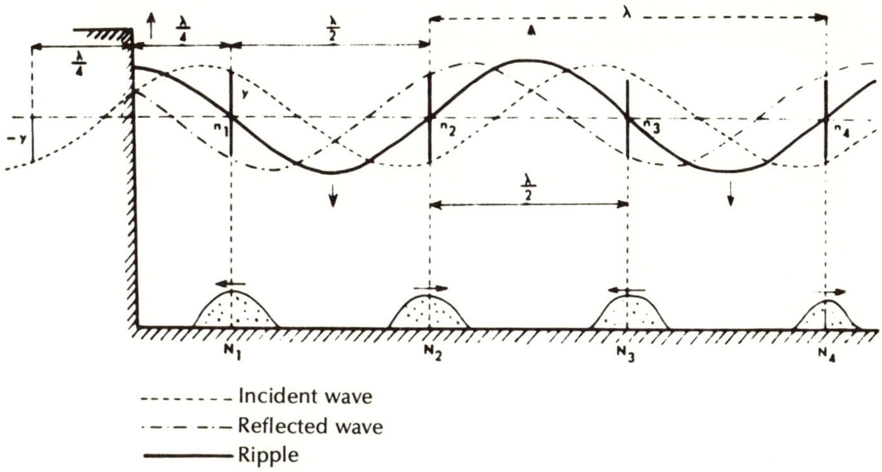

Fig. 4. Reflection of a swell in constant depth.

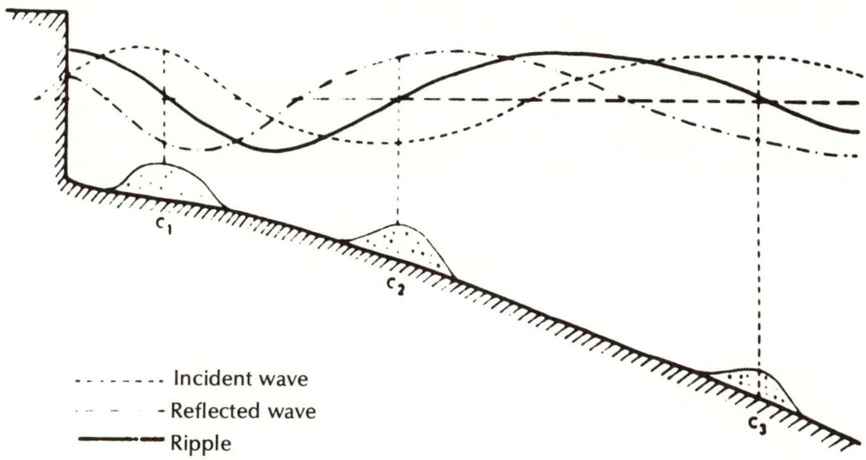

Fig. 5. Reflection of a swell in increasing depth.

by the undulation during the time *T*, *T* being the period of oscillation. It follows that the unit of length λ decreases when approaching a beach with gently sloping bottom. This is the case represented by Figure 5: it can be seen that the spacing of the bars corresponds to the experimental law enunciated above. From these observations, the following law is deduced (A):

The offshore bars are formed by the vertical component of the nodes of a clapotis caused by the interference of the approaching swell and of the reflected swell.

An objection is presented: the stability of the offshore bars implies the constancy of the period and direction of the swell. This objection is presented by the observation made of the stability of the long swells and by the fact that, on the gentle slopes of sandy shores, the short swells do not reflect. The Irribaren formula applies:

$$T = \frac{2}{P}\sqrt{\frac{a}{g}}$$

where p is the slope, T is the period of the wave, a its amplitude, and g the acceleration of the mass. This gives, in effect, the shortest period of the wave susceptible to being reflected. One sees that the period increases if p decreases, which practically eliminates the effect of the short waves. The swells of local origin will not, therefore, produce ripples. Their elimination leaves to the distantly originated long waves, much more stable, the main role in the creation of permanent oscillations.

EXPERIMENTAL VERIFICATION

The preceding theory came from the correlation between the observation of a natural form on aerial photographs and a theoretical form. But if we admit provisionally that the depositional forms are set in place along the nodes of a clapotis, the only mechanical element through which these radial lines can manifest themselves is through their vertical component; the horizontal movement of the water having negligible effect. One can then replace enunciation (A) with (B):

In areas where the water movement is horizontal, the effects are negligible in the zones of accretion.

This could be deduced from the fact that a body in suspension with a falling speed in the water of v in those regions takes a very long time to reach the bottom. The absence of a vertical component makes its descent untroubled.

Enunciation (B) has the advantage of being verifiable on site. In effect, it suffices to identify on an aerial photograph a zone of deposits of light material and to study there the movement of the water. The point chosen was the strait separating the islands of Saint-Honorat and Sainte Marguerite (Photo I.G.N., Frejus-Cannes 1955, No. 139). One can see on the aerial view the numerous deposits of sand or gravel, amongst them a veritable dune with a ridge normal to the axis of the strait. The appearance immediately makes one think of a strait with a uninodal seiche.

A simple walk with a diving mask permitted the following observations:

1. Along the length of the Sainte Marguerite beach, perpendicular to the sand banks, are masses of posidonies. These plants form an immense submarine prairie clinging to the rocky bottom.

2. At the bottom, between the rock ridges, slightly oblique to the axis of the strait, sand banks with ripple marks on some spots occur. These sand banks occupy elongated shallow basins at the bottom of which one sees posidonies debris oscillating in place. One can also

find numerous balls (aegagropiles) formed by posidonies fiber, some of which have the dimensions and almost the shape of a bowling ball. Their formation starts with light material which the slightest current would move and requires that the ball must be rolled over the spot for months. This suffices to establish that within the zone of the deposit of light materials under study the movement of the water is horizontal; in other words, in the neighborhood of a nodal line of permanent oscillation.

THE EVOLUTION OF COASTS

When the erosion of a coast, instead of attacking dunes or offshore bars, threatens human installations, one proceeds to work on defense.

This defense must be deployed over more or less space according to the amount of time allowed to elapse.

In the case where erosion is caused by longshore transport, it is the result of the obliqueness of the swell on which a current can be superimposed. The two often occur well in advance of the receding shore. It is evident, in effect, that a sandy shore must have advanced at first, otherwise it would not have existed. For instance, it suffices to examine the chart at 1:100,000 of the Montpellier sheet to see the place of the old dike hooks; and the Aigues-Mortes was installed formerly on a beach where the inflection reflects, with weaker curvatures, the site of Grau du Roi.

The state of relative equilibrium results from an equality of erosion and sediment supply. For erosion, the influence of the swell is relatively stable. But human industry can influence the currents and the quantity of available sediment.

THE DIMINUTION OF SEDIMENT

It is commonplace to say that the sea does not manufacture the sediment but confines itself to setting in place the materials supplied by the rivers. These have, during the historical epoch, delivered enough materials so as to result here and there in the construction of new depositional features (Point at Larree, the more recent ridges of L'Espiguette, etc.). But one can also note numerous discrepancies.

A general cause for the diminution of the fluvial sediments of rivers in the industrialized countries is the intensive use of water courses for different purposes. This use is expressed through dams, channelization, and modification of the courses. But, if a water course is not very erosive during a period of medium flow or during low-water periods, it becomes suddenly erosive during the high flow period. Suppress the high flow and there will be less sediment. In addition, the clays in suspension are deposited in the impoundments behind the dams.

Efforts at protection by channelization produce the same result. At the same time as the canal beds of Paillon are regulated and deepened to prevent damage during high flow, the point is reached where not very many pebbles are being fed to the beaches of the Bay of Anges.

In the same way, the great works made on the Durance and the Rhone have transformed a "working" river into a river with flow and speed almost constant at its mouth (about 2.10 meters per second at the mouth of the Grand Rhone). Moreover, since gravel and sand are not permitted either in turbines or in irrigation canals, the mud that the river used to supply to the sea is now reduced to a small amount. This results in the suppression of fluvial sediment at the start of They de Roustan. What we have said about its coastal progress affords the understanding that the situation has not been modified so much near the Grau du Roi.

For a long time the point of the Espiguette, in effect, will be still supplied by the sand eroded on the coast situated to the east. This is not a new situation: the majority of the supplies of the Rhone have been used for a long time (involuntarily) to fill up the gulf of Fos.

Whatever the case may be, this coast is profiting right now for the first time in its history from well-defined permanent conditions: current, supply, and density of known sediments at the estuary of the Rhone; the Coriolis current, as a consequence, has become constant. Measurements taken at any moment were found to be valid for almost all of the year. It is then possible to gather the given facts necessary for a study and for serious forecasts.

It is furthermore indicated in proceeding with such a study that the coast as far as Sete is in a relative state of stability. Let us point out that on such a coast work performed at one point can deprive other points downcurrent of sediment supply. The relative equilibrium is now disrupted, without conditions having changed in appearance. This situation has already been produced; it does not seem that there has been research to anticipate the consequences.

METHODS FOR PROTECTION OF SHORES

When a mobile shore is being attacked by the sea, there comes a time when the damage is so evident as to endanger property along the shore. The beach becomes smaller to the point of almost ceasing to exist. It is attacked by waves during high tide, and schemes for protective works are immediately planned. These consist of longitudinal walls (dikes or walls directly opposing the impact of waves) and transverse walls (wharves or jetties). Their behavior and strength are very well summarized in the study of Hommeril and Larsonneur published in May 1963.

The mechanism of destruction of these structures is evident. They are required to protect what is behind them. This requires the stopping of the waves, and, therefore, their reflection. This reflection (see Figure 6) creates an oscillation bulge in front of the dike, causes the washing away of its foundation, and eventually the collapse of the dike.

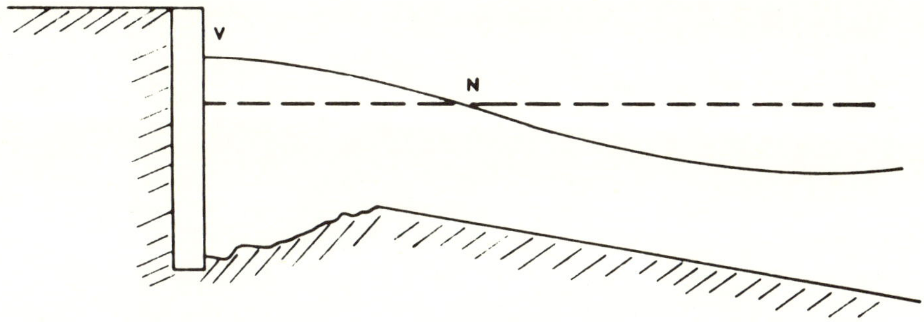

Fig. 6. Undermining of a dike foundation in the region corresponding to maximum oscillation near the shore.

Logically, there is only one possible remedy for this. That is to avoid this sinking by means of reinforcement of the soil. To be effective, concrete work should cover the greater part of the area. Waves upon which the stationary vertical oscillation in front of the dike develops, being 25 meters for a wave of 100 meters between crests in the vicinity of the shore. This is part of what has been done in certain places through the placing at random of big blocks in front of the dike.

A troublesome temptation is to incline the dike a little to decrease reflection. This half-measure is illustrated by the photograph of the Containville dike, published by Mm. Hommeril and Larsonneur. The nonreflected water has risen, excavating behind the dike, and the reflected water has uncovered the wall. Finally, the dike on one side has been partially sunk and uncovered on another, and did not accomplish its purpose.

THE ROLE OF WHARVES

Wharves are built to absorb the shock of impact. We have seen in connection with the Lewis law, that there are two types of stable coasts: those parallel to the crest of the swell and those perpendicular to the same crests. Wharves must be perpendicular to the waves. Those who build them must, then, forecast where the greatest concentration of waves will come from. The effect of a wharf thus conceived will be null on a stormy day. This is the time for dike defenses, with the reservations made above. In fact, one should avoid the situation where this defense is necessary: if it is, the area should be evacuated.

The function of the wharf is to regain ground by accumulation of sediment during periods of calm weather. Its mechanism is quite simple: it forms an artificial gulf in the outline of the shore. The sediments stop at the angle formed and, if the shore is fed, it advances to a line parallel to the wave passing by the point of the wharf. The Sete-Grau du Roi 1963 photograph, no. 12, showing this phenomenon, is particularly legible because there is not another wharf in three kilometers. It can be seen that this proceeds up to a small outlet of the pond, situated 1 kilometer east. The jetty protecting the mouth of the Vidourle at

Fig. 7. Effect of a wharf which is located here as a groin in front of the nursing home of Aigues-Marines, west of Grau du Roi. The presence of this wharf changed the outline of the coast over nearly 1 kilometer by causing it to advance seaward. (Sete-Le Grau du Roi, 1963, no. 12)

Grau du Roi plays the same role and already has caused a deposit 160 meters long by 50 meters wide whose border is exactly parallel with the occurrence of the swell crests (Figure 7).

The shore formed by the wharf is itself parallel along all its length to the visible surf. The wharf has caused an average advance of 20 meters on an 800-meter front. Also to be noted on this photograph, is the good performance of the wharf when the incidence of the wave approaches the perpendicular. It can be especially noted that the swell is composed of two crossed waves, the mixing of which produces at the border an unique surf that is parallel to the shore. This shows at which point it is prudent to introduce given facts in the calculation of a "wave plan." The reality often escapes calculation.

Another remark of practical interest is that near Grau du Roi, the sediments move from west to east. They are not then affected by the diminishing of supplies from the Rhone. The photograph, showing the action of the wharf, gives us this supplementary piece of information.

From all of this, it can be concluded that:

1. It is not necessary in envisioning the protection of a coast to be forced to protect it through dikes.
2. One must aim at helping the process of shore advance during calm weather rather than at defending it from storms.
3. The spacing of wharves and their dimensions must be calculated in each case in terms of the incidence of the swell.

POTENTIALITIES OF AERIAL PHOTOGRAPHS

The preceding study was made possible due to the fact that aerial photographs of the nearshore zone sometimes allow one to distinguish the bottom and, sometimes, the detail of the swell. One sees, as a general rule, the swell and all that occurs on the surface in the region covered by the photograph, which corresponds to the reflection of the sun or the sky. On the other hand, the bottom is visible within a zone centered on the point where the shadow of the plane is projected. This does not really occur when the plane is flying low. In seas where the water is transparent, the photographs provide enough details so that the topography of the bottom can be drawn. Thus, for anything having reference to offshore bars, festooned or not, we have precisely the topographic knowledge of the nearshore zone conveniently furnished by these aerial photographs (see Figure 8).

The uses of aerial photographs that we have mentioned are still far from exhausting their possibilities. They also furnish us with documentation of lesser known phenomena either put aside as too difficult to observe or as unsuitable for mathematical description. Photographs make it possible to arrive at descriptions that are not obtainable by simplifying hypotheses. These serve, then, to help us understand a phenomenon but cannot give an acceptable calculation in a real case. The calculation of the trajectory of a swell of known period has many points in common with an optics calculation; but where optics are concerned dispersion is a considerably stronger function of wave length and diffraction is almost never negligible.

The role that vertical photographs have played, particularly those of the I.G.N. collection, was to inform us directly in cases where the calculations would be too complicated and, sometimes, too uncertain to be of any interest.

Let us take, for example, the point of the Island of Ré, visible on the Langueville-Marans 1957 photograph, no. 175. It is enough to look at the 1:500,000 chart to see that it is the result of refractions of the Atlantic swell upon the shoals extending over 35 km to the northwest of the point, combined with the reflection of a portion of this swell upon the shore included between the Aiguille point and the Groulin point, 35 km long. None of this has a simple form. Calculations and graphics promise to be laborious: the photograph gives the plan of the waves as complete and exact as only experience can offer. Let us note on passing that all of this complication is translated for the shore north of

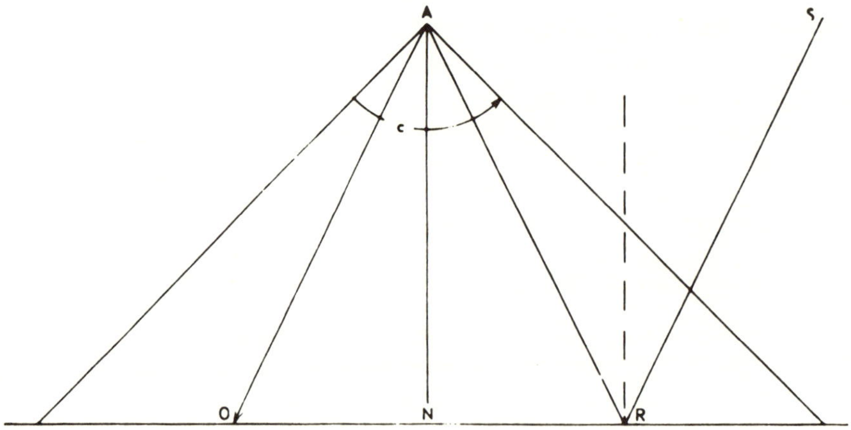

Fig. 8. Visibility zones of the surface and the bottom.
A: airplane and photographic equipment
c: photographic range
S: direction of the sun
R: reflection, center of the visibility zone of the swell
O: shadow of the plane, maximum visibility of the bottom
N: nadir

the point into a wave surf parallel to the beach, showing that the Lewis law, on our Atlantic shores, applies in general to the swell existing at a given moment.

Another "wave plan" on a large scale, resulting from a photographic mission at 1:5,000 scale makes visible a whole complex of swells with very different wave lengths. One can define the shadow of the plane, symmetrical in relation to the center of the plate, by the clearer reflection. All this complexity is reduced to a simple phenomenon: the intersection of two waves at a certain point. This intersection is very dangerous for a small boat. The examination of coastal photographs permits recognition of the places where this is possible.

In this photograph (AO 1960-61 AO 378-50, no. 53) details of the swell are apparent over the extent of visible water. This is due to the position of the sun near zenith and also to the diffused light of tropical skies (Figure 9).

The phenomena of diffraction appear to be subject to computation. But this computation must take into account the wave length and the curious mutation of a wave impacting against an obstacle. The photograph AO 378-50 AO 1960-61, no. 13 (Fig. 10), has furnished such a spectacular example as to make this indisputable.

Let us note in passing that the emitted secondary waves have a shorter period than the incidental waves; this is the opposite of what takes place in optics where, for instance, an incidental violet radiation brings on a red fluorescence. This transformation of a long wave into a short wave after impact upon rocks is a frequent case. It has the char-

Fig. 9. Intersection of two waves of a swell localized at a point in a cove near Dakar. (AO 1960-61 AO 378-50, No. 53)

acteristic that the wave produced depends little on the incidental wave, and this is found remarkably constantly, in spite of the irregularity of the waves that produce the phenomenon. The photographs of the Bay of Tamatave (Madagascar 1957, V45-V46, no. 007) (Fig. 12) represent such a diffracted swell induced in the bay. The aerial photograph clearly shows the dead angle on which the port is situated, sheltered from the waves.

It may be remarked that the necessary details are furnished for measurements on reduced scale models. In effect, the problem is to find out the given facts to be introduced under the heading "swell coming from the open sea." Let us assume that the topographic model is established in the form of a bathymetric relief map. The diffraction is, as in optics, proportional to the wave length. This then, must be reduced

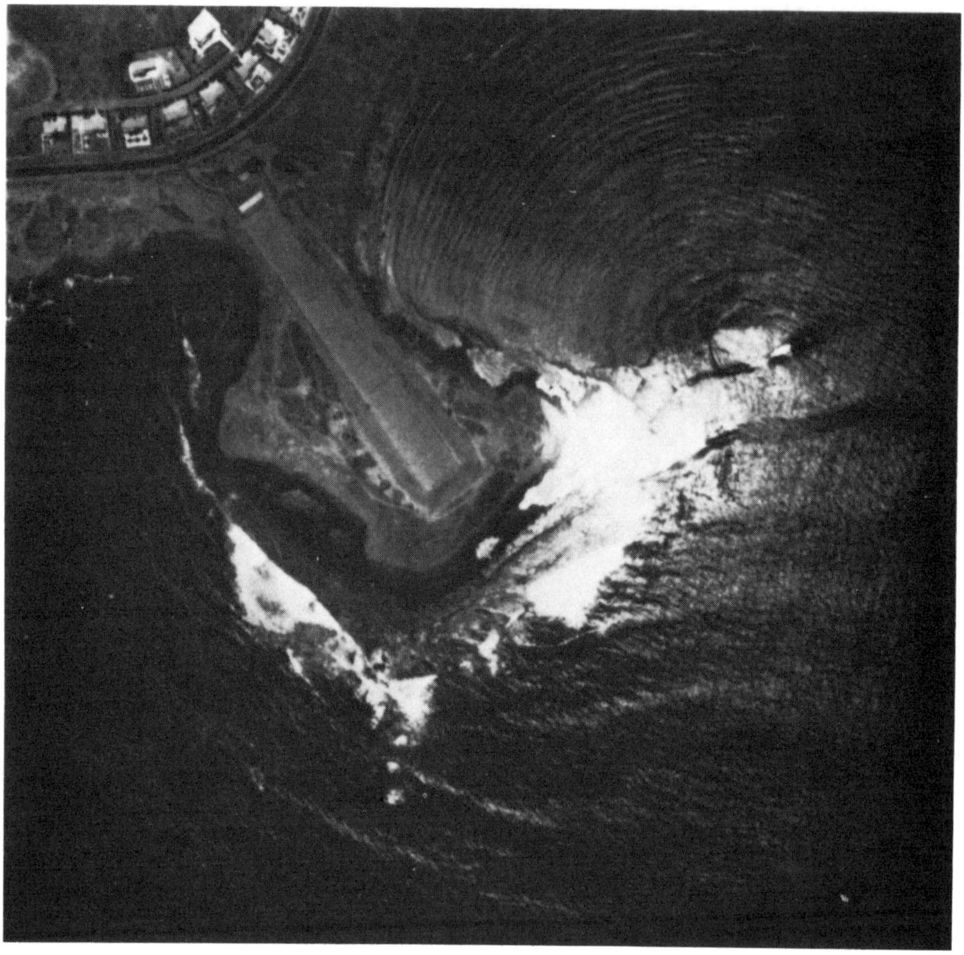

Fig. 10. Example of wave diffraction.

according to the range of the scale. If the plan scale is 1:1,000, a wave with 100 m between crests will measure 10 cm on the reduced model.

Another phenomenon, which also escapes calculation as does the preceding one, is the reflection of a wave by a current. This phenomenon, which occurs in the neighbourhoods of estuaries, is revealed by aerial photographs. We give as an example the photograph AO 1960-61 AO 364-100, No. 315 (Fig. 11), which represents a wave reflected at a right angle by the current coming from a lagoon at the Côte d'Ivoire. Here again, there is borne from the impact a very short resultant wave, whose relations with the approaching swell seem very difficult to foresee. All the given facts appearing in the Schou laws are the same for both cases, at least in appearance. In this case, the photograph offers us an already made "wave plan" that we would not know how to make.

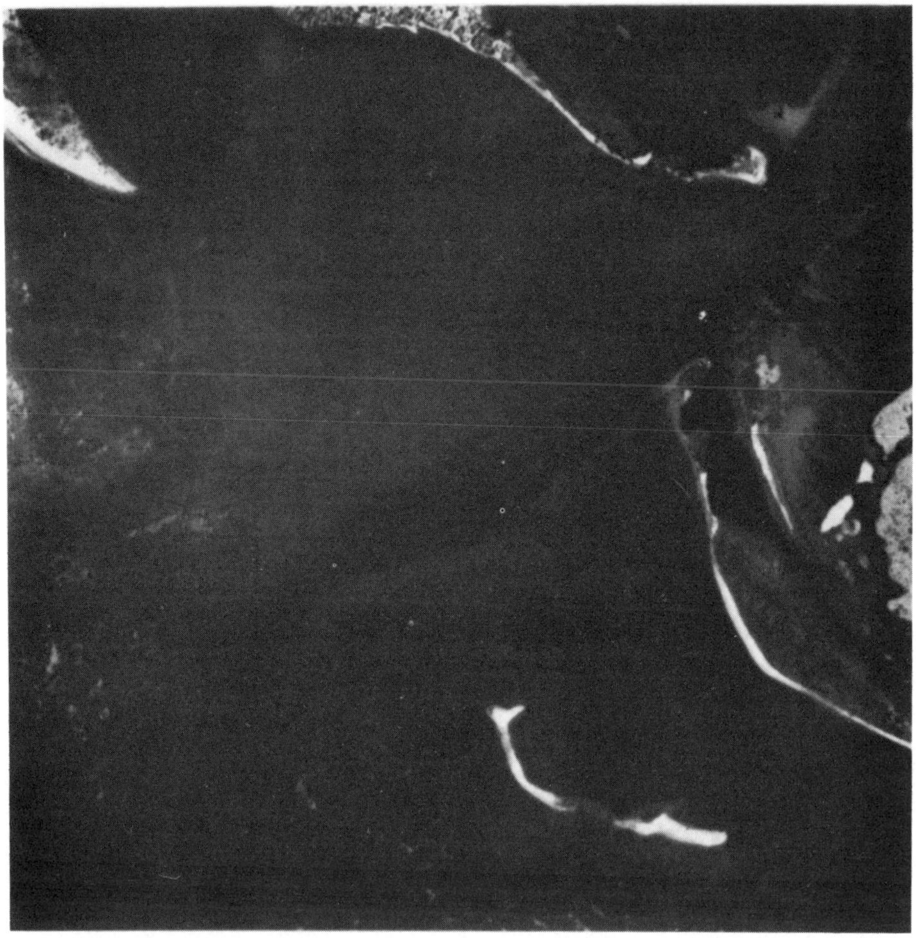

Fig. 11. Reflection of a swell by a current. (AO 1960-61 364-400, No. 315.)

Those who have observed the sea near an estuary with a swift current, such as the Rhone, or the Var, may have noticed very stable lines where the water seems animated with rapid movements with weak amplitude. These lines correspond to the boundary between two masses of water of different density. (Cf. article of Max Guy in No. 6 of *Photo-Interprétation*.)

This feature is also found at the interface of two sheets of water of different origin and density. The photographs, The Pieux-Saint-Vaast-la-Hougue 1955, nos. 136–37–38, are an example (see Fig. 13). One does not see the dioptric effect pointed out by M. Guy for the Rhone estuary, where the fresh-water sheet floats over the denser salt water, producing a freckled effect and the appearance of fringes. The sheets of salt water in contact near the point of Neville do not produce the same effect. On the contrary, the boundary line is uniform. Small waves can

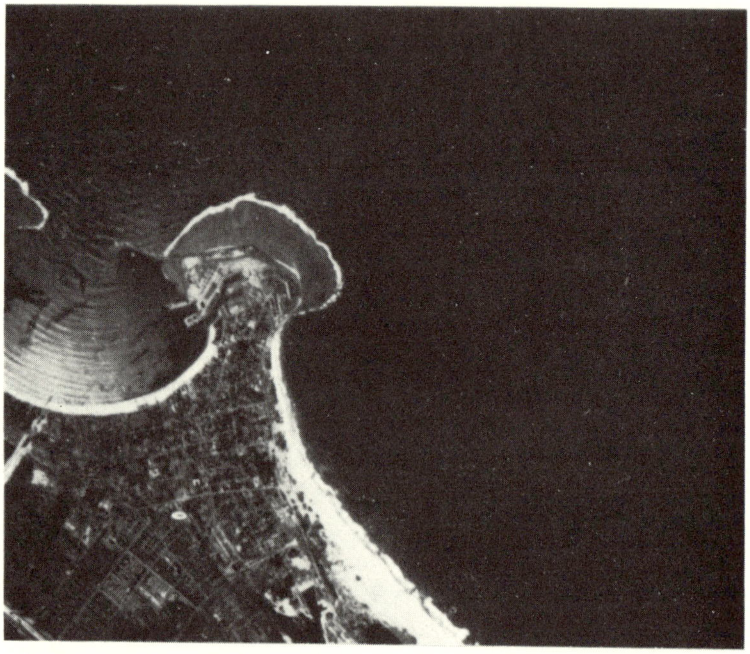

Fig. 12. Bay of Tamatave. Zone of a diffracted swell.

Fig. 13. Boundary lines of two sheets of water of different densities.

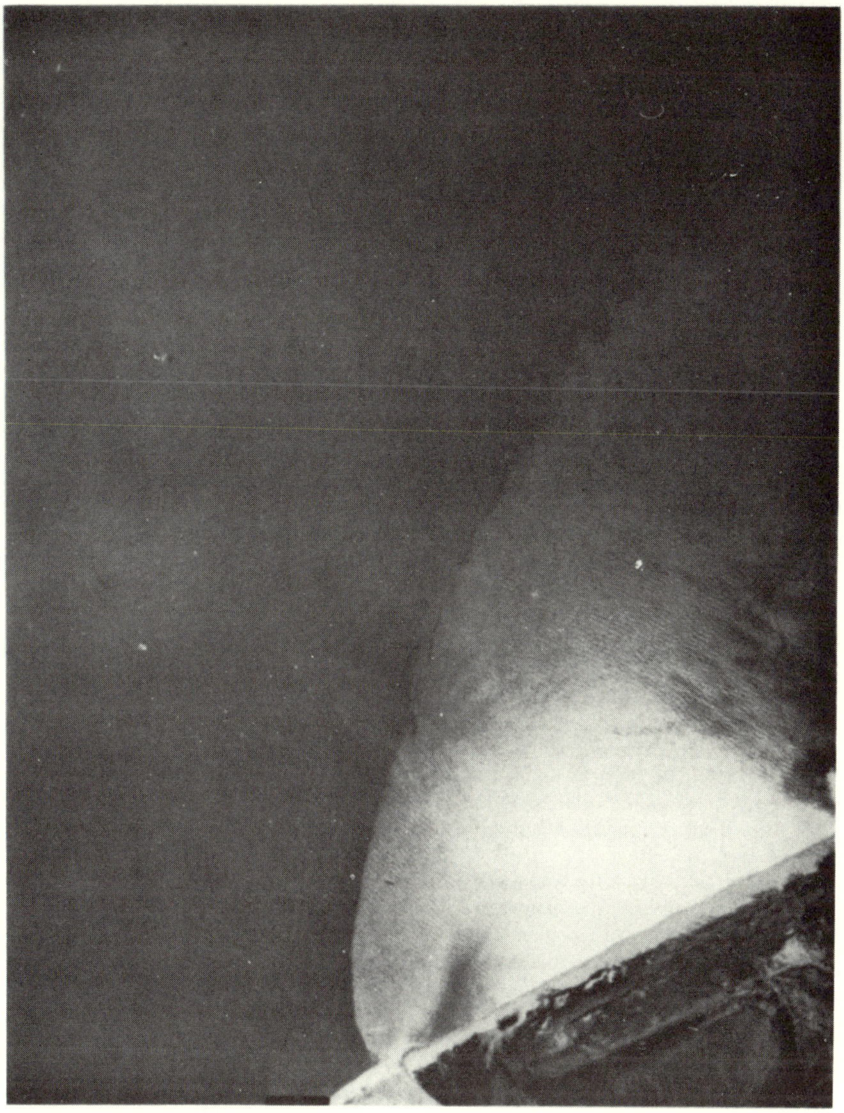

Fig. 14. Parabolic form of muddy water flow in rough weather. This form gives evidence of a Coriolis current.

be noticed, and their stability is sufficient that they may be seen by stereoscopy. Understanding of all this, and location of position of sheet boundaries, depends for the time being, on aerial observation.

The salt water current of a coastal torrent in rough weather flowing into the Mediterranean (Algerie 12–15, no. 001) (Fig. 14), supplies a remarkable example of the Coriolis effect, where we can see the boundary assumes the theoretical form of a parabola. This example demonstrates the possibility of acquisition of knowledge on coastal currents.

The examples presented in this chapter, together with those used in the preceding ones, are enough to show what resources can be obtained from the collection of photos of the National Geographic Institute for the study of the continental margin.

BIBLIOGRAPHY

Berthois. *Construction d'une carte des vagues* — Revue de Géomorphologie dynamique — N° 2 — 1954, p. 69 à 75.

F. Biesel et B. Le Méhauté. *Généralités sur les seiches et les ondes de seiches "La Houille Blanche".* Revue de l'Ingénieur hydraulicien — N° 5 — Grenoble — Oct. Nov. 1955.

Bouasse. *Houles, rides, seiches et marées* (Paris, Delagrave, 1924).

Caloi Pietro. *Sulle sesse del lago di Garda.* Archivio di oceanografia e limnologia — Anno V, fascicolo I–III — 1948, p. 3 à 24.

Clos-Arceduc. *Etude sur les vues aériennes, des alluvions littorales d'allure périodique, cordons littoraux et festons* (Société Française de Photogrammétrie, bulletin N° 4, p. 12 à 22).

Clos-Arceduc. *Etude simultanée des seiches et de l'alluvionnement* (Cahiers Océanographiques — 1963).

H. G. Gierloff-Emden. *Nehrungen und Lagunen* — Petermanns Geographische Mitteilungen — 105 N° 3 — 1961, p. 161 à 176.

Gougenheim. *Les seiches marines — Etat actuel de la question* — Comité Central d'Océanographie et d'Étude des Côtes (C.O.E.C.), N° 2 — 1955, p. 79 à 90.

A. Guilcher. *Morphologie littorale et sous-marine* (Presses Universitaires, 1954).

D. Lee Harris. *Wind tide and seiches in the great lakes* — Coastal Engineering published by Council of Waves Research, The Engineering Foundation — 1954, p. 25 à 51.

Jussy et A. Guilcher. *Les cordons littoraux entre la presqu'île de Quiberon et l'estuaire de la Vilaine* — Cahiers Océanographiques, Sept. Octobre 1962, p. 543 à 572.

H. Lacombe. *Cours de dynamique des mers — Circulation générale, houles et vagues* (Service Central Hydrographique).

Polli Silvio. *Le sesse del Laghi di Caldonazzo e di Levico* — (Estrato da: I laghi di Caldonazzo e di Levico, Bologna-1952).

Polli Silvio. *Le sesse nel porti di Livorno e Imperia.* Geofisica e meteorologica, vol III (1955) N° 4–5 — Genova, 1955.

J. Poncet. *Existence de formes festonnées prélittorales en certains points de la côte algérienne* — (Cahiers Océanographiques — N° I, Janvier 1961, p. 32–37).

A. Rivière, F. Arbey et Mlle Vernhet. *Sinuosités et festons de plage* — C.R. des séances de l'Académie des Sciences, 30 Avril 1962, p. 3230).

Romanovsky. *La seiche du port de Bonifacio* — C.O.E.C. N° 3 — 1956, p. 129.

B. De Saint-Guily. *Note au sujet de la seiche de l'anse de Saint-Roch à Antibes* — (C.O.E.C. — N° 7 — 1954, p. 289–290).

Sinclair. *Hydraulic Laboratory New South Wales*: Scale model investigations... The Dock and Harbour Authority — N° 407 — Sept. 1954, p. 146–148.

Basil W. Wilson. *Generation of long-period seiches in Table-Bay, Cape Town, by barometric oscillations* — Transactions American Geophysical Union — Vol. 35, N° 5, Oct. 1954.

Basil W. Wilson. *The Mechanism of seiches in Table Bay Harbor, Cape Town* — p. 52 à 76 à Coastal Engineering, 1954, p. 52 à 78.

Whittle. *The statistical analysis of a seiche record* — Journal of marine research volume 13, 1954 — number I (Oct. 1954 — New Haven, Conn. U.S.A.).

3

Reprinted from *Shore and Beach* **37**(1):23–27 (1969), journal of the American Shore and Beach Preservation Association

HISTORY of SHORE GROWTH *from* AERIAL PHOTOGRAPHS

By Francis H. Moffitt
Professor of Civil Engineering
University of California, Berkeley, California

Occasionally it becomes necessary to estimate the position of the shore line of a body of water at any particular time in recent past. If the subject of interest happens to be the ocean shore, there is every likelihood that the area has been photographed from the air repeatedly by various governmental agencies and certain industries. Beach erosion control, harbor improvement, topographic mapping, hydrographic charting, oil and mineral exploration, real estate development, and boundary surveys are among many endeavors which contribute to an interest in the ocean shore, and which require aerial photographic coverage from time to time.

If a time-history of the position of the shore is to be established, the first task is to research the sources of aerial photographic coverage, inquiring of every conceivable agency which is likely to be interested in the shore line. The results of the investigation will be a heterogeneous mixture of dated aerial photographs of various scales, taken from different flying heights, and with aerial cameras of different focal lengths. High oblique, low oblique and vertical photography will likely be included.

The next step requires that the general area of the shore to be documented be defined on each photograph of the collection. A very careful picture point selection must be made and correlated among all the photographs. This is extremely exacting work, because the appearance of natural and man-made cultural features appearing on aerial photographs change with time. Also, different scales and oblique-vs-vertical views add to the difficulty of picture point correlation. The ground elevations of the picture points selected for correlation should be all nearly the same. Stereoscopic examination of the photographs, and/or study of a good topographic map of the vicinity can be employed to good advantage. A minimum of four picture points which can be identified between two photographs are required for proper correlation of the two photos. However because of measuring errors, random relief displacements, film and paper shrinkage, and a host of other perturbations, correlation is enhanced by the selection of anywhere between six and twelve points.

Two systems of correlation are employed to fix the position of a shore line with respect to cultural features. The most direct system is that which relates the shore line directly to some horizontal datum such as a state plane coordinate system. Let it be supposed that in Fig. 1, points A, B, C, and D have known state plane coordinates, and that the differences in elevation between these points are fairly moderate. These known points appear on Photograph I which also contains that portion of the shore line of interest. The problem is to relate points along the shore to the state coordinate system.

The photographic coordinates of the four points are measured as x_A, y_A, . . . x_D, y_D, as well as those of

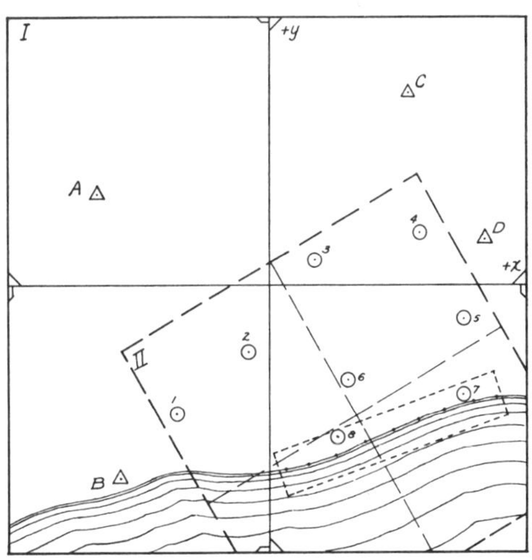

Fig. 1. Systems of points used to: (a) relate points on Photograph I to the State Coordinate System; (b) relate Photograph II to Photograph I; (c) define points on shore line.

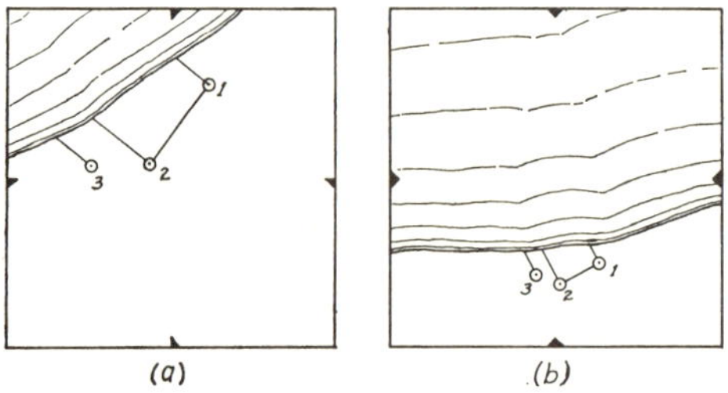

Fig. 2. Correlated points used to scale individual photographs and to define reference points from which distances to the shore line are measured.

the points on the shore, x_i, y_i. By the projectivity equations for a plane:[1]

$$X = \frac{a_1x + b_1y + d_1}{a_4x + b_4y + 1}$$

$$Y = \frac{a_2x + b_2y + d_2}{a_4x + b_4y + 1}$$

four pairs of equations can be written, one pair per point. In these equations, X and Y are the state plane coordinates, while x and y are the photographic coordinates. The eight unknown coefficients a_1, a_2, . . . d_2 can be solved by the four pairs of equations.

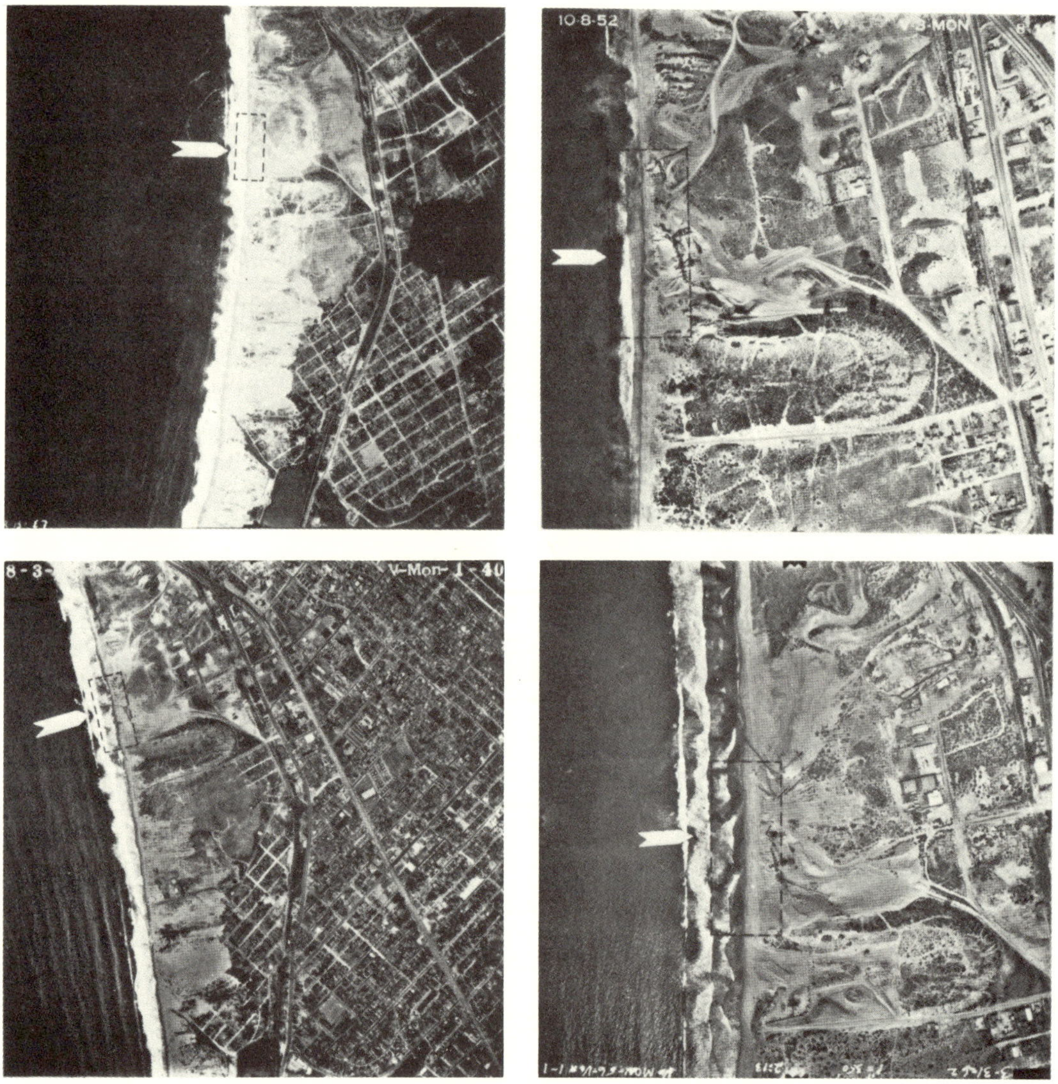

Fig. 3. Photographs used to measure change in position of shore line. Upper left photo taken March 5, 1946; upper right photo taken October 8, 1952; lower left photo taken August 3, 1955; lower right photo taken March 31, 1962.

The State Plane coordinates of the shore line points X_i, Y_i can now be computed by substituting the photographic coordinates x_i, y_i into the projectivity equations in which the coefficients are now known.

If the photograph is not badly tilted, a minimum of two known points such as A and B of Fig. 1 can be used in the transformation equations:[2]

$$X = a_1 + a_3x - a_4y$$
$$Y = a_2 + a_3y + a_4x$$

to solve for the four unknown coefficients $a_1 \ldots a_4$ by means of a pair of equations for each of the two points. It is advisable, however, to attempt to correct the measured photographic coordinates for relief displacement.[3] When the coefficients have been determined, the X and Y coordinates of the shore line points can be computed.

Photograph II of Fig. 1 is presumed to be a photograph taken at a different time and at a scale different (in this case larger) from Photograph I. The problem is to relate II to I. A group of picture points are selected which can be correlated between the two photographs. These are shown as points 1 through 8. The photographic coordinates of these points are measured on both photographs. The State Plane coordinates X_1, Y_1, . . . X_8, Y_8 are determined by either the projectivity equations or the transformation equations from the measurements on Photo I. Photo II can then be oriented to the State Coordinate System by means of these eight points in the same manner that Photo I was oriented using points A through D.

Finally, with the coefficients for Photo II known, the photographic coordinates of shore points measured on Photo II are used to solve for the State Plane coordinates. Thus the shore line as it appears at the time Photo I was taken can be compared to its position at the time Photo II was taken.

The above process is repeated for each selected photograph in the time series. Unless the correlation points between photos are all fairly equal in elevation, some attempt must be made to correct the photographic positions for the effect of relief displacement. If this is done, then low and high obliques can be accommodated by the projectivity equations just as well as vertical photographs.

If the absolute position of the shore line is not required, a system employing direct scaling from correlated features to the shore line can be used. One or more ground distances must be known in order to scale each of the photographs to make possible the calculation of distances to the shore line. In Fig. 2(a), three picture points 1, 2, and 3 appearing on Photo III are used for correlation with Photo IV of Fig. 2(b) and possibly many other vertical photographs. Assuming the ground length of line 1-2 is known, the series of photographs on which this line appears can be individually scaled. It is more desirable to have more than one ground length by which to scale, with at least one such line generally paralleling the direction of the lines measured to the shore line. Also, the shore line distances will be more reliable if the scale lines lie close to the vicinity of the shore. This criterion is very difficult to satisfy on sandy beaches because of cultural and topographic changes which occur with time. The next phase is the correction of the position of the shore line to some standard such as mean tide or mean sea level. This phase requires the three parameters of time of photography, stage of the tide, and slope of the shore. The researcher will find that some photography contains the time of day automatically recorded or else printed on the negative (and consequently appearing on the photograph). If this information is missing, the local apparent time can be approximated by examination of the direction of shadows with respect to true north. This can be converted to standard time for correlation with the tidal stage.

TABLE 1.

(1) Photo Date	(2) Time	(3) Scale	(4) Meas. Dist.	(5) Tide Stage (Sea Level Datum)	(6) Dist. Reduced to S. L.
			(ft)	(ft)	(ft)
1929	?*	1″ = 417′	181	?	?
11-22-41	?*	1″ = 408′	200	?	?
11- 7-44	16:15	1″ = 970′	166	+0.9	175
12-30-44	14:38	1″ = 944′	171	−1.6	155
1-22-45	11:25	1″ = 960′	208	−2.3	185
2-24-45	10:15	1″ = 928′	148	+1.5	163
7-19-45	12:12	1″ = 983′	176	−0.3	173
1-31-46	12:00**	1″ = 973′	184	−0.3	181
2-28-46	12:15**	1″ = 941′	254	−1.9	235
3- 5-46	12:00**	1″ = 954′	107	+1.9	126
5-24-61	11:00**	1″ = 1072	141	−1.7	124
3-31-62	14:13	1″ = 106′	136	−3.1	105
10- 4-67	12:00	1″ = 362′	78	+1.9	98

* Shadows not distinct enough to determine time of day.
** Estimated from shadows.

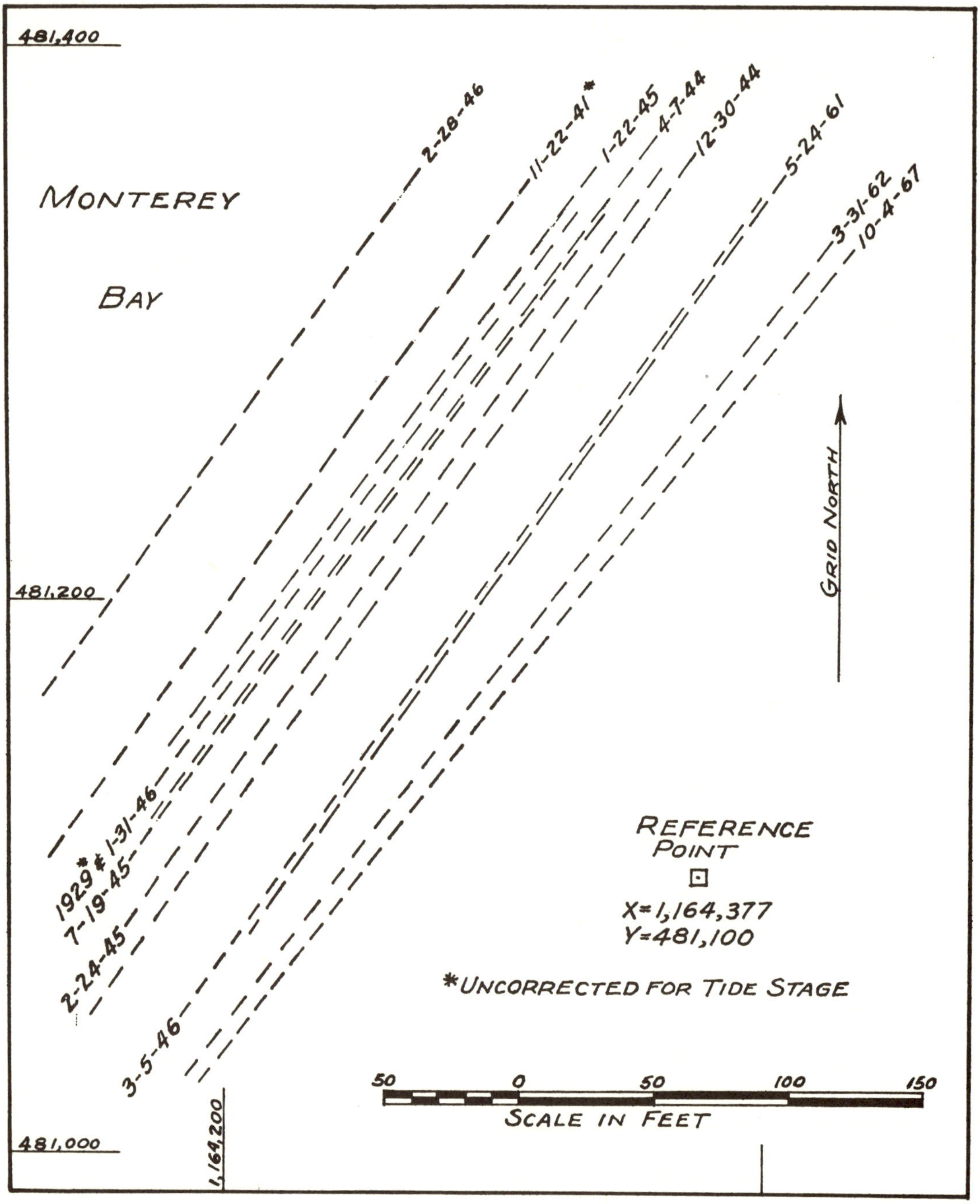

Fig. 4. Position of shore line of a portion of Monterey Bay, California, on twelve different dates as determined from aerial photographs.

An uncertainty is introduced in assuming that the beach profile in the subject area measured at any one point in time can be extrapolated to any other date to allow for shifting the plan position of the shore line to a standard datum. This, of course, is more serious for large tidal ranges. Probably the best that can be done is to find as many samples of the beach profile as possible over a period of time, and rely on the judgment of the researcher and his knowledge of local conditions regarding profile changes.

Fig. 3 shows four aerial photographs taken at different dates and from different flying heights. The area of interest is blocked off on each photograph. Note the large difference in scale exhibited by these four photographs. Two of these photos are part of a series of thirteen used to study the history of the shore line at Monterey Bay on the West Coast of California. The pertinent data are given in the table on page 25.

A 10:1 slope, determined from profiles measured in 1945, 1966 and 1967, was used to compute the corrections needed to reduce the position of the shore line to Sea Level Datum. The distances referred to in Columns (4) and (6) are obtained by measurement from a distinct point on shore to the water's edge. This point did not appear after the 3-5-46 photography and had to be reconstructed on the last three photographs by means of the projectivity equations. This reconstruction was expressed as a set of coordinates on the California Coordinate System, Zone 4.

The plan position of the shore line with respect to the point on shore for different dates is shown in Fig. 4. The coordinates of the reference point are given. This is an admittedly sketchy and discontinuous chronology, with the best portion lying between November of 1944 and March of 1946. However, even with the lack of completeness, a few interesting observations can be made. First of all, there is very little long-time change in this beach. The position of the beach in 1946 was essentially what it was in 1929. Secondly, the damage undergone by the beach in a violent storm which occurred in the first few days of March of 1946 is obvious. Compare the position of the beach on January 31, then on February 28, and finally on March 5. Between the latter two dates, the shore line receded somewhat over 100 ft.

Finally, if the position of the beach after March 5, 1946 is assumed to have recovered following the storm, it can then be noted that the beach has been gradually eroded in recent years. This observation is not conclusive unless the state of the sea is known to have been normal at about the times of the last three dates.

REFERENCES

1. HALLERT, B., *Photogrammetry* (New York: McGraw-Hill Book Co., 1960), Sec. 1-2.
2. MOFFITT, F. H., *Photogrammetry* 2nd Ed. (Scranton, Pa.: International Textbook Co., 1967), Sec. 12-15.
3. *Ibid.*, Sec. 3-7.

4

Reprinted from *Photogramm. Eng.* **14**(4):454–462 (1948)

UNDERWATER DEPTH DETERMINATION BY AERIAL PHOTOGRAPHY*

A. C. Lundahl, U. S. Navy Photographic Interpretation Center

INTRODUCTION

SOLUTIONS to the problem of underwater depth determination by aerial photography have been proposed by numerous military and civilian researchers during World War II, and many of their suggestions have been tested under operating conditions. The present paper describes some of the most interesting developments and concludes that they are vital in war where cost is unimportant but are of questionable value in peacetime when indifferent degrees of accuracy must be balanced against expensive photography. Where coastline aerial photography has been taken for other purposes, it is well worth while to attempt to extract hydrographic data by the techniques described.

Aerial photographic depth determination methods may be divided into two general groups. The first is based on evaluations of water depth from measurements of horizontal parallax of sea surface and bottom detail photo images, or from measurements of photo emulsion densities obtained by simultaneous exposures with various filter combinations.

In either case, the sea water must be relatively transparent to the light wavelengths for which the film used is most selective or sensitive. Dietz (1) states that oceanic water is most transparent to blue light with a wave length of 0.48μ, whereas coastal water is most transparent to green light of 0.53μ or higher. The second group of methods is based on evaluation of water depths and bottom gradients from timed photographic records of sea surface wave behavior. An unclassified though very useful method described by Williams (2) is based on aerial shoreline photography taken at carefully recorded times. From the time record, and appropriate tables the tide height is computed. By this method, the waterline shown on each photograph is transferred to a mosaic or master map and Section profiles may be constructed.

The importance of determining the gradient of a beach on which forces are to be landed is obvious. As Williams (2) questions—How far to the seaward will craft ground? To what extent will it be necessary to waterproof the vehicles to be unloaded? What kind of craft will be used? Will special equipment such as pontoons be needed? Will men be in danger of drowning in deeper water inshore

* Published with permission of the Commanding Officer, U. S. Navy Photographic Center. Subject material is based on unofficial investigations of the author and represents no recommendations or conclusions of the Navy Department.

of where the craft ground? Scores of these and related problems arise, varying slightly between tidal and non-tidal water. Based on the minimum draft requirements for amphibious craft, and with some exceptions, the 2-fathom line was generally considered the lower limit for beach gradient determinations. Few maps or charts available before World War II set out in detail all the beach gradient information required. Outside the limitations of scale and the changes which beaches undergo with the passage of time, both maps and charts showed a tendency to regard beaches as a kind of "no man's land" (2—p. 76) with which neither is primarily concerned. This statement surely needs qualification. Where large scale hydrographic surveys of beach areas had been made, the results were excellent; but such surveys were rare and the areas covered were small. Usually the beach was regarded as the border to the sea bed being sounded. Hydrographic work becomes difficult in shoal water, since boats may ground or swamp and positions are difficult to resect because of the poor geometry of survey triangles. Further, the common user of charts is more concerned with keeping off the shore than grounding on it, and it becomes clear that charts cannot be expected to always give detailed information about beaches. Thus it can be seen how important the problem of determining beach gradients appeared to a few officers in 1941 who were beginning to consider invasion problems.

DESCRIPTION OF METHODS DEVELOPED

PARALLAX METHODS

Initial attempts to use the conventional methods of height determination from parallax measurements of stereo pairs of aerial photos were largely unsuccessful and had to be abandoned. High reconnaissance flying altitudes would not permit parallax equation to give heights correct to 1 or 2 feet with the laboratory equipment available. Also, small scale photography and loss of beach texture would not permit accurate placement of the floating dot in parallax evaluations. It was equally difficult to derive depth results from low altitude oblique photographs despite their value in determining the nature of beach materials, obstacles, etc. Navy technicians had been experimenting (3) with the Sonne continuous strip stereo camera† on underwater depth determination problems following the earlier camera developmental efforts of the Army Air Force specialists at Wright Field. The principle of the Sonne Camera is related to that of the old-time panoramic view camera and is simple to outline. The film, travelling opposite to the direction of flight, is made to pass before a continuously open slit which acts as a shutter and in effect the image is literally "painted" on a continuous strip of $9\frac{1}{4}''$ film 200 feet in length. Modified to produce a continuous stereo strip, the Sonne camera was supplied with two lenses, instead of one, which were mounted on the camera and made to rotate slightly fore or aft of the neutral position. For more details on the Sonne Camera it is suggested that the reference to Mr. P. Kistler's work (4) be consulted. Mounted in fast fighter aircraft, the Sonne Camera could be reasonably expected to provide coverage at a scale of 1/500, or 10 times larger than with other known methods and still maintain pilot safety according to Mansfield (3—p. 126).

The operation of the Sonne camera was made completely automatic by designing an electronic system that would synchronize the optical system and film speed of advance with the ground motion of water surface or beach as the plane passes over it.

† Now developed and sold by the Chicago Aerial Survey Co.

Experimental tests were made over sounded ranges in clear water reef areas near Cat Cay, Bahama Islands. Bottom detail was found necessary to provide reference for measurement. Also, it was found essential to have ripples or slight waves on the water in order to record the surface image on the film. The Sonne Stereo Comparator is used to measure the distance between the surface of the water and the bottom and is so constructed as to use either negatives or prints. Cat Cay tests result showed an average error of six inches with a maximum error of slightly over one foot in depths measured. Operational tests were equally heartening in handling such questions as the following: "I want to land heavy artillery from an LST on Gree Beach on Love Day at H plus one hour. How far do you figure I can get her in before she goes aground?" (3—p. 124). The Sonne derived answer of 810 feet offshore was almost perfect as the subject LST actually grounded 800 feet offshore (3—p. 232). During the pre-invasion preparations for the Okinawa campaign, as Katz (12—p. 606) points out, measurements were made from stereo-strip films of the sea-wall back of one invasion beach. The Sonne estimated height of 6 feet differed from the post-invasion field mesurement by only 2 inches. The general averages for height and depth determinations by the Sonne technique during the Okinawa Campaign were slightly over one foot. While there is usually no objection to compiling results whose accuracy is greater than that actually required, there is a degree of precision beyond which refinement becomes pointless. Lateral accidental height changes of 1 foot along a beach are of frequent occurrence. Therefore, an accuracy of 1 foot in height was considered sufficient for prediction of how craft would ground.

In the Normandy invasion, it can be assumed, according to Thoren (13—p. 7), that the Germans were ignorant of the Sonne strip camera which made it possible for the Allied Air Force to photograph the beach at low tide from 50 to 100 meter altitude and thereby acquire information concerning landing craft obstructions along the entire French coast.

FILM DENSITY OR TRANSPARENCY METHOD

British research workers in 1942 noted the correlation between density of the photo image and depth of water while examining aerial photos of an area which happened to include a swimming pool. This condition might have been expected since more light will be reflected through a thin layer of water than through a thick one. While attempting to deduce the depth of sea water from aerial photographs, the British workers noted certain exceptions in the density-depth correlation. For example, when the sea was ruffled by waves or ripples, the photo image came from the water surface and not from the bottom. As might be expected, no results were obtained in such cases. Intermediate cases showed part of the photo image to be formed by surface reflection and part of the reflection to come from the sea bed. In other situations, the density of the photographic image was not a simple function of the water, as settled or suspended mud, sand, weeds, shell or foreign material exerted a positive though unevaluated influence. It was early decided that the Density Method was unreliable if not dangerous, and it was generally abandoned. More recently, other organizations have evaluated the method and some remarkable results have been obtained. However, none of the later tests appear to have been carried out under operating conditions and it is doubtful whether the method can be of great value quantitatively. Qualitatively, there is no question as to the usefulness of the method. For example, the principle was successfully used in 1943 to plot the deep water channel in the Seine below Rouen (2).

WAVE METHODS

It seemed possible that quantitative data on water depths might be derived from timed aerial photographs of lines of breakers or of wave patterns since it was known that both react to water depths and sea bed configuration among other things. A mathematical relationship between wave length, wave velocity and water depth is expressed in Stokes First Approximation as follows:

$$C^2 = \frac{g\lambda}{2\pi} \tan h \frac{2\pi d}{\lambda} \qquad [1]$$

where C = wave velocity; λ = wave length; d = depth of water; g = gravitational constant and tan h is the hyperbolic tangent. Where the depth is less than 0.1 of the wave length, a simplified expression of this equation can be used:

$$C^2 = gd \quad \text{or} \quad d = \frac{C^2}{g}\cdot \qquad [2]$$

The theoretical considerations which form the basis of the indicated computation are found in classical hydrodynamics. The mathematical expressions apply specifically to surface waves of low steepness—that is, waves whose height is less than one-half per cent of their length. The difficulty in using the wave methods results from computations that are based on somewhat uncertain theoretical relationships of wave properties such as length, height, period, and steepness as the surface waves advance from deep water across increasingly shallow sloping bottoms. For theoretical discussions related to this problem, the works of Miche (5), Lewy (6), Stoker (7) and Friedrichs (8) might be consulted. Despite the difficulties encountered with the wave methods, various procedures have been developed and applied.

1. Wave Velocity Method

British researchers obtained highly encouraging results from the application

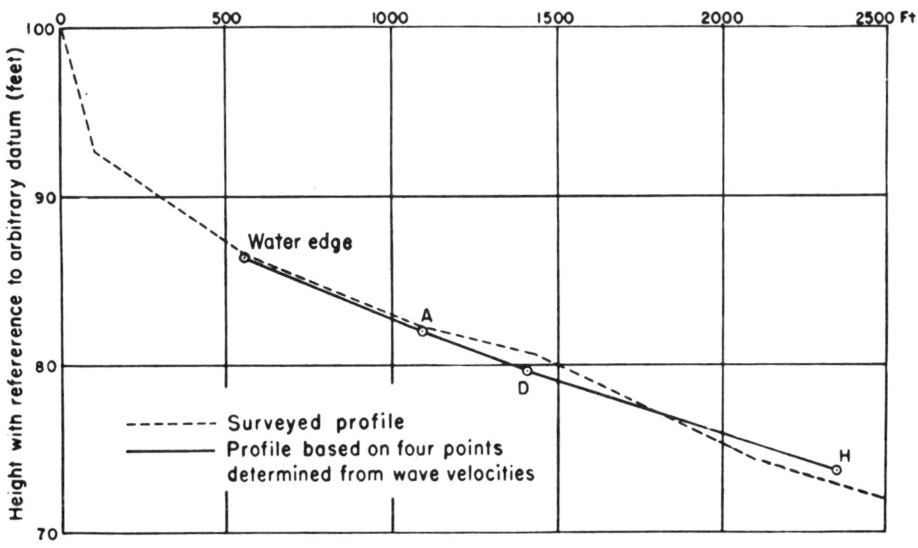

FIG. 1. Pendine Beach profile as determined from air photographs and ground surveys.

of the simplified formula [2] to some excellent aerial photographs of Pendine beach (2—p. 81). A comparison of the depth profiles obtained by ground survey and the wave velocity method is shown in Fig. 1. Unfortunately, these results turned out to be misleading as such accurate results were only rarely obtained. The assumption that depths are less than 10 per cent of the wave length is not always justifiable, and formula [1] must then be applied without simplification. In order to make this repeated operation as simple as possible, it has been found convenient to draw curves of water depth against wave velocity for several selected wave lengths and to interpolate where necessary. The curves have been plotted in Fig. 2. An examination of the curves shows that long waves travel

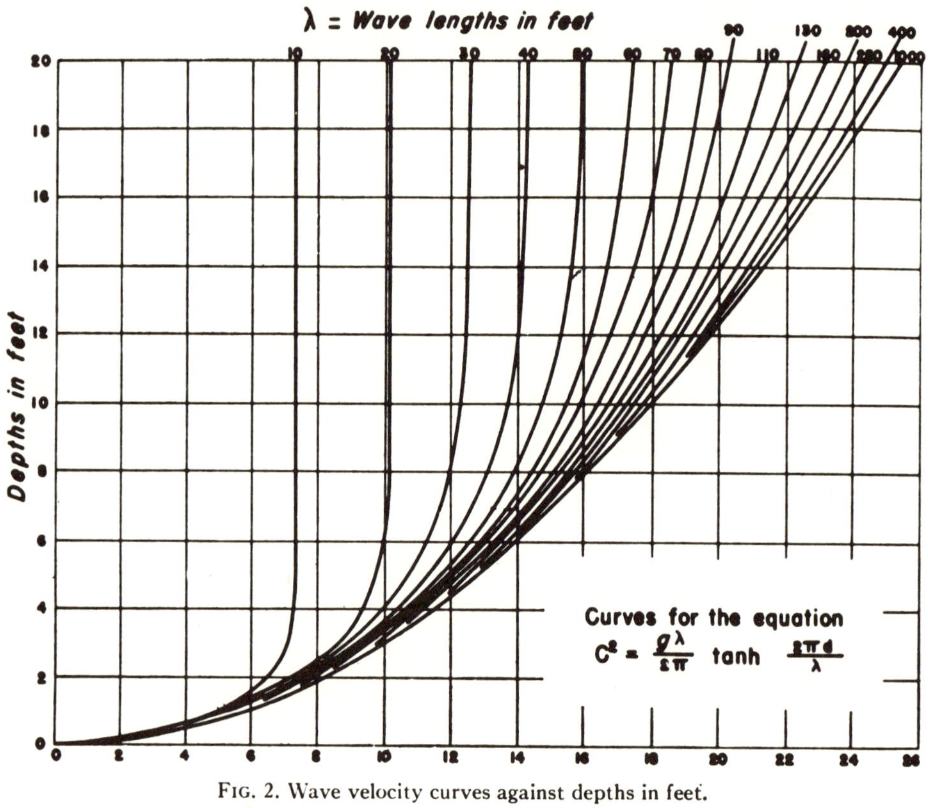

FIG. 2. Wave velocity curves against depths in feet.

faster than short ones. Second, a given wave travels in deep water at a speed dependent upon its length, but as it enters water of less depth than its length, it is retarded and the velocity decreases with decrease in depth. It is apparent then that for a given wave system, the wave velocity in shoal water is a function of both wave length and the water depth. Inasmuch as methods for determining underwater depths from aerial photographs were scheduled to play a prominent role in World War II Olympic Operation according to Sciwell (9), the Army 6211 Beach Intelligence Unit was reassembled at Oahu after cessation of hostilities to investigate military oceanographic methods. The purpose was to evaluate characteristics, requirements and limitations of the wave pattern methods, among others and to formulate lines of future research. The organization of the

6211 Beach Intelligence Unit for research is outlined by Seiwell (9—p. 209).

Other examples of how waves are retarded by a shoaling sea are shown by Williams (2—p. 83). The procedure for determining water depths (d) from aerial photographs by means of the wave velocity formulas is relatively simple if the aerial camera is equipped with a clock bearing a second hand. In order to compute *d*, it is necessary to measure wave velocity (c) and wave length (λ). Wave lengths are simply measured off the photographs once the scale is known. Velocity is determined by dividing the magnitude of advance of any wave crest between two consecutive photographs by the time interval elapsed. Corrections for tilt must be taken into account as in normal aerial survey practice. The depth determined from Fig. 2 is the still-water depth at the mean point of the wave length. Depths so computed are below the tide level at the time of photography. The tide level must be determined in order that all depths can be converted to chart datum.

Formula [1] neglects the effect of wave height. It is known that, for any given wave length, a steep wave travels faster than a flat one. Stokes' Third Approximation takes this into account:

$$C^2=\frac{g\lambda}{2\pi}\tan h\,\frac{2\pi d}{\lambda}\left\{\frac{1+e^{(8\pi d)/\lambda}+e^{(-8\pi d)/\lambda}+2(e^{(4\pi d)/\lambda}-e^{(-4\pi d)/\lambda})+12}{(e^{(2\pi d)/\lambda}-e^{(-2\pi d)/\lambda})_4}\;\frac{\pi^2h^2}{\lambda^2}\right\} \qquad [3]$$

where λ, *C*, and *d* are defined as before and h = wave height. Inasmuch as this formula is more cumbersome that [1] and it is often difficult or impossible to measure wave heights from operational photography, among other reasons, the formula was not widely used. In December 1944, the Beach Erosion Board, Washington, D. C. made its laboratory available for investigating the effect of wave height in connection with this particular application. It was concluded among other things that neglect of wave height by using equation [1] can result in computed depths being exaggerated by 10 per cent for the steepest waves. As a rule of thumb then, it has been the practice when using the curves in Fig. 2 to reduce the depths obtained by 10 per cent if the photographs indicate very steep wave, or by 5 per cent for flatter waves.

To yield to the temptation of computing water depths and then adding a large safety factor by reporting gradients or clearances less favorable than the actual determinations is to work contrary to an unwritten rule of assault landing, namely to beach as many of the largest craft as soon after "H" hour as possible.

2. The Wave Period Method

In view of the fact that not all aerial cameras are equipped with internal clocks, the wave period procedure was developed to make use of wave photography taken during reconnaissance flights. A fundamental equation of wave motion is

$$\lambda = CT \qquad [4]$$

where λ and *C* are defined as before and *T* is the wave period. Rewriting [4] as $C=\lambda/T$, and substituting in [1] we obtain:

$$T^2 = \frac{2\pi\lambda}{g}\cot h\,\frac{2\pi d}{\lambda}\cdot \qquad [5]$$

Curves for this equation are plotted in Fig. 3. Some objections to this method have been raised by observers who maintain that periods of ocean waves are

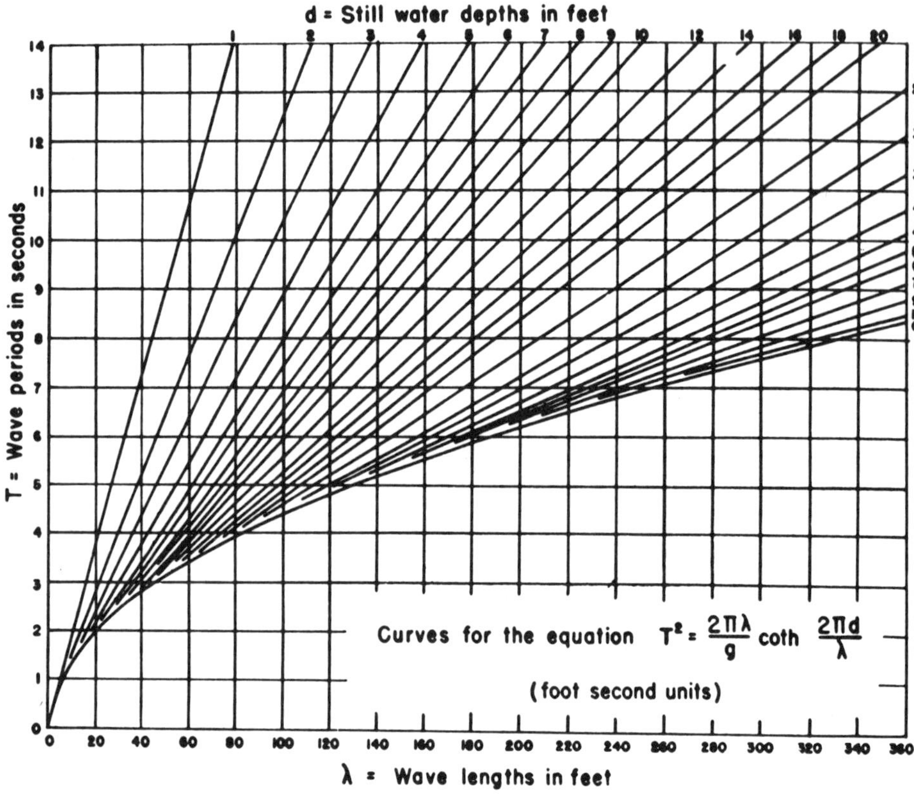

FIG. 3

not constant for a given time and locality. These criticisms appear to be based on shore observations of wave breaks or crests passing a fixed mark. It is true that such observations can be quite erratic, but inshore wave crests rarely result from one wave system alone but usually from a combination of several systems, as Williams points out (2—p.86).

After good photographs have been selected, the method of applying the wave period is as follows. First, the aerial photographs are scaled and the wave lengths are measured, including waves as far to the seaward as possible. The latter are termed deep water waves. The length of ordinary waves on a photograph whose scale is 1/10,000 is so small that gross errors might occur if single wave measurements were used for evaluation of λ. It is usually best to group the waves into threes or fours and to use the mean length of each group. The offshore distances of all waves measured are essential in order to plot the depths on charts or section paper. Using the best hydrographic chart available, the depths of the sea at the positions of the deep water waves are read off. These waves must be carefully selected for sharp definition of crests and great water depth. Finally, the curves in Fig. 3 are entered with the measured wave length and corresponding depth and the wave period is determined. Once having established T, the curves may be reentered with the same period and successively shorter wave lengths approaching the shoreline and the various depths may be read off quickly. Determination of the correct value for T, is all important and several computations should be made to check the value used. Accuracy further depends on definition of the wave crests and the precision with which wave lengths are measured. Other representative profiles illustrating the reliability of the method under operational conditions are shown by Williams (2—p.89).

In applying the wave methods, the greatest difficulties are traced to the absence of a camera clock, determination of the scale of wave photography being analyzed, dual systems of waves giving complex patterns, and lack of wave definition.

3. Wave Refraction Method

The last wave method to be noted is based upon the continually decreasing angle of incidence of deep water waves with a long straight beach as they proceed inshore. This refraction effect is shown diagramatically in Fig. 4 where the angles θ_1, θ_2, θ_3 are obviously functions of the water depth at points P_1, P_2,

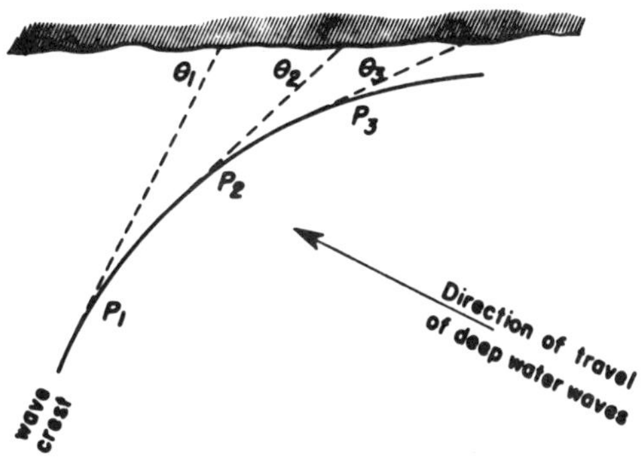

FIG. 4. Wave refraction method.

P_3. Unfortunately photographic records of this type of phenomenon are not too abundant. It would appear that excellent results might be obtained where a large angle of wave incidence exists initially and varies rapidly as the shore is approached. A recommended procedure consists of determining the depth, wave length and angle of incidence of the waves at a deep water point and then reducing the depth at points inshore as functions of θ_2, θ_3, etc.
The basic formula is:

$$\frac{\sin \alpha}{\sin \alpha_0} = \frac{C}{C_0} \qquad [6]$$

where $\alpha_0 =$ the wave incidence angle offshore, $\alpha =$ the wave incidence angle at any given point inshore, and C_0 and C are the corresponding wave velocities. For tables relating the offshore water depths and initial wave incidence angles, the work of Bigelow and Edmondson (10—p.155) should be examined.

The Waterline Method

Reliable beach gradients may be determined from conventional aerial photographs taken at carefully recorded times over certain types of coastlines. With accurate time data and appropriate H.O. Tables, the tide height can be computed. On each photograph the waterline is traced, and thus from a series of photos taken at different times, a succession of contours can be transferred to a mosaic or master map and profiles may be constructed. This procedure was first carried out under wartime conditions during 1914–18 when the "Dover Patrol" (11) described how beaches of the French coast were photographed at scheduled intervals from an aircraft while the tidal rise and fall was observed from a submarine offshore.

During World War II, the accuracy of this method was assessed in a series of trials carried out at Porthkerry Beach near Barry, South Wales. The conditions were made to resemble those under operational conditions as closely as possible. Photographs at 1:10,000 scale were used and Admiralty Tables were used to compute tide heights. Despite the problems encountered in scaling sea photographs, in computing tide heights, in correcting for "run-up" of waves giving a false waterline or in seeing the waterline on photographs, the waterline method had an obvious application because of its simplicity. Most amphibious operations undertaken in tidal waters during World War II were planned on beach gradients derived substantially from the waterline method. The first of these operations was the Dieppe landing.

The most common sources of error lay in the inaccurate photographic time records, changes in the beach gradients as a result of wave action after profiles were drawn, or difficulty in interpreting the waterline. Despite certain specific advantages of infra-red film, it was found that panchromatic film generally gave a clearer waterline. Further, the method had certain limitations as it was useless for gradients below the level of low spring tides and had no application to tideless waters.

REFERENCES

(1) Dietz, Robert S., "Aerial Photographs in the Geological Study of Shore Features and Processes," PHOTOGRAMMETRIC ENGINEERING, Vol. XIII, No. 4, 1947, pp. 537–545.

(2) Williams, W. W., "The Determination of Gradients on Enemy-Held Beaches," *The Geographical Journal,* July 1946, pp. 76–93.

(3) Snyder, C. C. and Frame, R. R., "How Deep is the Ocean," *Popular Mechanics,* May 1946, pp. 124–127.

(4) Kistler, P. S., "Continuous Strip Aerial Photography," PHOTOGRAMMETRIC ENGINEERING, Vol. XII, No. 2, June 1946, pp. 219–223.

(5) Miche, A., "Mouvements ondulatories de la mer en profondeur constante on decroissante," *Annales des ponts et chansees,* Tome 114, pp. 25–78, 131–164, 270–292, 369, 406, 1944.

(6) Lewy, H., "Water Waves on Sloping Beaches," *Bulletin of the American Mathematical Society,* Vol. 52, No. 9, pp. 737–775, 1946.

(7) Stoker, J. J., "Surface Waves in Water of Variable Depth," *Applied Mathematics Group, New York University No. 153; Quarterly of Applied Mathematics,* Vol. 5, No. 1, pp. 1–54, April 1947.

(8) Friedrichs, K. O., "Water Waves on a Shallow Sloping Beach," *Communications on Applied Mathematics, Institute for Mathematics and Mechanics,* Vol. 1, No. 2, pp. 109–134, May 1948.

(9) Seiwell, H. R., "Military Oceanography in World War II," *The Military Engineer,* Vol. XXXIX, No. 259, pp. 202–210, May 1947.

(10) Bigelow, H. B. and Edmondson, W. T., "Wind Waves at Sea, Breakers and Surf," *H. O. Publication* No. 602, 1947.

(11) Bacon, Sir Reginald, "The Concise Story of the Dover Patrol," pp. 186–187, London, 1932.

(12) Katz, Amron H., "Aerial Photographic Equipment and Applications to Reconnaissance," *Journal of the Optical Society of America,* Vol. 38, No. 7, pp. 604–610, July 1948.

(13) Thoren, Ragnar, "Photographic Intelligence and Photo Interpretation," *The Swedish Society of Photogrammetry to the International Society for Photogrammetry, Sixth International Congress,* The Hague, 1–10 September 1948.

5

Reprinted from pp. 3–13, 16–17, 23–50, 54–55 of *Technical Report Number 6, Volume 5*, School of Civil Engineering, Cornell University, Ithaca, N.Y., 1954, 74 pp.

A METHOD FOR ESTIMATING BEACH TRAFFICABILITY FROM AERIAL PHOTOGRAPHS

D. R. Lueder

DISCUSSION OF THE TERM "TRAFFICABILITY"

The term "trafficability" is used frequently in this volume. This term refers to the ability of a beach to support the movement of military vehicles. When the term is used it should be understood as referring to "virgin trafficability," i.e., the ability of the beach to support the movement of a "first" vehicle. No considerations of "multiple-pass" trafficability are included. The effect of multiple passes on beach conditions is a function of so many variables, that it can be estimated best in terms of the landing plan for a specific operation on a specific beach.

It is important to remember that trafficability is a rather vague term which covers the combined effect of variables such as slope, bearing capacity, and tractive capacity. It is also affected significantly by outside factors such as vehicle-types, driver-abilities, speed of approach, tire-pressures, etc.

The trafficability—or "virgin" trafficability—discussed is considered to be a function of two factors:

1. The slope(s) of the beach
2. The penetration(s) of the beach

Beaches can be classified, on the basis of these variables, into a number of classes. A possible classification is illustrated by Figure 1. However, THIS DIAGRAM IS INDICATIVE ONLY and should not be used without verification or modification in the light of current operational equipment and techniques.

The effect of slopes on beach trafficability is obvious. The significance of penetrations is not immediately clear. Briefly, penetra-

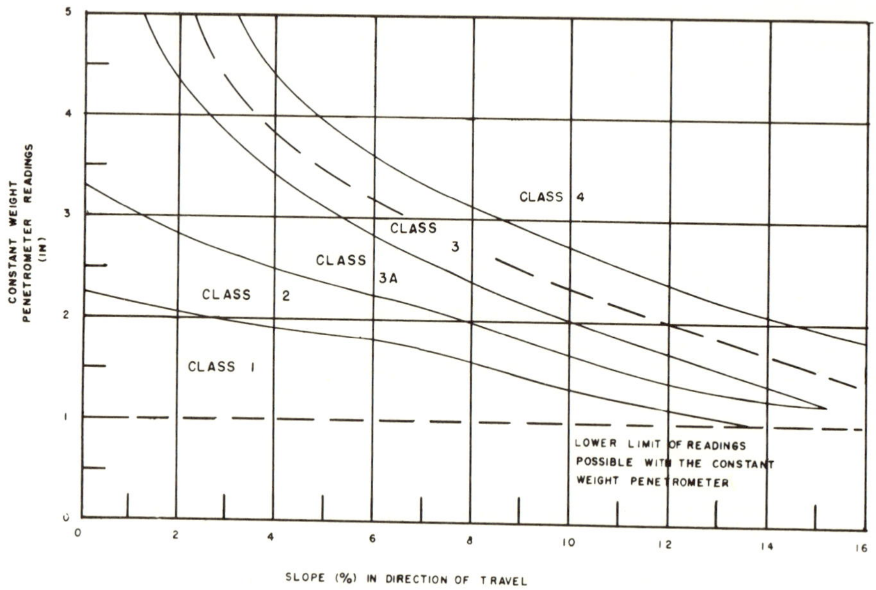

Fig. 1. Areas and classes of beach trafficability, 2½ ton truck (combat loaded), low range, 2α gear, standing start, standard (1951) tires at 40 p.s.i.

tions (constant weight type) are believed to provide a reasonable and acceptable *index* of the bearing capacity of a beach.[a] Consequently, the determination of penetration-values (together with slopes) is the ultimate objective of the method described in this volume. Once these factors have been determined, it is possible, using a chart similar to Figure 1, to classify the beach for operational purposes. Again, however, it is desirable to emphasize that the RELATION BETWEEN VARIOUS VALUES OF THE INDEX (i.e., penetration-values) AND ACTUAL OPERATING CONDITIONS MUST BE DETERMINED BY USING AGENCIES. For example, it is possible for a given beach to have a slope of 8% and a range of constant weight penetrations of 1.75 inches to 2.75 inches. It still remains for the using agency to determine whether such conditions should be described as good, fair or poor for the landing plan proposed. However, this is a relatively simple matter and should cause little difficulty.

DISCUSSION OF PROBLEM

During the process of completing prior investigations, it became apparent that beaches are exceptionally complex natural features whose major characteristic is variability. It became further

[a] See Volume IV of this report.

evident that an estimation of beach trafficability is one of the more severe tasks that can be assigned to the technique of aerial photographic analysis. This opinion is still maintained. The determination of beach trafficability by means of aerial photographic analysis is no simple job. However, it can be done if the method presented in this volume is applied carefully and intelligently.

The first four volumes of this report discussed various aspects of the problem. This volume, while based largely upon the results presented in the preceding four, is concerned with the practical technique. Consequently, it has little room for discussions of the experimental and survey results presented in the other volumes. A knowledge of certain generalities is, however, so important to the successful operation of the method, that they are listed again. They may be considered as essential introductory elements of the recommended method. These generalities follow:

1. General

a. The trafficability of a beach is reflected by numerous details of the beach and its environment. These details are related to:
   (1) Characteristics of the physical profile (width, slopes, cusps, scarps, current ripple marks).
   (2) Characteristics of the beach materials (grain-size distribution and shape).
   (3) Characteristics of the surf zone (number of breakers, spacing of breakers, frequency and height of waves, location and number of bars, turbidity).
   (4) Characteristics of the environment (offshore and onshore protection, river mouths, sources of fresh material supply, modes of littoral current movement).

b. These details, as well as trafficability itself, vary significantly with time. The variation, and therefore the details, result from the interaction of numerous causative variables such as winds, tides, waves and currents on the materials composing the beach and the materials in the immediate environment that may be deposited on the beach.

c. It is difficult to relate specific beach details to the variable or combination of variables that produce them. It is possible, however, to relate details of the beach and its environment to trafficability conditions in terms of an index of supporting capacity (penetrations).

2. Correlations

a. Definite relations have been found between penetrations and the following details of the beach:

(1) Average foreshore slope.
(2) Median grain-size of the foreshore material.
(3) Grain-size class of the foreshore material.
(4) Foreshore mean-sea-level width.
(5) Foreshore current-ripple marks.
(6) Cusps

b. Relations have been indicated, but not proven, between penetrations and the following beach details:

(1) Number of breaker lines.
(2) Number of bars.
(3) Turbidity.
(4) Relative height and frequency of waves.
(5) Grain size shape.
(6) Offshore protection.
(7) The existence of nearby river mouths.
(8) Sources of fresh rock erosion.

3. Airphoto Potentiality

a. Most of the details listed under 4 and 5 can be measured, or evaluated, either directly or indirectly, by careful analysis of properly executed aerial photography.

b. To complete the airphoto analysis efficiently and practically, it is important that the interpreter have some control over the photographic specifications and that the interpretation be accomplished in a logical manner.

4. Variable Nature of Beaches

a. One of the limitations of the method is the time lag that must inevitably result between the execution of photography and the completion of analysis. This lag incorporates an unavoidable element of uncertainty into every analysis regardless of the care with which it is made. This is due to the essentially variable nature of beaches.[b]

b. Generally, beach conditions are characterized by fluctuations with respect to time, approaching static conditions only for comparatively short periods. (However, wide,

[b] For a more comprehensive discussion, see Volume II of this report.

gently sloping beaches composed of fine sands are essentially static for long periods.)

c. The features of each beach fluctuate daily, for long terms, within the limits of a "normal" range. See Figure 2. These daily fluctuations may be significant with regard to trafficability conditions.

d. Occasionally, each beach will undergo changes such that the midpoint of the daily range raises or lowers with respect to the normal midpoint. These different ranges may persist for weeks or months, but compared to the normal range, are of short duration. See Figure 2. Changes of this nature are definitely significant with regard to trafficability conditions.

e. The normal 24-hour changes in beach features are generally quite small. Consequently, significant changes of relatively long duration *usually* occur over a period of weeks[c] being accompanied by increased instability of beach features. However, significant changes *may* occur within 24 hours or less.

f. Beaches, whose "norm" consists of low slopes, broad widths and fine-sand compositions are more stable than other beaches.

g. It is probable, but unproven, that seasonal changes in wave attack, and currents are responsible for seasonal variations in beach conditions.

5. Major Factors Affecting Beach Conditions

a. Firmness, slope and width are primarily related to material characteristics, but are also related, within the range dictated by material characteristics, to variations in wave attack.

RECOMMENDED SPECIFICATIONS

The following specifications are recommended for aerial photography that is to be used in making estimates of beach trafficability:

1. SCALE 1:2500 or larger.[d]

[c] Exceptional weather conditions not taken into account.

[d] 1:5000 is acceptable if absolutely necessary. However, the accuracy of estimate may be expected to decrease.

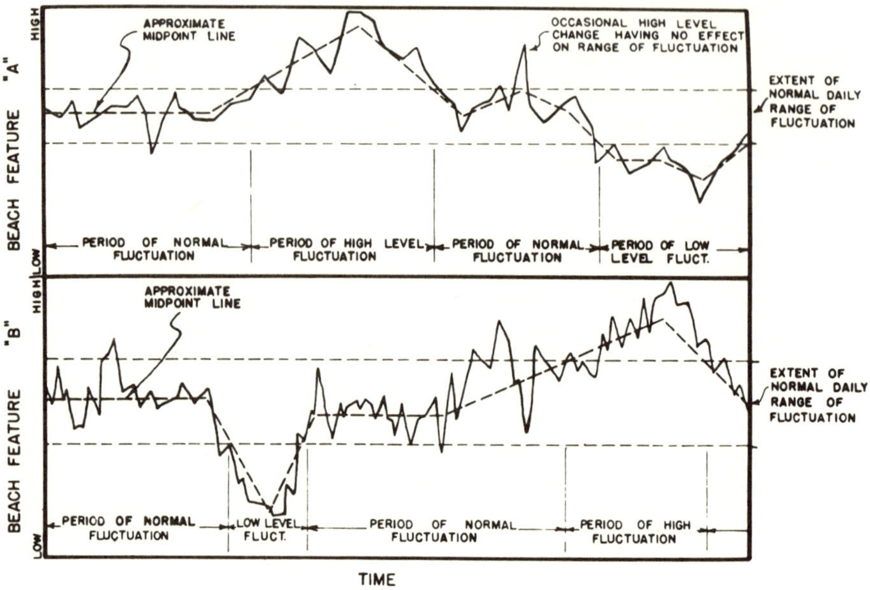

Fig. 2. Schematic representation of beach changes.

2. DIRECTION OF FLIGHT	Parallel to waterline.
3. TYPE OF PHOTOGRAPHY	Conventional vertical with standard stereoscopic overlap *or* Sonne vertical with stereoscopic coverage (only if it is believed that slopes can be estimated with sufficient accuracy.)
4. COVERAGE	Two thirds land and one third water with principal point of photograph on the beach face.
5. TIME OF PHOTOGRAPHY	During a period two hours before to two hours after low tide.
6. FILM-TYPE AND FINISH	Black and white, Super XX, Tri-X (or equivalent) glossy finish.
7. FILTERS	Standard.
8. FREQUENCY	See Section III.

9. MARGINAL DATA — Date, time and scale *must be accurately* recorded on the margin.

The most important of these specifications is the one dealing with the time of photography. If the photos are taken at or near high tide, THEY ARE VIRTUALLY USELESS INSOFAR AS THE METHOD DESCRIBED IN THIS REPORT IS CONCERNED. Every reasonable effort should be made to obtain them at or near low tide.

SYSTEMATIC ANALYSIS OF PHOTOGRAPHY

The systematic analysis of photography is accomplished in accordance with analysis methods one to ten. These methods are described as follows:

Analysis Method 1
Division of Beach into Zones

The first step in an estimation of beach trafficability is a division of the beach into zones.

The beach is an area extending landward from the mean low water line to a marked change in material or land-form, or to a permanent vegetation line. It may be divided into four sub-areas which may be designated as:

1. The wetted foreshore.
2. The drying foreshore.
3. The backshore.
4. The forward dune apron.

The wetted foreshore is arbitrarily defined as that part of the beach which is alternately but frequently covered and uncovered by shallow water swash at any given tide stage. The wetted foreshore is saturated or almost completely saturated at all times. Its boundaries are indefinite. Naturally, since it is a function of tide stage, it moves up and down the beach face.

The drying foreshore is that part of the beach lying between the landward boundary of the wetted foreshore and the average point of maximum uprush at high tide. At any given tide stage, the drying foreshore is completely exposed to the air and has moisture contents ranging from nearly saturated at the wetted foreshore boundary to nearly dry at its inland boundary. The width of the drying foreshore varies from a maximum at low tide to zero at high tide. THE DRYING FORESHORE ZONE IS THE MOST VALUABLE INDICATOR OF BEACH CONDITIONS.

The backshore extends inland from the average point of maximum uprush at high tide to a marked change in material or landform, or to the line of permanent vegetation. The backshore, by definition, is beyond the zone of wave action. Consequently, it is composed of loose sand with very low surface moisture contents. It is of little value in predicting beach conditions.

The forward dune apron is an area of transition between the backshore and the dune zone—if dunes are present. It is composed of soft, dry sand in practically every case. It is of little value in predicting beach conditions.

The various zones are depicted schematically on Figure 3. It is emphasized that the relative width and shape of all zones may change within a very short time, if the surf characteristics change. This variation is also indicated on Figure 3.

The delineation of the various zones is not always easy on an aerial photograph. However, in general, they will appear as follows:

1. Wetted foreshore—an even-toned dark area close to the surf zone. May often be partially or completely obscured by uprushing swash water. At low tide the zone is narrow in comparison to the remainder of the foreshore.
2. Backshore—a light toned area, flat, or sloping, extending from a point near the dunes, seaward to a definite zone or line where the tone becomes visibly darker. This area *may be* coincident with a line of cusps, a scarp or a slope change (berm). It is not always coincident, however. The zone of darkening is often coincident with a zone of debris and seaweed accumulation.
3. Drying foreshore—that area of the beach between wetted foreshore and backshore. It has tones which are intermediate between those of the backshore and those of the wetted foreshore. The tones may be uniform or may vary due to slope changes, low spots, debris accumulations or increasing surface moisture contents.

Further information concerning the delineation of the various beach zones will be found under "Method 5—Estimation of Material Size." Illustrations of tone patterns typical of beach zones appear in this section.

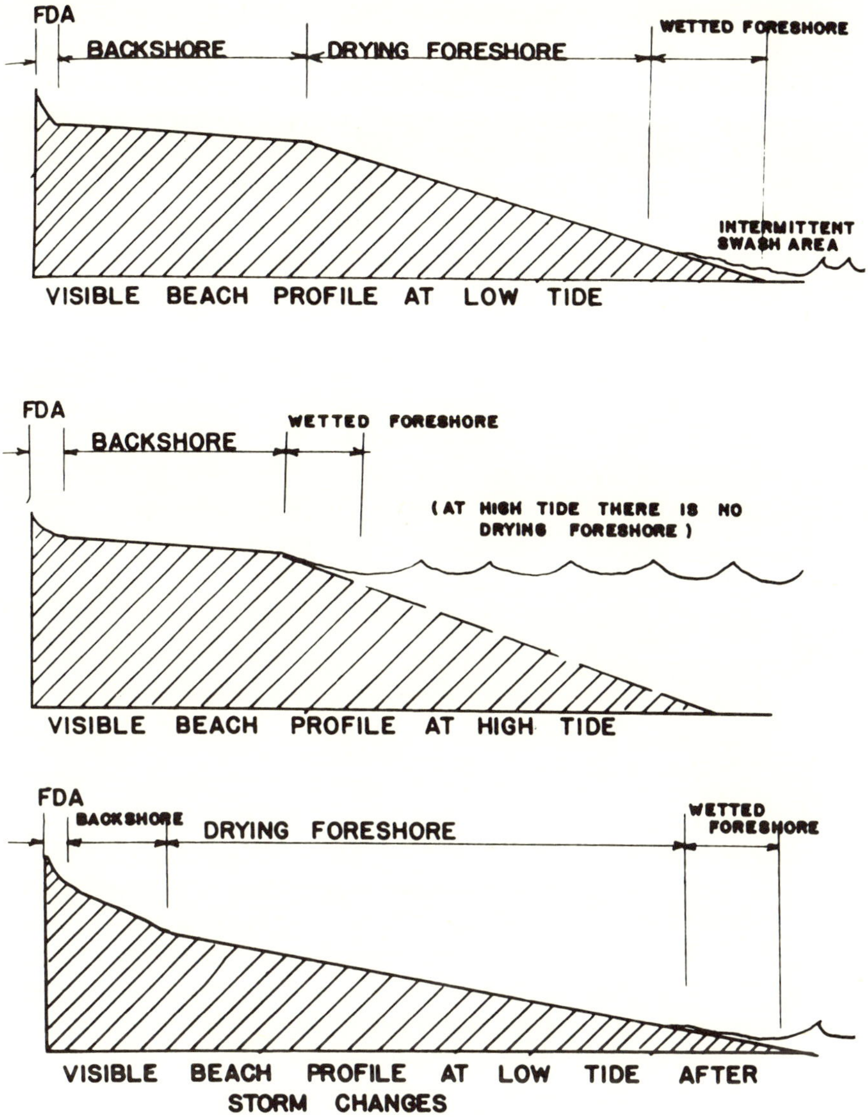

Fig. 3. Illustrations of beach zones.

Analysis Method 2
Determination of Average Foreshore Slope

The determination of the average *foreshore* slope (AFS) is one of the most important steps in completing an estimate of beach conditions. Unfortunately, it is also one of the hardest. The difficulty is not related to the concept itself, but rather to its physical determination from aerial photographs.

It will be shown in later sections that there are several ways to make an estimate of beach conditions. If the method to be used is that of substitution into the proper statistical formula (METHOD C) it is essential that the determination of average foreshore slope be accurate to ±1%. If the method involves several separate judgements based upon the indications of several features, one of which is slope, (METHOD B) it is *desirable* that the slope be determined to ±1% but it is *allowable* to make a rougher estimate.

A few recommendations, mainly of a sepcification nature are re-emphasized.

1. Photography should be taken at a scale of 1:2500 or larger. (Sonne photography is acceptable, if satisfactory slope determinations can be obtained from it.)
2. If vertical photography is specified, the principal points of the photographs must appear on the beach face.

If a rough estimate of slope is allowable, it is acceptable to place the average foreshore slope into one of five categories:

1. Less than 3%.
2. Less than 5%.
3. Between 5% and 10%.
4. Between 10% and 15%.
5. Greater than 15%.

If the average foreshore slope cannot be definitely placed in one of these categories, it is best to neglect the slope indication in formulating an estimate of beach conditions.

The determination of slope is further complicated—in many instances—because it is an *average* slope. The foreshores of many beaches often display more than one slope. The average foreshore slope represents their rough graphical average. The graphical average is best obtained by plotting the slopes and drawing an average line through them by eye. The method is illustrated by Figure 4.

Rough foreshore slopes can often be estimated in terms of other visible features, such as breaker patterns, tone patterns, width etc. Additional information concerning such estimations is included in the last discussion of this section entitled "Estimate of Trafficability Conditions."

Analysis Method 3
Determination of the Foreshore Mean-Sea-Level Width

The foreshore mean-sea-level (Fs MSLW) width is determined as follows:

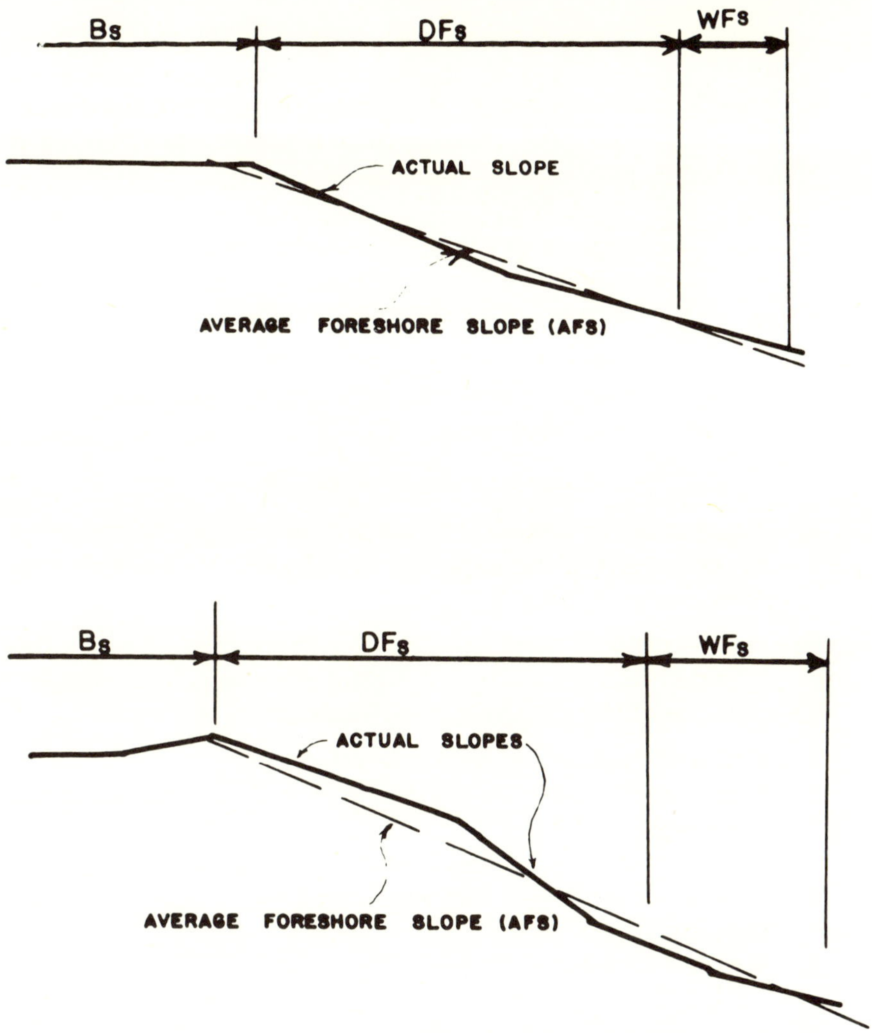

Fig. 4. Method of graphically estimating the average foreshore slope (AFS).

1. Measurement of the horizontal extent of the exposed foreshore (including the visible portion of the wetted foreshore) and its conversion to feet.
2. Determination of time of photography from the marginal data on the aerial photograph.
3. Determination, from available tidal data, of the vertical distance in feet between the tide stage at the time of photography and mean-sea-level. If the tide stage is below mean-sea-level, the vertical distance is negative. If the tide

stage is above mean-sea-level, the vertical distance is positive.

4. Substitution of average foreshore slope, determined according to Method 2, and vertical distance, determined as in Step 3, in the following equation:

   $x = \frac{100V}{S}$ where:

   x = a horizontal distance (feet)

   V = the vertical difference between tide stage at time of photography and mean-sea-level (feet)

   S = average foreshore slope (%)

5. Application of distance x, according to sign, to the horizontal extent of the exposed foreshore determined in accordance with Step 1.

The result of these computations may be accepted as a sufficiently accurate measure of the foreshore mean-sea-level width.

An example of the computations follows:

1. Width of exposed foreshore= 200 feet
2. Tide stage at time of photography= 2 feet above mean low water
3. Mean-sea-level= 3.5 feet above mean low water
4. Difference between tide stage and mean-sea-level=−1.5 feet
5. Average foreshore slope = 3%
6. $x = \frac{(100)(-1.5)}{3} = -50$ feet
7. MSLW = 200 − 50 = 150 feet

Analysis Method 4
Estimation of Material Size

An estimation of material size is best made in terms of average foreshore slope (AFS), foreshore mean-sea-level width (Fs MSLW) and foreshore gray tone pattern.

If accurate values of AFS are available, the correlation shown in Figure 5[e] may be used to estimate the *median grain size* of the

[e] For further discussion, see Volume I of this report.

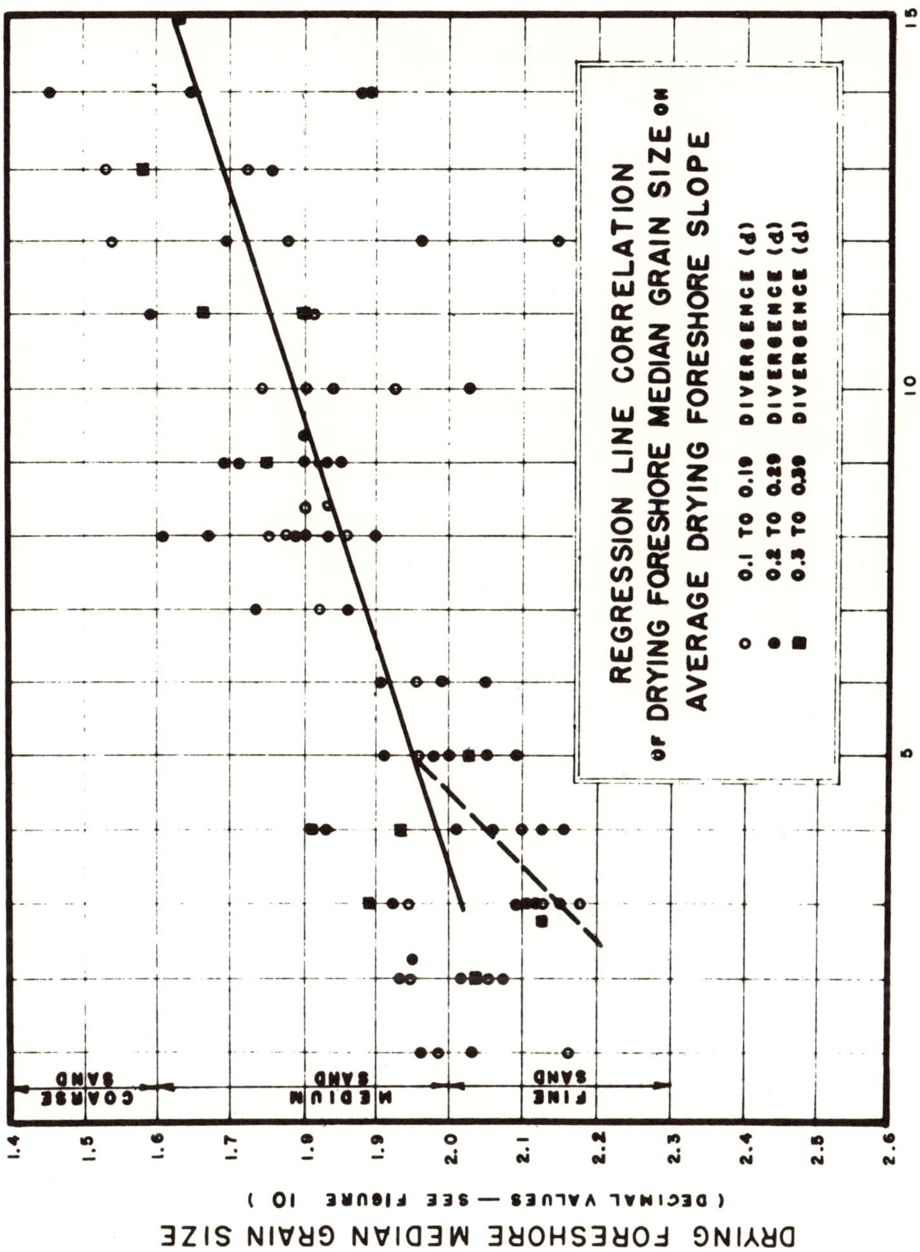

Fig. 5. Average drying foreshore slope (%).

foreshore. It is apparent that, from an inspection of the point spread on either side of the lines, that the estimate is quite rough. In the slope range below 5%, the dashed rather than the solid line should be used.

If a value of the Fs MSLW is available, the correlation shown in Figure 6[e] may be used to provide another estimate of median grain size. The point spread also indicates that the estimate will be quite rough. For widths less than 60 feet, the dotted rather than the dashed line should be used.

Gray tone patterns provide an excellent means for distinguishing fine sand beaches from coarse and coarse-medium sand beaches. The analysis of gray tones should be accomplished in accordance with the following statements:

1. Gray-tones of the dry backshore (0–10% saturation) and the wetted foreshore (80–100% saturation) may be assumed essentially constant for all beaches. They may be considered as upper and lower limits of the range of tones appearing on the drying foreshore.
2. By eliminating the gray tones of the backshore and wetted foreshore, the gray tones appearing on the drying foreshore are the final indicators of predominant grain-sizes composing the beach.
3. The drying foreshores of beaches composed predominantly of fine materials ($D_{50} < 0.2$ mm) have gray-tone patterns characterized by:
    a. A band of rapid darkening adjacent to the backshore boundary.
    b. A relatively broad expanse composed of narrow bands whose tones becomes gradually darker as the waterline is approached (the gradual darkening is not always discernible by eye.)
    c. An abrupt darkening or leveling of tone near the waterline.

This pattern is typically shown by Figure 7a.

4. The drying foreshores of sand beaches composed predominantly of coarse materials ($D_{50} < 0.5$ mm) have gray-tone patterns characterized by:
    a. A band of rapid darkening adjacent to the backshore boundary.
    b. A relatively broad band, of fairly uniform tone, only slightly darker than the backshore.

[e] For further discussion, see Volume I of this report.

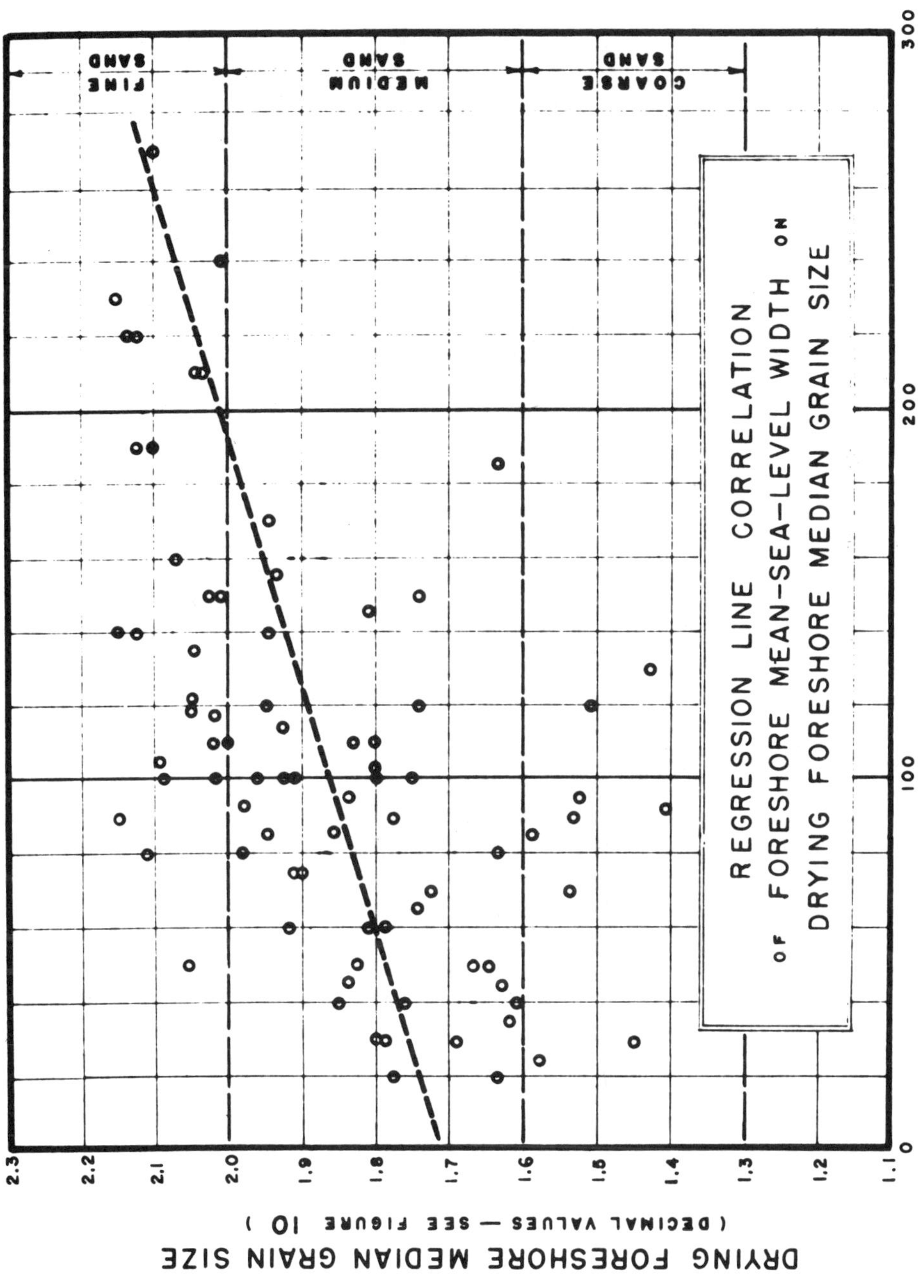

Fig. 6. Foreshore mean-sea-level width (ft.).

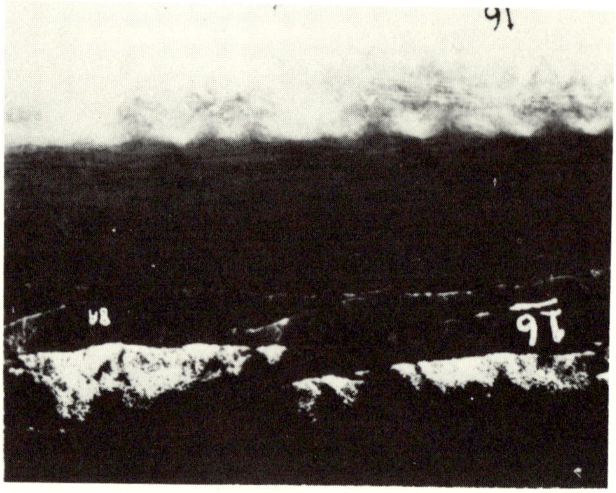

Fig. 7a. Typical gray tone pattern of a fine sand beach.

Fig. 7b. Typical gray tone pattern of a coarse sand beach.

c. A narrow band or abrupt darkening of tone close to the waterline.

The typical pattern is shown schematically by Figure 7b.

5. Gray tone patterns of beaches composed of the lower medium sizes (D_{50} = 0.2 to 0.3 mm) will resemble the patterns of fine sand beaches.
6. Patterns of beaches composed of coarser materials (D_{50} = 0.3 to 0.5 mm) will resemble those for coarse sand beaches.

7. The gray-tone pattern on beaches composed of coarse materials is practically independent of physical slope.
8. The tone pattern on the drying foreshores of beaches composed of fine materials is affected by the physical slope, the rate of darkening increasing in direct proportion to the degree of slope.
9. When analyzing gray-tone patterns, it is important that:
   a. The photography show the beach when the tide stage is below its mid-height.
   b. The backshore and wetted foreshore boundaries be carefully located.
   c. The local accumulations of seaweed and debris as well as other erratics be eliminated from the tone analysis.
10. The gray-tone pattern may be modified near the waterline by the uprush due to high waves frequently interspersed among low waves. This uprush may cause a two-toned pattern for the wetted foreshore. (See Figure 8.)

A final estimate of foreshore grain sizes should be obtained by combining all the estimates derived from indications based upon slope width and gray-tone pattern.

Analysis Method 5
Investigation of Special Features

Special beach features of significance are cusps, current ripple marks and scarps.

Cusps are roughly triangular erosion features, always created at the foreshore-backshore boundary. In some instances, old cusps will be found some distance within the backshore while new ones will be apparent at the backshore boundary. The recognition of cusps is easy. They appear as shown in Figures 9a and 9b. The definition, and spacing of cusps should be described. Figure 9a shows well-defined closely spaced cusps. Figure 9b shows ill-defined, broadly spaced cusps. Ground views of cusps, appear in Volume I of this report.

Current ripple marks are very faint depressions, roughly parallel to each other and to the waterline. They are found only on the foreshore. They are extremely difficult to recognize on aerial photographs. However, if definitely identified they are a positive indication of good trafficability. The significant marks should not be confused with wind ripples on dry sand or current ripples formed in the frequent current depressions found on beaches. These have

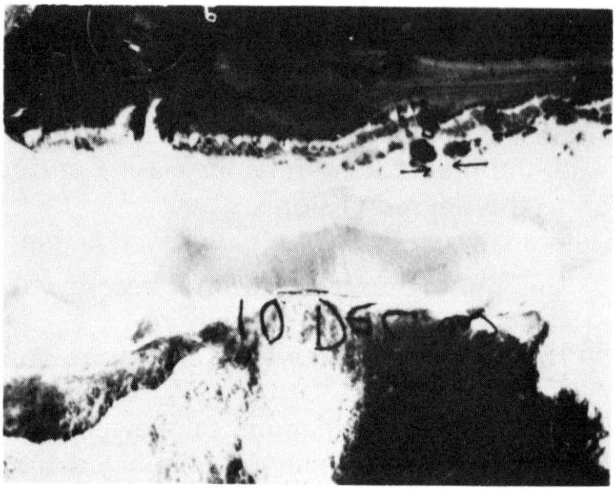

Fig. 8. Two-toned pattern for wetted foreshore caused by uprush.

no apparent significance. Various types of ripples are illustrated on page 48 of Volume I of this report.

Scarps refer to erosion features of a clifflike nature. They have only slight significance with respect to trafficability condition. However, they create a serious obstacle to vehicle movement. Consequently their height should be noted.

Analysis Method 6
Determination of Backshore Data

Data concerning the backshore is of little value in estimating the general trafficability condition of the beach foreshore. The backshores of all beaches, regardless of foreshore conditions, are characterized by an essentially constant high range of penetrations.[f] However, the backshore has a trafficability condition of its own. If the zone is steep and/or long, its dry loose condition may impede vehicle progress to such an extent that it governs the suitability of the beach for landing operations. Consequently, it is essential to determine the width and approximate slope of the backshore. This can be done according to standard methods.

Analysis Method 7
Division of Surf into Areas

It is convenient to divide the surf zone into a nearshore and offshore area. The surf zone includes the entire area seaward of the waterline, in which breaking waves appear. The nearshore area

[f] Unless crusted or recently wetted.

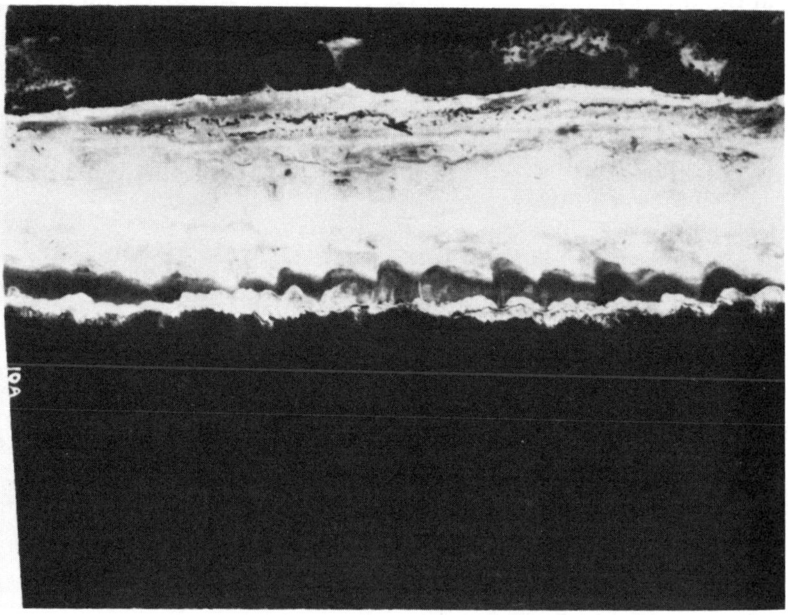

Fig. 9a. Well-defined closely spaced cusps.

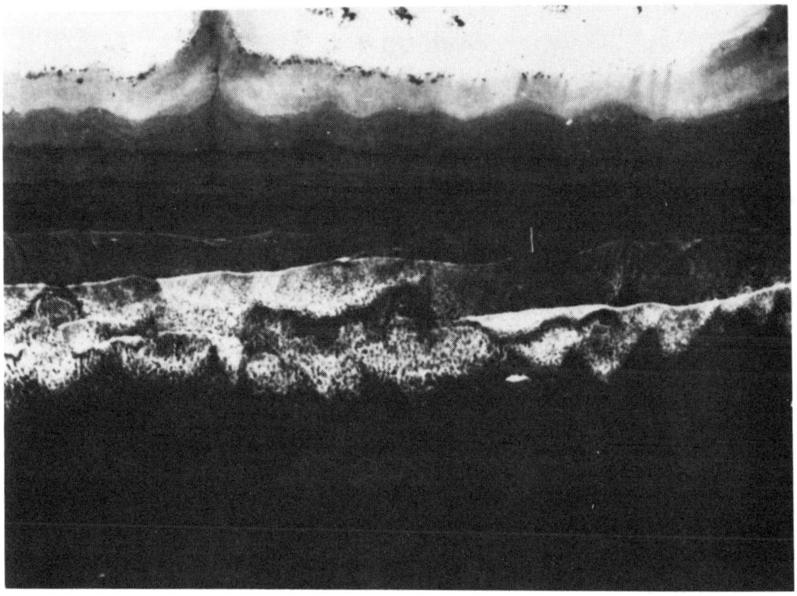

Fig. 9b. Ill-defined broadly spaced cusps.

of the surf zone is that portion of surf nearest the waterline where wave breaking is caused merely by shoaling. Often, the nearshore area is the only area present. Occasionally, the nearshore area of surf will be separated by an expanse of normal water surface, from a line, lines, or groups of breaks at some distance seaward. These breakers constitute the offshore surf zone and are indicative of offshore shoaling conditions caused by bars, reefs or rocks.

Figure 10 illustrates both the nearshore and offshore areas of the surf zone.

Method 8
Analysis of the Nearshore Surf Zone

The analysis of the nearshore surf zone is quite simple. Three items should be noted. These are:

1. The number and spacing of roughly continuous breaker lines adjacent to the foreshore.
2. The width of the nearshore surf zone.
3. The turbidity of the zone, indicated by a waterstain of uniform gray tone.

Items one and two are rough indicators of slope. Item three is indicative of the beach stability at the time of photography.

Method 9
Analysis of the Offshore Surf Zone

The analysis of the offshore zone consists of an inspection and evaluation of underwater obstacles at some distance from the shoreline. The obstacles may be reefs, rocks or bars. They may be detected by changes in gray tone or by lines of breakers. Their detection and classification as to type should offer little difficulty to an experienced interpreter.

Method 10
Division of Surroundings into Types

The area surrounding a beach should be examined for various features which can have an effect upon beach conditions. These features may be divided into three classes, or types:

1. Features providing protection from wave action.
2. Possible sources of contributory beach materials.
3. Possible sources of changes in longshore currents.

The features which provide protection from wave action are exclusive from bars and reefs, which are considered as part of the offshore surf zone. They consist of headlands and offshore islands.

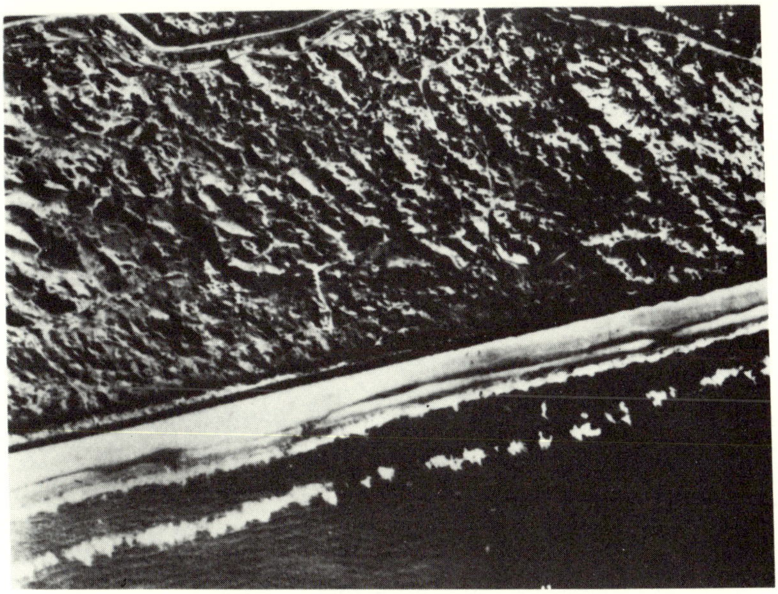

Fig. 10. Nearshore and offshore areas of the surf zone.

ESTIMATE OF TRAFFICABILITY CONDITIONS

As mentioned in the introduction, the trafficability condition of a beach can be estimated in terms of a slope-penetration combination. Figure 1 shows a *possible* resolution of slope-penetration combinations into beach classes.

The primary objectives of the trafficability estimate are determinations of the significant slopes and penetrations that may be considered representative of the beach.

The various slopes of beach foreshore and backshore are determined directly, using instruments, or are estimated indirectly, using low oblique photographs, breaker patterns, widths and graytone patterns.

The representative penetrations of the beach are determined in accordance with quantitative statistical correlations and qualitative observational records.

There are three conceivable methods of estimating the representative penetrations:

1. Method A—A single estimate based upon a single beach feature.
2. Method B—Several single estimates based upon several beach features leading to a single final judgement.
3. Method C—A single final judgement based upon the substitution of quantities in a statistical formula and modification in terms of qualitative information.

6

Reprinted from *Geol. Soc. America Bull.* **82**(2):345–357 (1971)

Mapping and Landform Analysis of Coastal Regions with Radar

H. C. MACDONALD *Department of Geology, Georgia Southern College, Statesboro, Georgia 30458*
A. J. LEWIS *Department of Geography and Anthropology, Louisiana State University, Baton Rouge, Louisiana 70803*
R. S. WING *Princeton Geological Group, Continental Oil Company, Princeton, New Jersey 08540*

ABSTRACT

Although aerial photographs still constitute an important source of information for the coastal geomorphologist, data acquisition is severely limited in cloud-shrouded coastal environments. Sequential radar mapping of remote coastlines, however, will provide a practical and rapid method of updating maps and monitoring coastal processes. Contrasts in signal return from the land-water boundary produce a striking interface on radar imagery which is advantageous for the delineation and mapping of coastal features. Along the Pacific and Caribbean coasts of eastern Panama and northwestern Colombia, marked discrepancies exist between available maps and the actual configuration as revealed by radar.

In addition to the revision of coastal maps, radar imagery also provides the geomorphologist with most of the information interpretable from aerial photographs. Mud flats, beach ridges, natural levees, backswamp basins, beach features, and zones of breaking surf are examples of the observable features. Mangrove swamps, which abound on tropical shores throughout the world, provide a readily recognizable vegetation type which can be unmistakably interpreted from the imagery. Although offshore sediment transport cannot be directly interpreted, inferred trends can be obtained through sequential radar coverage. The Atrato River Delta of northwestern Colombia is cited as an example.

Radar imagery interpretation has indicated marked landform differences between the Pacific and Caribbean coasts of eastern Panama and northwestern Colombia. The partially submerged mountainous terrain of the southeastern coast is contrasted with the mangrove swamps, beaches, and barrier reefs of east-central Panama.

INTRODUCTION

During the past two decades, aerial photographs have been used increasingly for coastal geomorphic studies. Generally, more detail of the coastal configuration and near-shore features are observable on the photographs than can be recorded in field studies. Sequential aerial photography is an excellent tool for monitoring shoreline changes; however, in some regions of the world, photographic acquisition is extremely limited by daylight or weather conditions.

The use of radar imagery for identifying landforms, interpreting processes, and delimiting physiographic regions is a relatively new tool for the geomorphologist. Radar as a tool for the coastal geomorphologist has several interesting and advantageous characteristics, including the ability to scan a broad band of terrain (40 mi wide) with a single pass, and presenting the imaged area on a continuous strip of film. Continuous strip presentation of a 40-mi band of terrain of several hundreds of miles is of special importance to the coastal geomorphologist who is interested in a long, but relatively narrow, area.

Radar imaging systems have a near all-weather, 24-hr imaging capability, and this advantage is of special importance for coastal studies of areas masked by darkness as in polar regions, or by clouds in tropical environments. Coastal changes, which are greatest during high winds, waves and tides are best monitored during the period of maximum erosion and not after the storm. The ability of certain radar wavelengths to penetrate all but thick precipitating clouds may enable monitoring near maximum erosional conditions and, therefore, should result in a better understanding of the cause and process of coastal erosion.

Except for a few cursory studies, the potential of radar as a tool for coastal geomorphology has not been documented. It was the purpose of this study to define both the capabilities and limitations of radar for coastal geomorphic studies; specifically to determine the coastal landforms that are identifiable.

RADAR PARAMETERS[1]

For those readers not familiar with radar imagery it is important to note, that although the imagery is recorded on photographic film, we have recorded relative differences in signal return from the microwave portion of the electromagnetic spectrum. For example, Figure 1 shows an area being imaged by a typical SLAR system. The transmitter generates short bursts or pulses of radio frequency (RF) energy. These pulses are propagated into space by means of a directional antenna (A), and radiate from the antenna as a block of energy (B) at the velocity of light (3×10^8m/sec). If a terrain feature capable of intercepting RF energy is irradiated at point (a), a fraction of the transmitted energy will be reradiated back to the antenna (A). An object at (b) will also reradiate energy back at a later time when it is illuminated by the pulse packet of RF energy. The same is true for features at (c) and (d). This portion of energy returned to the antenna from the terrain features (a), (b), (c), and (d) is converted to a video signal by the receiver. The signal return from these features is displaced from the origin as a function of the range from the antenna to the target, whereas the amplitude of each return is a function of the scattering properties of the terrain.

The antenna (A) is repositioned laterally at the velocity of the aircraft (V_a). Each transmitted RF pulse (B) returns signals from the targets within the beamwidth. These target returns are converted to a time/amplitude video signal (C) which is displayed on a cathode-ray tube (CRT) and then recorded as a single line (E) on photographic film (F). Returns from subsequently transmitted pulses are displayed on the CRT at the same position (D) as the previous scan lines. By moving the photographic film past the CRT display line

[1] For a more detailed discussion of radar systems operation, the reader is referred to MacDonald (1969, p. 6–19).

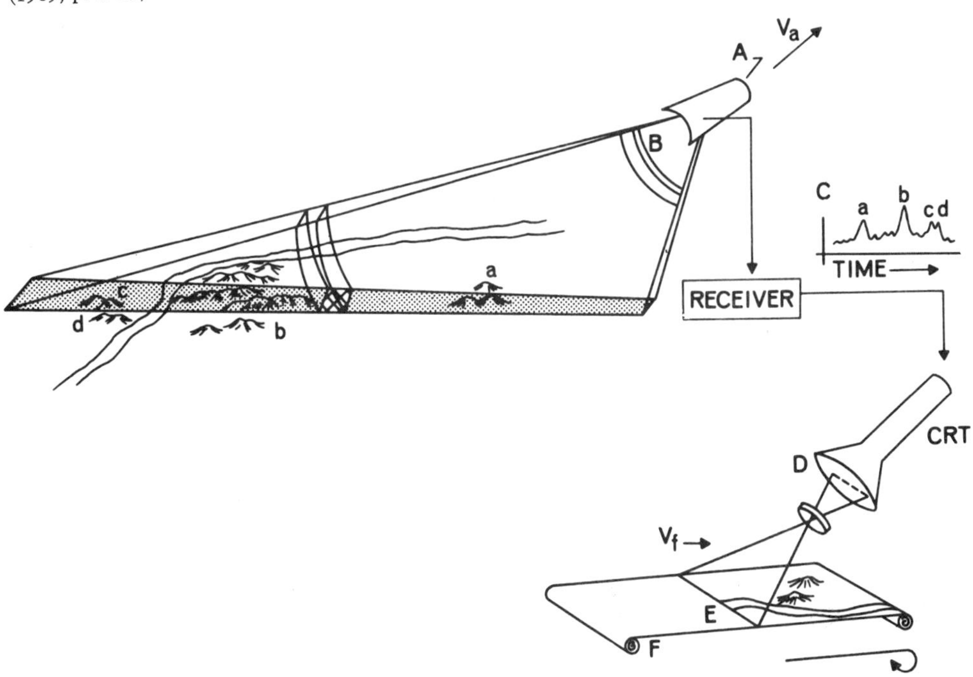

Figure 1. Operational schematic, typical side-looking radar system.

at a velocity (V_f) proportional to the velocity of the aircraft (V_a), an image of the terrain is recorded on the film (F) as a continuous strip.

SENSOR-TERRAIN INTERACTION

Depending on the relative surface roughness[2] of the terrain, the signal return from vegetation, rock, and soil is generally higher than the signal return from water. Thus contrasts in signal return from the land-water boundary produce a striking interface on the radar imagery which is advantageous for the delineation and mapping of coastal features. For example, in Figure 2 along the Texas Gulf Coast, the coastal waters are shown in black (low or no radar return) and the terrain in contrasting gray tones (relatively higher return). It is significant that the radar imagery shown in Figure 2 was taken with an obsolete, low-resolution, side-looking radar system, and even though we are using a relatively crude-resolution system, many coastal features are easily identifiable. For example, the distinctive pattern of a chenier plain can be located south of Sabine Lake. The distinctive asymmetrical shape provides information of past and present shoreline processes. Some of the more obvious features of Figure 2 include: (1) beach ridges or growing lines of a chenier plain, (2) double jetty to contain the river in a narrower channel (to increase its erosional power and maintain a deep channel), (3) old meander scar, and (4) offshore drilling platforms.

[2] Surface roughness is not an absolute roughness, but the relative roughness expressed in wavelength units. For a surface roughness much less than the wavelength ($\lambda/10$), the surface appears "smooth" to the imaging radar; for a surface roughness on the order of a wavelength or more, the surface appears "rough."

Figures 3 and 4 are more representative of the quality of radar imagery that can be expected with the present-day commercial imaging radar systems. In the vicinity of Seaside, Oregon, several coastal features have been illustrated on Figure 3. The bright offshore feature just south of South Jetty of the Columbia River designated as area (1), is a line of surface convergence probably caused by a combination of wave refraction and longshore current. An additional surface pattern has been defined in area (2). The surf zone is visible in area (3). The relatively bright return region (high signal) shown at area (4) represents vegetated foredunes. Beach ridges landward of the foredune can be delineated in area (5). These beach ridges are essentially parallel to the present shoreline and provide evidence of a prograding shoreline.

Figure 4 of San Diego, California, illustrates the ability of radar to detect moving sea-

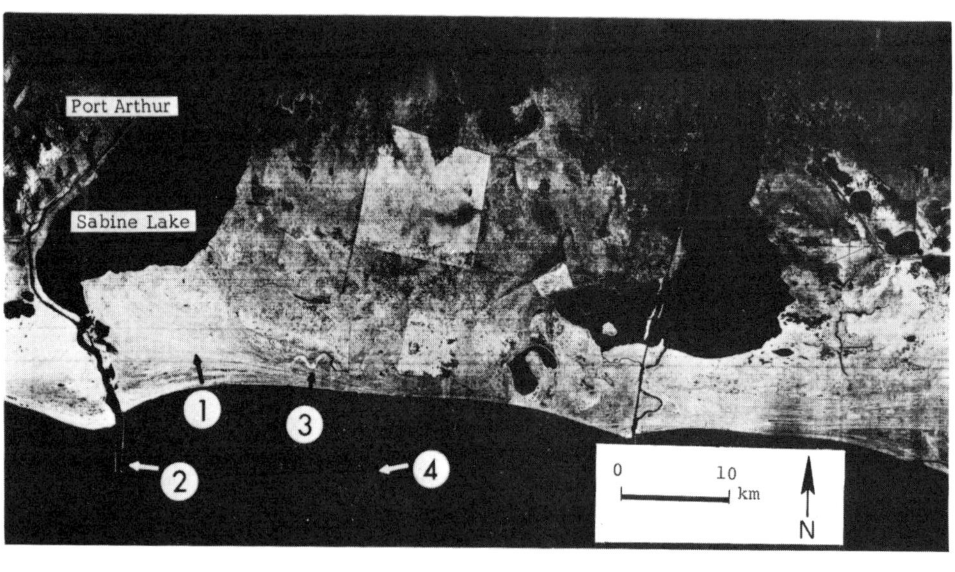

Figure 2. Radar imagery of the Gulf Coast area near Port Arthur, Texas. Photograph from CRES, University of Kansas.

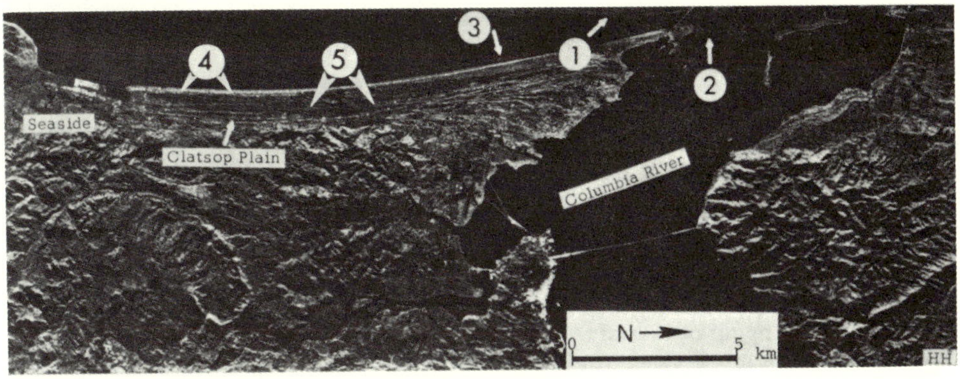

Figure 3. Radar imagery of the Seaside, Oregon area. Photograph courtesy of CRES, University of Kansas.

going vessels and their wakes (1), and kelp beds (2). The separation of sea state is also indicated where the open sea (3), was sufficiently rough to produce enough backscatter to be recorded, whereas in the more sheltered area (4), the water surface was smooth enough that the emitted radar signal was reflected away from the receiver thus lost and a no-return area recorded.

RADAR IMAGERY PRACTICALITY

Obviously, the three previous examples of SLAR imagery were obtained from geographic areas where aerial photography fully satisfies the reconnaissance needs of the coastal geomorphologist, and where radar provides only supplementary information. Consider now the role of an imaging radar in a coastal environment where aerial photography is difficult or impossible to obtain. For example, in east-central and southeastern Panama (Fig. 5), attempts to obtain complete aerial photographic coverage of this area have been notably unsuccessful for several decades because parts of this region are perpetually covered by clouds. Particularly in Darien Province, the area northwestward from the Panama-Colombia border to north of San Miguel Bay (Fig. 5), gathering topographic data has defeated photographers for several decades.

To exploit radar's capability to penetrate clouds, in 1967 the U.S. Army Topographic Command, Engineer Topographic Laboratories successfully collected radar topographic map data and imagery of southeastern Panama and northwestern Colombia. The data were gathered during six separate days of flight operation, and a portion of the imagery was used to construct a radar mosaic (Fig. 6) encompassing approximately 17,000 sq km. Although not without error, the radar mosaic provides a much more accurate configuration

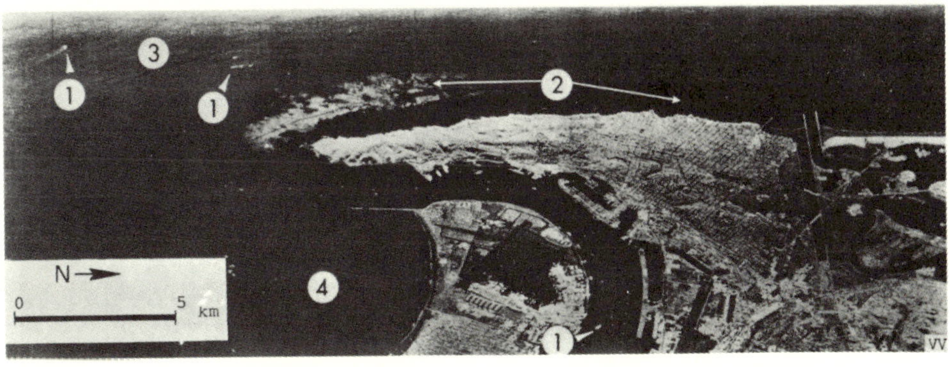

Figure 4. Radar imagery of San Diego, California. Photograph courtesy of CRES, University of Kansas.

of the Darien region than has been previously available and reveals numerous errors incorporated in earlier maps of the area. Figure 7 represents a reduction of the best available topographic map made prior to radar coverage in the area. Some of the more obvious discrepancies between the mapped features and the actual configuration as delineated on the radar include: (1) the meander pattern of the Rio Tuiria (upper center of the map), (2) the position of the Isla Mangle which lies upstream from where it was originally mapped, and (3) the incorrect orientation of the low hills east of the Rio Tuiria.

Because this study is concerned with coastal features, the selection of two areas along the Caribbean are of particular interest. There are several obvious discrepancies revealed on Figures 8A and 8B when the mapped coastline and the radar imagery are contrasted. These illustrations provide evidence that sequential mapping by radar, and ultimately the production of orthomaps from imagery would result in a practical and rapid method of updating coastal maps. Figure 5 can be used for location of Figures 8A and 8B.

RADAR IMAGERY INTERPRETATION

Tidal flats usually exhibit striking patterns on air photos. On radar imagery the pattern is also unique because of its offshore location, herringbone texture-drainage, high tonal contrasts with the adjacent water, and marked textural contrast with the land. The tidal flats on the Pacific coast of east-central Panama (Fig. 9) are generally nonorganic, fluviomarine accumulations occurring in shoal areas, and like most coastal areas where such features are found they are protected from

Figure 5. Physiographic sketch map, eastern Panama and northwestern Colombia. Photograph from CRES, University of Kansas.

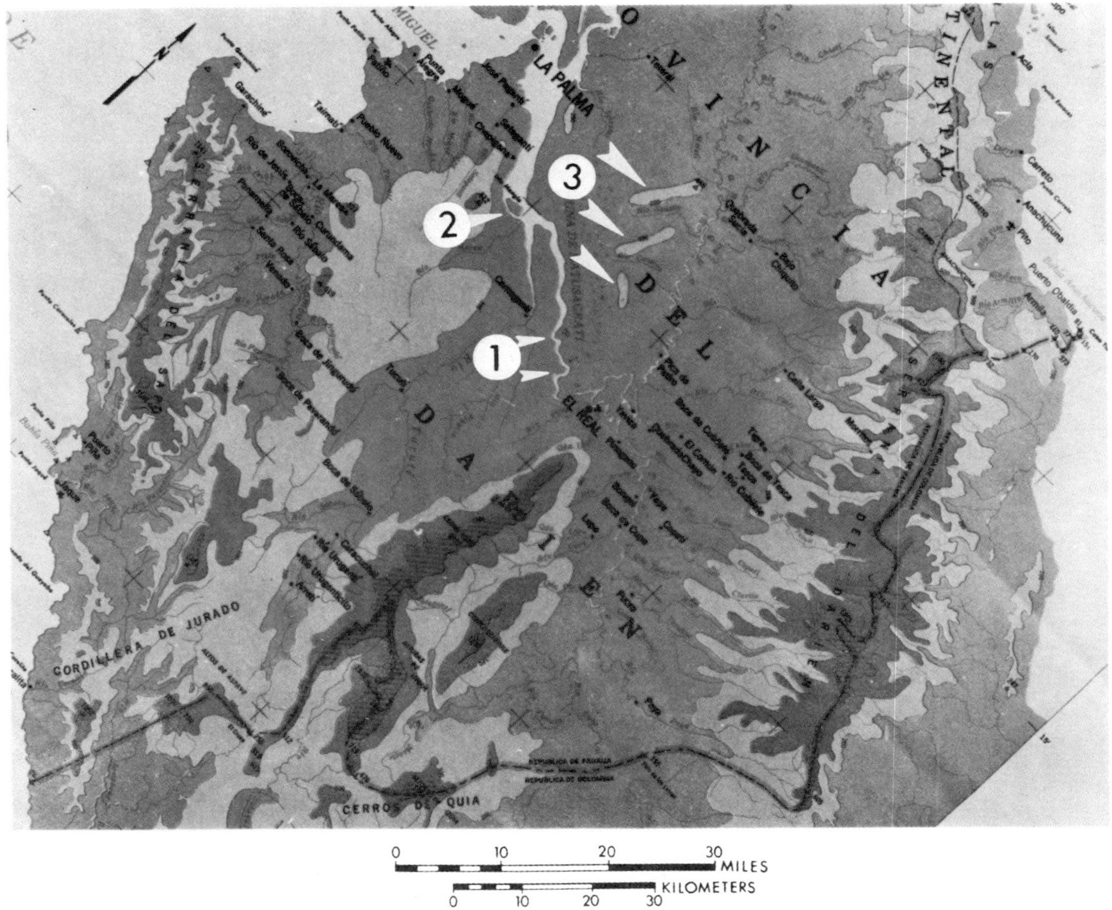

Figure 6. Radar mosaic, Darien Province, Panama and northwestern Colombia.

strong wind and current action. Tidal flats, such as those outlined by arrows on Figure 9, are believed to be the first stage in the formation of mangrove swamps and exhibit a slightly undulating surface devoid of vegetation. In this particular location, extensive mangrove swamps can be delineated shoreward of the tidal flats.

Mangrove swamps, which abound on tropical shores throughout the world, are a readily recognizable vegetation type unmistakably identifiable on radar imagery. On Figure 10A white arrows define the boundary between mangrove vegetation to the north and jungle vegetation to the south. Proximity to coastal waters and distinctive tone and texture pattern aid in delineation of the mangrove swamp limits.

The unique shape of estuarine meanders[3] are easily defined from radar imagery, and Figures 10A and 10B provide examples. The stippled areas on the silhouettes (traced from the imagery) of the major rivers outline the zones of estuarine meanders. These diagrams can be contrasted with the actual configuration on the radar imagery. Differing from either river or tidal meanders by their position in the fluvial-marine scheme of the hydrologic system, and by their relationship to a specific family of curves, the recognition of the estuarine meander form allows insight into the

[3] The identification of estuarine meanders is based primarily on planimetric shape. They were first described by Ahnert (1960) as a succession of oblong pools connected by narrow channels at the bends.

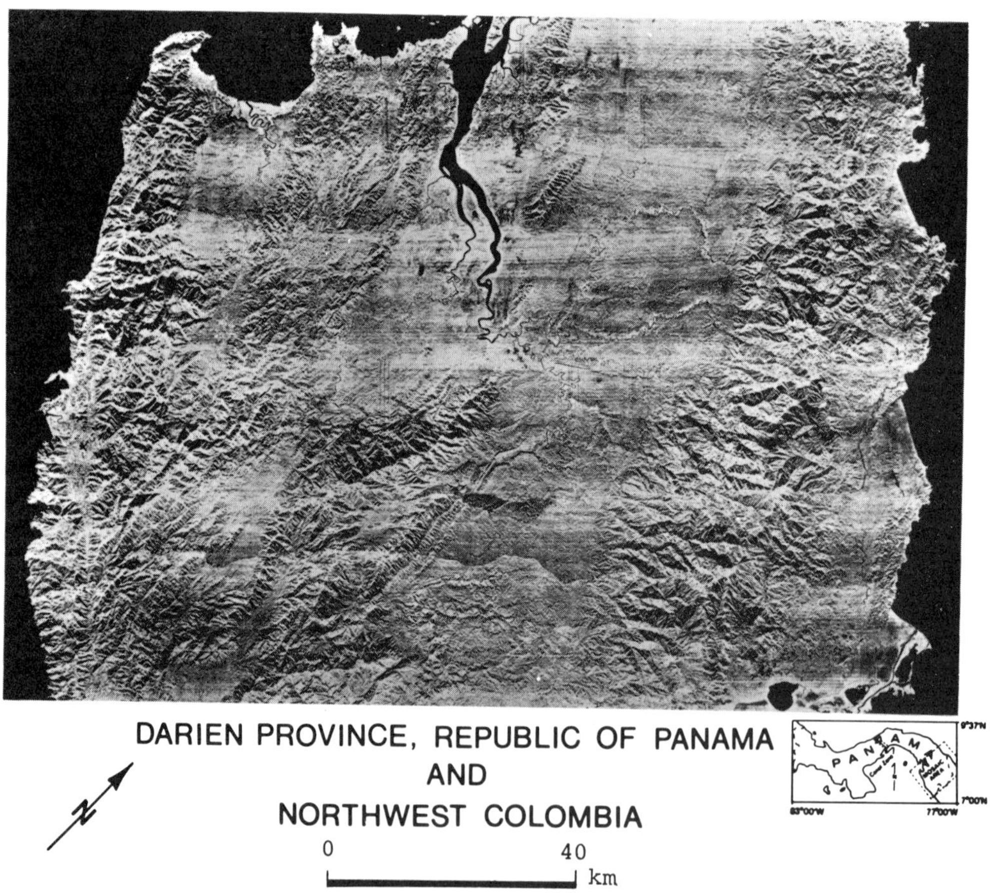

Figure 7. Topographic map, Special Map No. 2, sheet III, Army Map Service, USAR CARIB, 1:500,000, 1967.

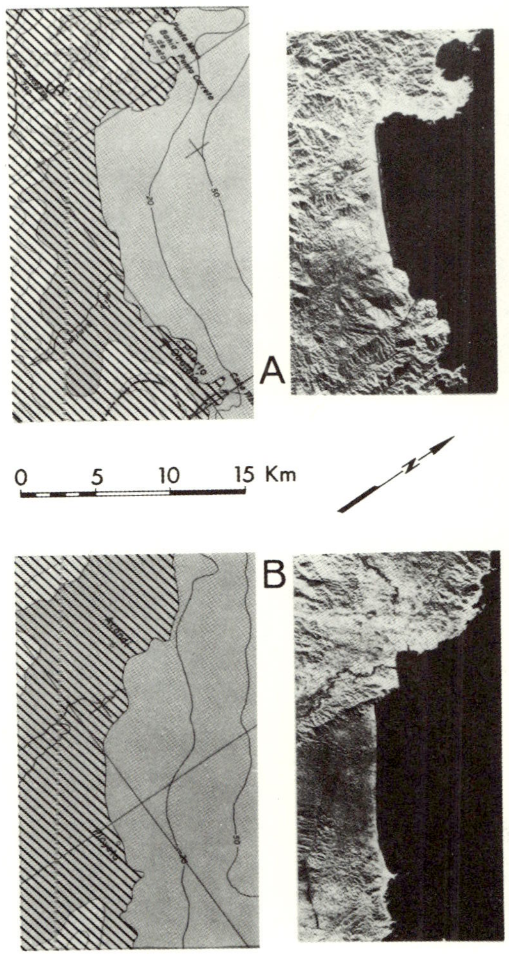

Figure 8. Comparison of radar imagery and topographic map, Instituto Cartografico Tommy Guardia, Sheet # 11, La Palma, Panama, 1:250,000, 1966; radar imagery, 1967.

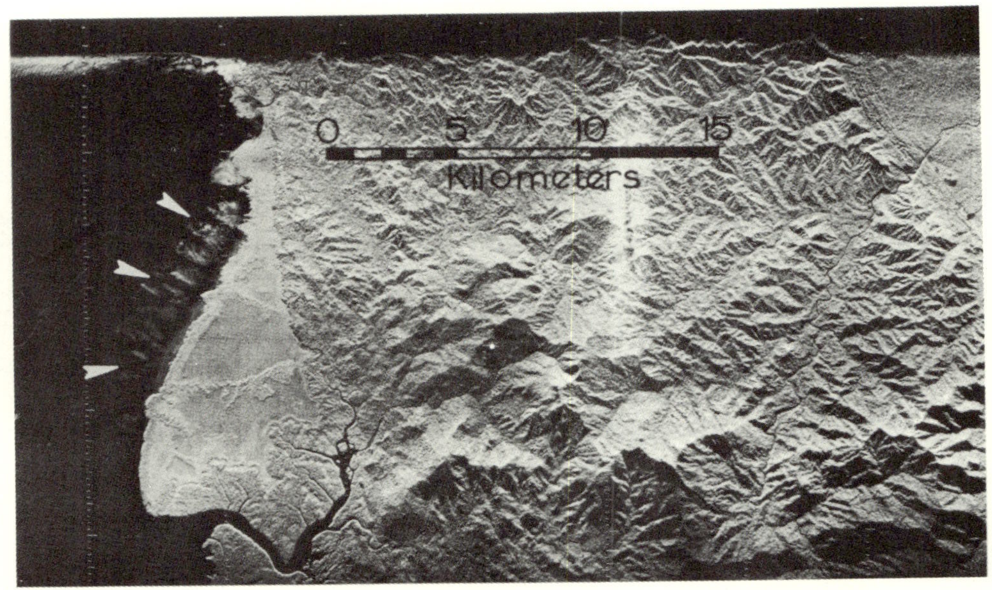

Figure 9. Radar imagery, tidal flats along Pacific Coast, east-central Panama, mouth of Rio La Maestra lower left of imagery.

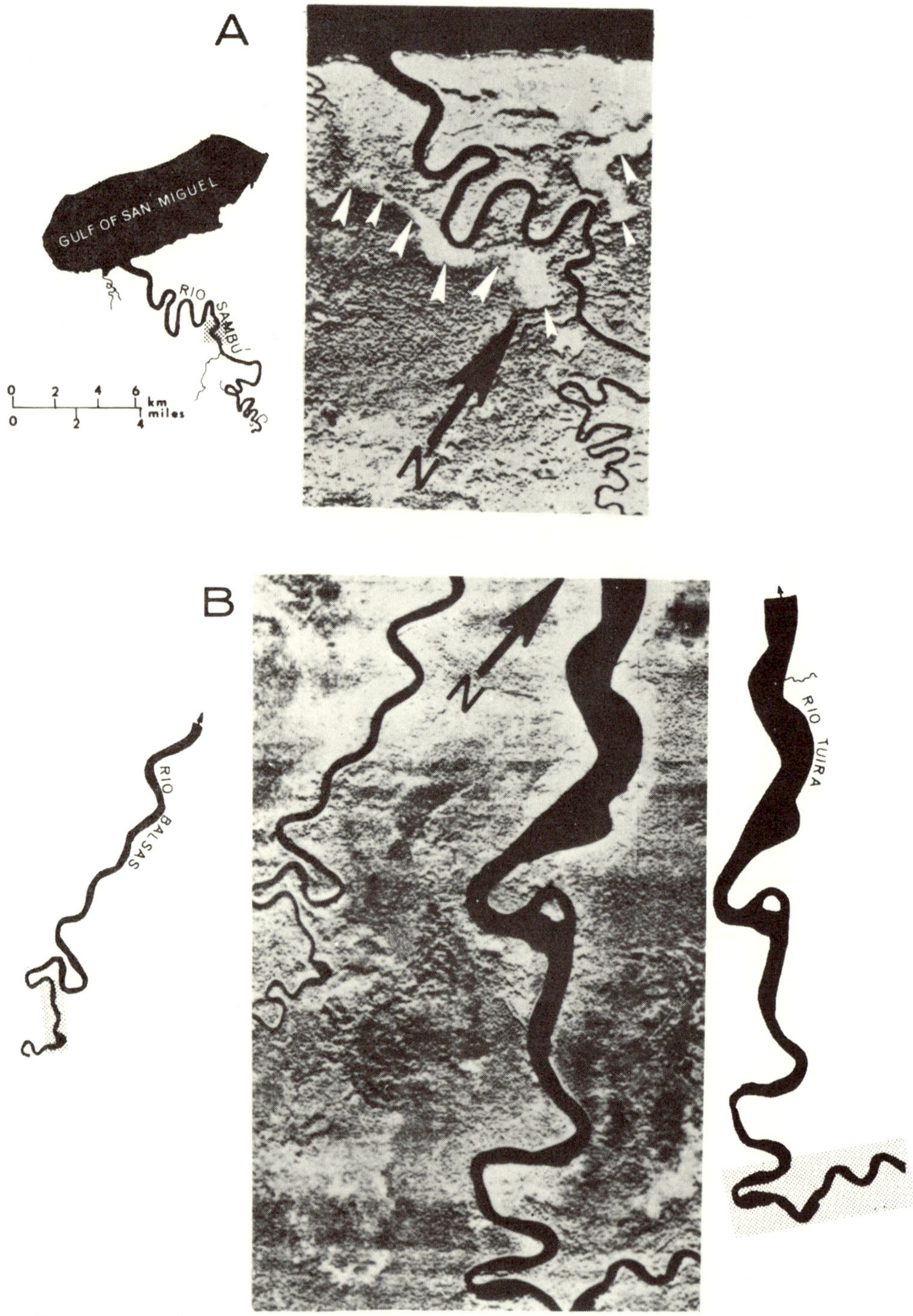

Figure 10. Radar imagery and silhouette diagrams, zones of estuarine meanders within stippled areas. A. White arrows define the boundary between mangrove vegetation to the north and jungle vegetation to the south.

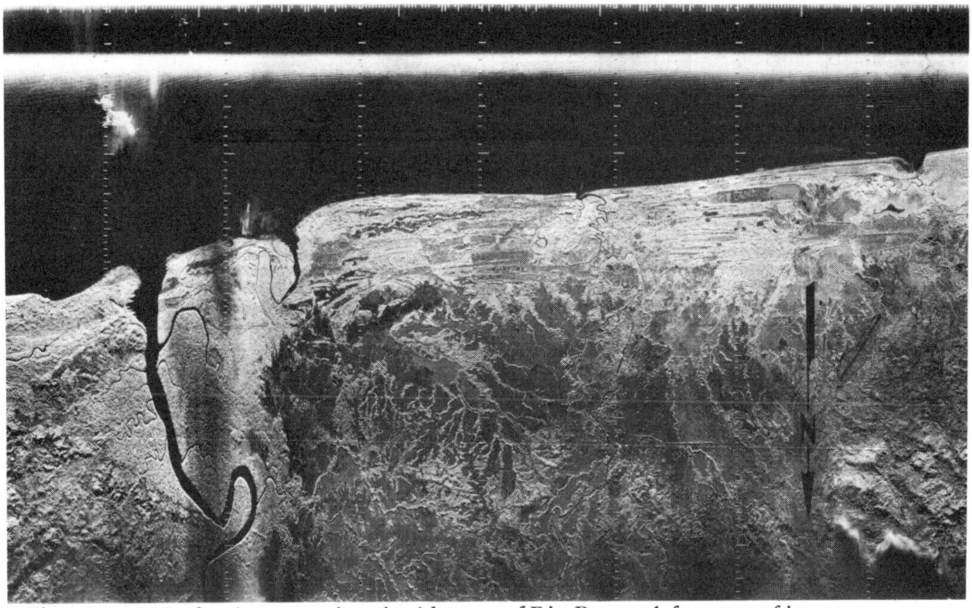

Figure 11. Radar imagery, beach ridges along Pacific coast, east-central Panama, mouth of Rio Bayano left center of imagery.

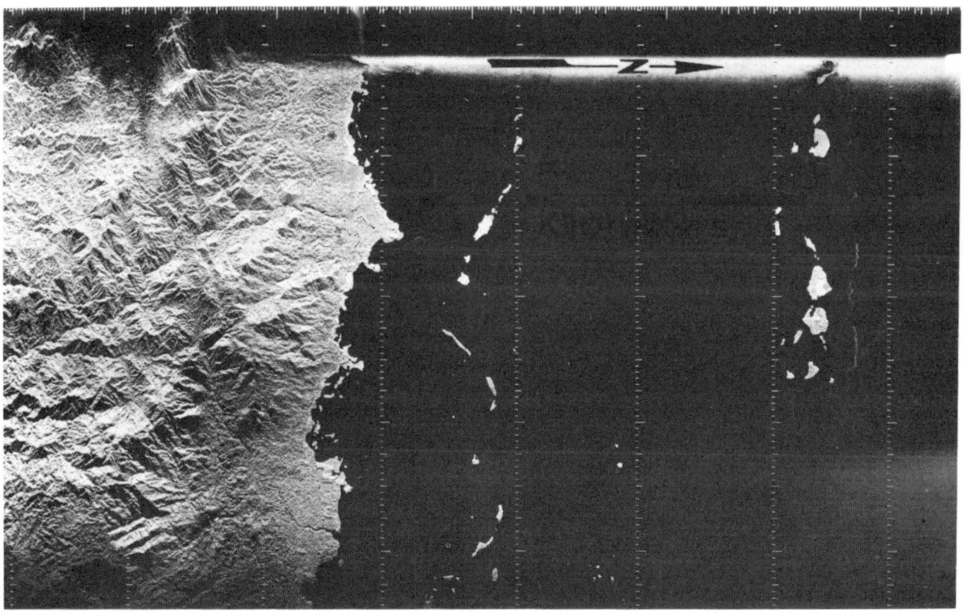

Figure 12. Coral sand islands on leeward slope of barrier reef, surf zone (arrows) outline windward part of reef.

type of coast, channel flow, availability of sediment, and increased knowledge of the balance between marine and fluvial processes. The estuarine meanders illustrated in Figures 10A and 10B had not been previously reported until acquisition of the radar imagery (Lewis and MacDonald, 1970).

Beach ridges provide evidence (Fig. 11) of progradation along the Pacific coast between the Rio Bayano Estuary and the Panama Canal (*see* Fig. 5 for location). Back swamp drainage is well defined north of the beach ridges; to the east the outline of a mangrove swamp is equally apparent. A wave refraction pattern can be delineated (black arrows) along the upper margin of the imagery.

On the Atlantic coast are the San Blas Islands (Fig. 12) which consist of coral sand that has collected on the leeward slopes of barrier reefs. The outline of the windward part of the reef is delineated by the surf zone (arrows, Fig. 12). Landward from the San Blas Islands, coral flats fringe the coast, but this cannot be determined from imagery interpretation; however, the occurrence of these dense clusters of dead corals at the present sea level was noted in a study by Tuan (1960, p. 24).

The delta of the Atrato River is located on the southwestern side of the Gulf of Uraba in northwestern Colombia (Fig. 5). The delta forms a coastal lowland belt bounded on three sides by mountains, thus the Atrato River receives large quantities of water from the mountain runoff. Vann (1959) has revealed that the Atrato delta exhibits physical features which permit recognition of recent changes by analyzing landform and vegetation. Landforms in the delta consist of mud flats, natural levees, backswamp basins, beach features, and round lakes.

Figure 13 represents a summary of Vann's investigation of the Atrato delta. Figure 14 provides a comparison between the 1954 coastal conditions constructed by Vann in the field and from aerial photographs, and the coastal configuration as recorded by radar imagery some 13 yrs later. Since 1954, the northern spit (area 1, Fig. 14) has been either eliminated by wave action or has been breached, then later connected from the south by a spitlike tombolo (2). Encroaching vegetation from the natural levee (3) in the main channel reflects a reduction in stream flow. The largest lake in the region (4) is now being reduced in area because of sedimentation, and

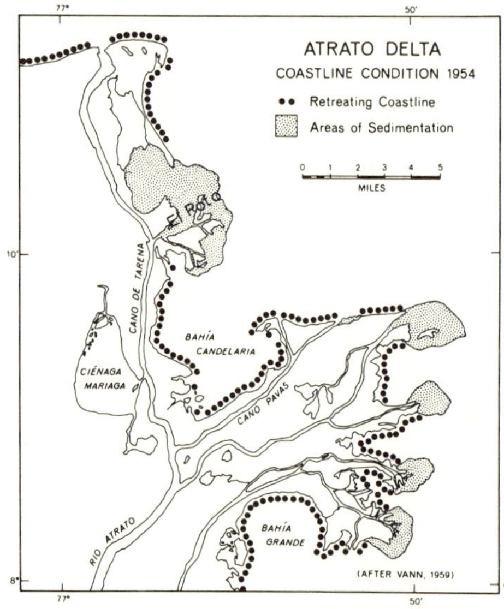

Figure 13. Atrato Delta coastline conditions, 1954.

appears to reflect the initial stages in the formation of an inland swamp. Contrasting Vann's study with that of the radar imagery suggests that Vann's projection of coastal retreat was correct.

COASTAL LANDFORM ANALYSIS

Interpretation of single strip radar imagery (Figs. 8, 9, 10, 11, 12, and 14) and the radar mosaic (Fig. 6) of the eastern half of Panama provides evidence of marked landform differences between the Pacific and Caribbean coasts. However, the present shorelines on both sides of the eastern Panamanian Isthmus are marked by over-all characteristics of submergence. The Pacific side of southeastern Panama and northwestern Colombia (Fig. 6), is dominated by a coastline of partially submerged mountainous terrain which extends from Colombia northwestward to San Miguel Bay. Although the Pacific coast of Darien Province is in an early youth stage of submergence, the Caribbean coast represents a more advanced stage of youthful submergence. For example, along the Darien mountain range (Fig. 5) partially submerged mountainous terrain is interrupted by straight stretches of sandy beaches (the bayhead beaches in Figs. 8A and 8B). To the southeast is the delta complex of the Atrato River.

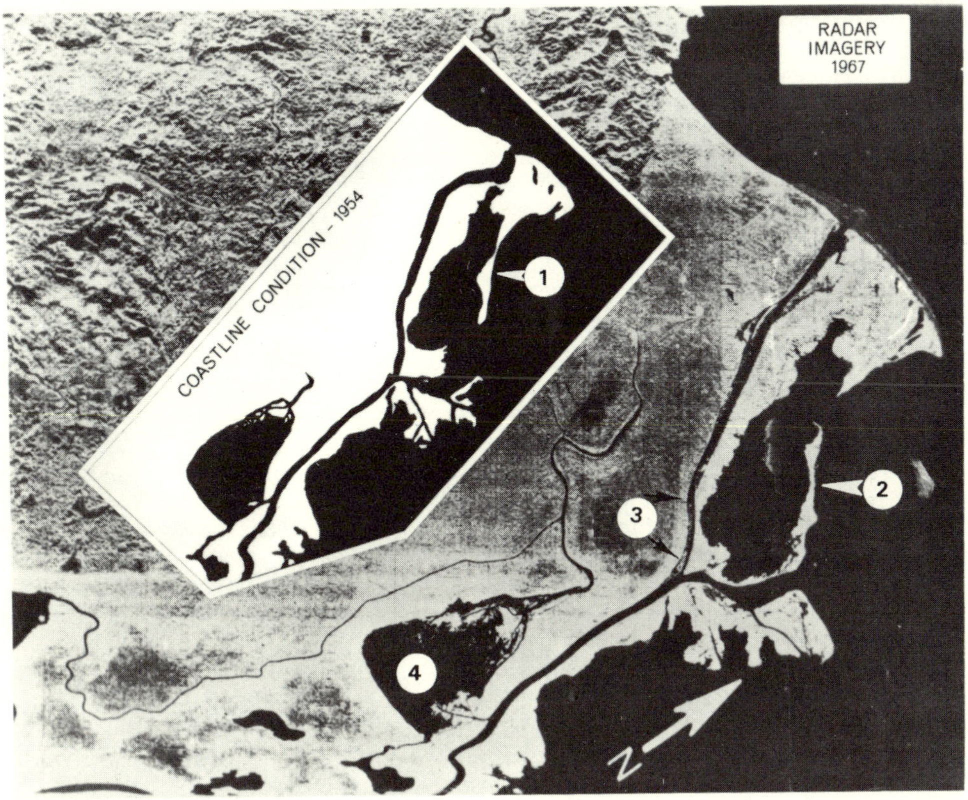

Figure 14. Comparison between radar imagery (1967) and coastline conditions constructed by Vann in the field in 1954.

The rise in sea level at the end of the Wisconsin glacial substage provided sufficient submergence to cover most of the narrow coastal plain sediments and to bring both the basement complex and sedimentary strata in contact with the sea. Along the submergent Pacific and Caribbean coasts of eastern Panama, the sedimentary strata of Eocene age or younger (MacDonald, 1969; Wing, 1970) provide a less rugged coastline than does the basement complex. Contrast, for example, the coastal terrain in Figure 8A and 8B, where the mountainous coastline (Punta Carreto and Puerto Obaldia of Fig. 8A) represents outcropping of the basement complex, but the sediments of the beach areas are juxtaposed with relatively less resistant strata of Eocene age.

The Bay of San Miguel (Fig. 5, 8°15′N, 78°20′W) is a compound estuary. Formerly it was the main trunk of a large river whose tributaries included all of the major rivers now supplying fluvial sediments to the estuary (Wing, 1970, p. 110). The shoreline of San Miguel Bay is submergent, but many bayhead beaches and especially bay deltas attest to a rapid fill-in process. This region contains the most extensive development of mangrove swamps in all of eastern Panama, and the entrapment of fluviomarine muds contributes considerably to the rapid sedimentation now taking place.

The coastlines of east-central Panama are more complex than those of Darien Province. The Caribbean shoreline from Punta Ailigandi westward to the Canal Zone (Fig. 5), appears to be both contraposed and compound. Here there is no continuous coastal plain, but Holocene sedimentation is much more prevalent than to the southeast. This fringe of Holocene sediments lends support to Terry's (1956) supposition of slight emergence of the Caribbean coast in post-Pleistocene time, but this subsequent emergence was relative to the submergence of the Pacific littoral. Nevertheless, numerous coral islands (Fig. 12) along the Caribbean coast of east-central Panama are not yet tied to land or each other by spits, bars, or tombolos. The net impression is that of a shoreline of contraposed submergence.

Along the Serrania De Maje (8°50′N, 78° 30′W) from San Miguel Bay to the mouth of the Rio Bayano (9°00′N, 79°05′W) a submerged coastline is interrupted by mangrove swamps and mudflats. Bay deltas building out from the mouths of nearly all rivers are filling in estuaries and forming extensive tidal flats. Along this coast, a shallow bottom extends far offshore protecting the deltas from wave erosion. This coastline appears to qualify for a neutral classification though net submergence is characteristic of the recent past.

Immediately east of the Canal Zone and extending to the mouth of the Rio Bayano a coastal plain has developed ranging in width from 4 to 25 km. The beach ridges previously illustrated in Figure 11 provide evidence of progradation, the causes of which are attributable to one or all three of the following possibilities (Wing, 1970, p. 46): (1) Acidic terrain drained by the Bayano tributaries has supplied an abundance of quartz sand debris to a locally regressive coastal segment; (2) slight progressive uplift (on the coastal side of a major fault) has resulted in regression and progressive abandonment of beach bars; and (3) a combination of high tides with anomalous wind and wave conditions provides unique sorting and sediment distribution action.

CONCLUSIONS

The spatial resolution capability of radar is less than that of aerial photogtaphy, and the coastal detail interpretable from radar imagery does not equal that obtainable from aerial photographs. For example, wave height and beach features such as dune shape, size, and orientation cannot always be observed; however, refinements in operational and system parameters such as gain-setting and resolution may improve certain aspects of data retrieval. Water depth information provided by color photography, and near-shore bottom detail interpretable from panchromatic aerial photography are not available with radar imagery.

Radar does, however, provide a rather precise basis for measuring the net changes in dimensions or shape by sequential imaging. A qualitative estimate of the amount of sediment deposited in the form of shoals, tidal deltas, and spits can be determined. Also, estuarine changes, delta formation, nearshore processes, and sea state can be observed. Thus in cloud shrouded environments where photography cannot be obtained, radar will provide an important source of information for the coastal geomorphologist and for updating coastal maps.

ACKNOWLEDGMENTS

Facilities for this study were provided by the Remote Sensing Laboratory, Center for Research in Engineering Science (CRES), University of Kansas. This study was sponsored by the Department of the Army, Engineer Topographic Laboratories under THEMIS Contract DAAK02-68-C-0089 and NASA Contract NAS9-10261.

REFERENCES CITED

Ahnert, F. 1960, Estuarine meanders in the Chesapeake Bay area: Geogr. Rev., Vol. 50, No. 3, p. 390–401, 1960.

Lewis, A. J., and MacDonald, H. C. Significance of estuarine meanders identified from radar imagery of eastern Panama and northwestern Colombia: Mod. Geol., Vol. 1, No. 3, p. 187–196, 1970.

MacDonald, H. C. Geologic evaluation of radar imagery from Darien Province, Panama: Mod. Geol., Vol. 1, p. 1–63, 1969.

Terry, R. A. A geological reconnaissance of Panama: Calif. Acad. Sci., Occas. Pap. 23, 91 p., 1956.

Tuan, Y. Coastal landforms of central Panama: Dept. Geogr. Univ. California, Berkeley, Contr. ONR–222(11) NR 388 067, AD No. 239 568, 30 p., 1960.

Vann, J. H. Landform-vegetation relationships in the Atrato Delta: Ann. Ass. Amer. Geogr., Vol. 49, No. 4, p. 345–360, 1959.

Wing, R. S. Structural analysis from radar imagery, eastern Panamanian Isthmus. Mod. Geol., 180 p., 1970, in press.

MANUSCRIPT RECEIVED BY THE SOCIETY JUNE 16, 1970

REVISED MANUSCRIPT RECEIVED SEPTEMBER 9, 1970

7

Reprinted from *NASA/Goddard Space Flight Center ERTS-1 Symp., 1973, Proc.,* pp. 108–116

USE OF ERTS-1 IN COASTAL STUDIES

Orville T. Magoon
Coastal Engineering Research Center
U.S. Army, Corps of Engineers

The strong private, public and federal interest in our nation's shorelines is readily apparent from material in the published press and in past and pending legislation. This presentation reflects the federal involvement in shore protection and planning and also reflects congressional and public recognition of beach and shore erosion as a local as well as a regional problem.

The physical processes which shape our shorelines are extremely complex and diverse. Both natural forces and man-induced changes affect the movement of and supply of sediments along a shoreline. The shores of the United States include practically all known land forms consisting of many materials and at various stages of geologic evolution. These land forms and materials have different vulnerabilities to wave action and their responses to the powerful forces which they oppose form a very broad spectrum. In order to effect realistic coastal planning and sound coastal engineering, it is necessary to quantifiably determine the movement of sedimentary material along the coastline under consideration. Although littoral processes are conceptually understood, it is not yet possible to adequately describe or quantify long or short term forecasts of the effects of improvements or modifications to the shoreline. In some instances, past shoreline changes have resulted in costly damage worth millions of dollars. Research in coastal processes at the Coastal Engineering Research Center is currently directed toward efforts to improve the state-of-the-art on qualitative and quantitative approaches to this problem. These efforts are based in part on tedious, extremely complex and short-lived in situ sediment transport measurements and experiments.

In addition to the examination of in situ sediment movement at specific locations, it is also necessary to describe and quantify the major coastal processes which occur over large physiographic reaches of our nation's shores. Of the 84,000 miles of the United States' ocean and Great Lakes shorelines, about 25% is undergoing significant erosion. Almost 3,000 miles of the nation's shoreline are undergoing critical erosion. The cost of remedial measures to halt such erosion, if desired, would be about two billion dollars plus major annual costs. From a practical standpoint, it is impossible to make detailed in situ measurements to describe and quantify the coastal phenomena in all these areas sufficiently to enable the planner and engineer to provide economically and environmentally viable solutions of these problems.

One thus looks for new tools that advance our capabilities in coastal studies. This material will summarize a few of the exciting possibilities of the use of ERTS-1 imagery in coastal studies. The material presented here is very preliminary and is a result of the synergistic contributions of personnel of the NASA-Goddard Space Flight Center and the Coastal Engineering Research Center.

ERTS-1 SATELLITE

The ERTS-1 satellite and its user oriented applications are described in NASA-Goddard Space Flight Center Data Users Handbook, Earth Resources Technology Satellite, Document 71SE4249 with appropriate revisions (1). Analysis of multispectral images is given in reference (2). The nation's shorelines are discussed by Sheppard and Wanless in reference (3) and in the Report on the National Shoreline Study (4). Coastal applications of the ERTS-1 satellite are given in (5), (6) and (7). General coastal processes are contained in the Shore Protection Manual (9).

REGIONAL VIEWS

An obvious application of the ERTS-1 imagery is in obtaining regional views of extended coastal areas. An example of this is shown in Figures XI-1 and XI-2 which show multispectral scanner (MSS) bands 4,5,6 and 7 and a color composite from Orbit 0333, NASA E-1024-15073 and NASA E-1024-15071 and NASA E-1024-15080. Note that a clear line of demarcation along the New Jersey coastal areas is seen on the color composite view. Black and white views show a similar effect. Although readily compiled from ERTS imagery; mosaics made from aerial photography are often extremely difficult to match due to problems inherent with variation of sun angle and camera geometry. This is especially true when comparing the large, well-defined sediment plumes seen in ERTS with attempts to prepare mosaics of areas over water. We thus see in ERTS for the first time the structure of the large coastal and lacustrine plumes associated with sediment producing areas in many parts of the world. The four channels of the multispectral scanner show increasing sediment with decreasing MSS channel number. As shown on Figure XI-3 taken near Admiralty Island, Alaska (Orbit 0266 NASA E-1019-19430) and consistently shown in other coastal locations, minimum sediment is shown on MSS-7 gradually increasing to maximum sediment on MSS-4. In ERTS imagery where atmospheric haze is a problem on MSS-4, good sediment patterns are shown on MSS-5. Well-defined sediment sources indicate the direction of coastal currents away from the source. A large number of types of sediment plumes have been discussed in the literature. Considerable work on this subject has been supported by the Office of Naval Research and in particular, the Geography Branch. Material in this regard is available from Evelyn Pruitt, Chief of the ONR Geography Branch. The major sediment sources that have been seen in ERTS imagery appear to be derived from river discharges and to a lesser extent from cliff erosion. Sediment plumes are also seen in the Gulf of California near Mazatlán, Mexico (Orbit 0111, NASA E-1008-17062, not shown).

COASTAL CONFIGURATION

Information on the predominant direction of littoral movement is often derived from a number of specific shoreline configurations or changes. Particular work in this regard has

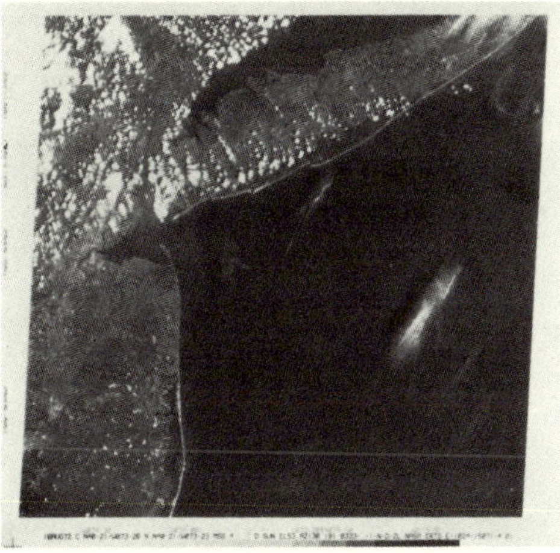

MSS 4

MSS 5

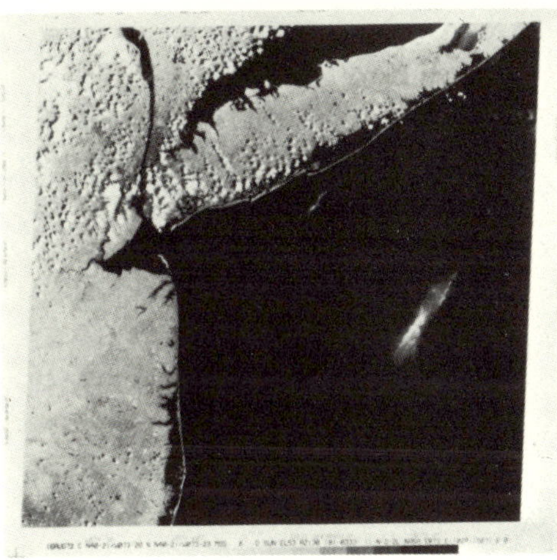

MSS 6

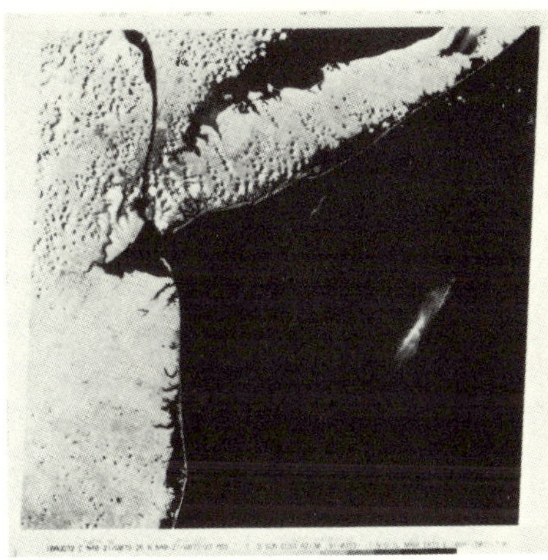

MSS 7

Figure XI-1. ERTS-1 images of the New York–Long Island area showing all MSS bands.

Figure XI-2. Color composite mosaic of the New York–New Jersey coastal zones made by combining consecutive ERTS images.

MSS 4

MSS 5

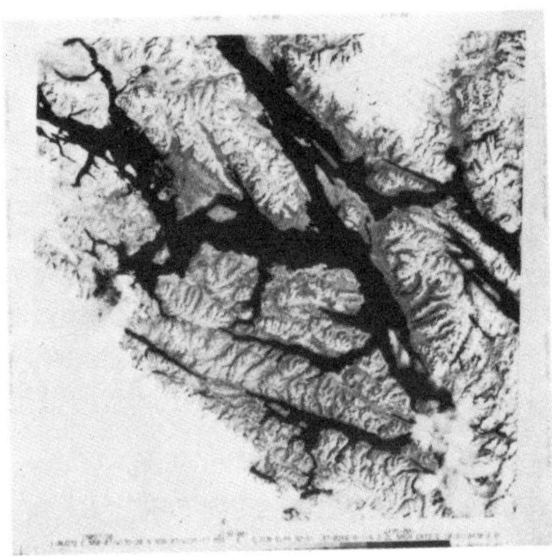

MSS 6

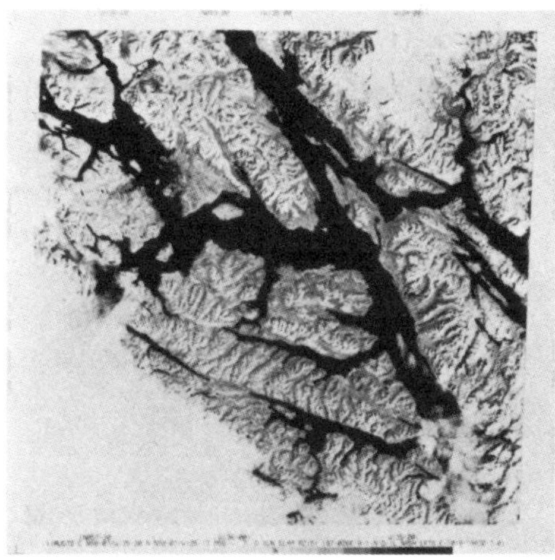

MSS 7

Figure XI-3. ERTS-1 images of the Admiralty Island–Lynn Canal area near Juneau, Alaska showing all MSS bands.

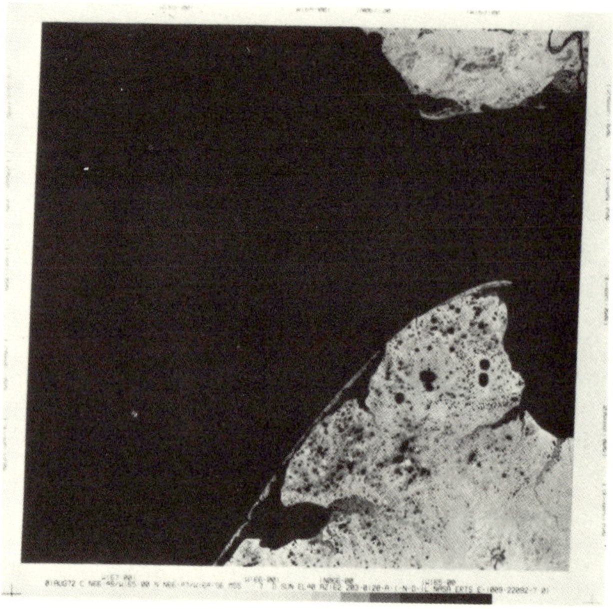

Figure XI-4. ERTS-1 MSS Band 7 showing the coastal zone west of Kotzubue Sound in the Seward Peninsula, Alaska.

Figure XI-5. ERTS-1 MSS image of Lake Marangua area on the southeastern African coast just south of the Tropic of Capricorn.

been published by Galvin regarding coastal inlets (8), and also in the Shore Protection Manual (9). Note particularly on the barrier beaches seaward of Long Island, the strong overlapping entrances indicating drift towards New York Harbor (Figure XI-1 and Figure XI-2). Note the similarly occurring spit pointing toward the entrance to New York Harbor. Although shown in the figures, and more easily seen on the 70 mm originals, the pronounced offset at some coastal inlets is an example of an indication of the direction of littoral drift.

BARRIER ISLANDS

One of the major types of natural protection to the shore and thus of great interest to the coastal planner and the coastal engineer are the barrier islands. Due to the large extent of these islands, it is relatively difficult to obtain good synoptic coverage of major portions of the islands after severe storms. One important use of ERTS will be in studying the changes in barrier beaches and barrier islands after severe storms or hurricanes. The precise location capabilities of ERTS will be particularly useful in the measurement of the changes in these islands. Some idea of the capability of ERTS to resolve small features in barrier beaches was shown on a random comparison of a section of ERTS imagery with the Army Map Service scale 1:250,000 sheets. An example is Shishmaref and Kotzebue, Alaska (Figure XI-4). The particular area (Orbit 0128) shown is located between Arctic Lagoon and Cape Essenberg, roughly latitude 66°, 30′ north, longitude 165° west. Note that very small features in the smallest channels are clearly seen. A new inlet or closed inlet would be easily detected. I would estimate that the inlet channels are in the range of three to four hundred feet wide. The interface between land and water is, of course, particularly enhanced by the infrared bands.

UNDERWATER PENETRATION

During examination of the ERTS imagery, it is clear that considerable underwater penetration is possible under the appropriate conditions. This will be particularly useful in detecting, for example, changes in underwater sand formations such as those located between Florida and the Bahamas. Considerable penetration was also observed in ERTS imagery of tropic waters where the comparison between various bands of the multispectral scanner gives an indication whether the phenomena being observed is at the surface or under water. Analysis of results from specific test sites will be required before detailed results can be presented. An interesting fringing reef which shows up well on RBV1 around Poeloe Besar, Flores, Indonesia (see HO Chart 3090) (Orbit 46, NASA E-1004-01241, not shown).

COASTAL WAVES

In viewing ERTS imagery of coastal areas, particular attention was paid to areas where coastal waves might be present. Breaking waves were identified on the coast and around a coastal bay mouth bar located along the Spanish-Sahara coast (Orbit 0205, NASA E-1015-01523, not shown). Waves are in the series of small dots along the coast and out over the bay mouth bar. Breaking waves along a bar ten miles in length are seen on the coast of

Mozambique (Orbit 0119, NASA E-1009-07064) (Figure XI-5). Breaking waves also appear on the north side of bars along the New Jersey coast (Orbit 0333, NASA E-1024-15073) (Figure XI-2). Additional analysis will undoubtedly reveal other areas of breaking wave identification. In areas of high waves in deep water with appropriate sun angles, it is expected that long straight deep water waves as seen in the U-2 simulation flights will be observed in ERTS imagery.

CONCLUSIONS

Only a few samples of the ERTS-1 imagery to coastal studies have been presented. Even with the limited inspection possible of available imagery, a number of clearly demonstrated applications of ERTS to coastal studies have been described.

ACKNOWLEDGEMENTS

Acknowledgement is gratefully made to the personnel of NASA/Goddard, particularly Messrs. Bill Nordberg and Tom Ragland for their support. This presentation would not have been possible without the assistance of Dick Holmes, Jack Palgen in Building 23 and Clair Hill and others who kept the ERTS Browse Room open many evenings while I was scanning ERTS imagery. A special thanks is given to the sympathetic understanding of Miss Gail Blackmore in actually obtaining the ERTS imagery shown here. Acknowledgement is also made to the following CERC investigators who viewed the ERTS imagery and commented on its applications: Mr. Jim Balsillie, Mr. Dennis Berg, Mr. Charlie Chesnutt, Dr. Dave Duane, Dr. Craig Everts, Dr. Jerry Galvin, Dr. D. Lee Harris, Mike McClenan and Dr. Don Woodard. Acknowledgement is also made for Herb Bruder for mounting mosaics, Jim Dayton for producing the slides and Barbara Fletcher for typing the manuscript.

Data presented in this presentation, unless otherwise noted, were obtained from research conducted by the United States Army Coastal Engineering Research Center under the Civil Works research and development program of the United States Army Corps of Engineers. Permission of the Chief of Engineers to publish this information is appreciated. The findings of this paper are not to be construed as official Department of the Army position unless so designated by other authorized documents. The review and comments on the manuscript and continued support by LTC Don S. McCoy, Director, CERC are gratefully acknowledged.

REFERENCES

1. NASA/Goddard Space Flight Center Data Users Handbook, Earth Resources Technology Satellite, Document 71SE4249.

2. Short, Nicholas M. and MacLeod, Norman H., "Analysis of Multi-Spectral Images Simulating Earth's Observation", NASA/Goddard Publication X-430-72-118.

3. Sheppard, F.P. and Wanless, H.R., "Our Changing Coastlines", McGraw-Hill, 1971.

4. National Shorelines Study – Summary Report, Dept. of Army, Corps of Engineers, Washington, D.C., August 1971.

a. Shore Management Guidelines, Dept. of Army, Corps of Engineers, Washington, D.C., August 1971.

b. Shore Protection Guidelines, Dept. of Army, Corps of Engineers, Washington, D.C., 1971.

c. Regional Inventory Reports for Entire U.S. Coastline, published by various Corps of Engineers Division Offices.

5. Magoon, O.T., Jarman, J.W., Berg, D.W., "Use of Satellites in Coastal Engineering", Proceedings First International Conference on Port and Ocean Engineering Under Arctic Conditions, The Technical University of Norway, Trondheim, Norway, August 23-30, 1971.

6. Magoon, O.T., Jarman, J.W., Perry, Doug M., "Coastal Engineering Application of the ERTS-A Satellite", Proceedings 13th International Conference on Coastal Engineering, Vancouver, British Columbia, Canada (in press).

7. Magoon, O.T., Perry, Doug M., "Use of Remote Sensing in the Study of Coastal Processes", Proceedings 13th International Conference on Coastal Engineering, Vancouver, British Columbia, Canada, July 1972 (in press).

8. Galvin, Cyril J., Jr., "Wave Climate and Coastal Processes", The Water Environment and Human Needs, Massachusetts Institute of Technology, Symposium, October 1-2, 1970.

9. Shore Protection Manual, Coastal Engineering Research Center, Washington, D.C. (Replacement for Technical Report No. 4, Shore Protection Planning and Design) (in preparation).

10. Schubel, J.R., Carter, H.H., Schiemer, E.W., Whaley, R.C., "A Case Study of Littoral Drift Based on Long-Term Patterns of Erosion and Deposition", Chesapeake Science Vol. 13, No. 2, p. 80-86, June 1972.

11. Szekielda, K.H., Kupferman, S.L., Klemas, V. and Polis, D.F., "Element Enrichment in Organic Films and Foam Associated with Aquatic Frontal Systems", Journal of Geophysical Research, Vol. 77, No. 27, p. 5278, September 20, 1972.

Part II

PHOTOINTERPRETATION AND SHORELINE CHANGES

Air Photograph (1958) of Ocracoke Island, North Carolina, after hurricane Helene of 1958. (U.S.C. & G.S. photo)

Editor's Comments on Papers 8 Through 14

The position and configuration of a shoreline in a given coastal area is determined by the balance between sediment contribution and sediment removal. An imbalance between rates of accumulation and removal of sediments results in a prograding or a receding shoreline. Beach sediments are contributed to the shore zone by inland rivers, from erosion of coastal cliffs, by longshore currents, or from offshore. Removal of sediments takes place mainly by waves and currents, and in some cases by man through the construction of upstream dams and inlet jetties. Stability of a coastal zone does not infer a static condition. The coastal zone is a dynamic system in which sediment is in continuous motion. Stability refers only to conditions in which the amount of sediment removed is equal to the amount of sediment added. It also refers to a long-term condition where seasonal changes are more or less balanced. Kenney (1962) described the process of shoreline changes as follows: "The sea is one of the earth's most powerful forces, matched only by winds and earthquakes and running inland waters. It is whimsical—building one day, destroying its own handiwork the next. It can be brutal, it can be exceedingly subtle, but it can never be still."

Waves play a major role in determining coastal configuration and shoreline features. The capacity of waves for moving sediment along coastlines is determined by their height and length, as well as by the direction of their approach. The position of offshore shoals and the configuration of the sea bottom influence the amount of wave erosion from the beaches. Wave refraction may be responsible for alteration of bottom features by accumulation or removal of sediment (Munk and Traylor, 1947). In areas where a diversified submarine topography exists offshore, the nearshore circulation is ordinarily influenced to a large extent by wave divergence at the heads of submarine valleys and by wave convergence over submarine ridges (Shepard and Inman, 1950). Transportation of sediments in the nearshore zone is dependent mainly on longshore currents, rip (return) currents, and horizontal diffusion. The existing refraction patterns greatly influence all of these factors.

The presence of a wide, gently sloping, beach can effectively protect the coast against wave attack. Dissipation of wave energy takes place as a result of friction between the lower portion of the waves and the gradually sloping bottom.

Longshore bars are common along gently sloping sandy beaches. They offer protection to the beaches immediately behind them as waves tend to break and dissipate their energy against the bars. Beach erosion is usually greater behind gaps in bars or where bars are completely absent.

Barrier islands exist along many coasts. They are particularly common along slowly subsiding coastal plains and around deltas, especially along sections of the delta that are not actively growing. Typical barriers consist of three major divisions (Shepard, 1960a): (1) an outer beach with a broad berm; (2) a belt of dunes; and (3) an inner flat or marsh. Typical examples of barrier islands exist along the Atlantic and Gulf Coasts of the United States and portions of the southeast coast of Australia. Barrier islands migrate landward by washovers, which involves the transfer of sediment by storm waves from the oceansides of the islands to the lagoon sides.

Barrier islands are normally interrupted by inlets. Inlets develop as a result of waves washing over narrow and low portions of the barrier islands. When storm tides pile up water in a lagoon behind a barrier island inlets may form by washouts toward the sea. Tidal currents can widen inlets and keep them open. Inlets tend to migrate in the direction of littoral drift by sand deposition on the updrift side of an inlet and erosion on the downdrift side. They close as a result of a shift in the balance between littoral drift and tidal currents which results in shoaling inside the inlet and reduction of the tidal prism.

Prograding shorelines are usually marked by beach ridges. Beach ridges are low ridges of sand or shingle, alternating with slight depressions, which mark former positions of prograding shorelines. When a beach is eroded, the eroded face may develop at an angle to the beach growth direction, and the truncated beach ridges may help in restoring earlier beach outlines.

The configuration of coastlines and the rates of their progradation or recession are influenced to a large degree by the type of material that forms them. They may be made of loose material that is easily eroded and transported; they may be made of hard rock material with occasional pocket beaches; or they may be formed of unconsolidated and easily eroded cliffs.

Shepard and Grant (1947), summarized the factors that influence wave erosion on rocky coasts as follows: (1) hardness of the rock; (2) structural weakness; (3) coastal configuration (straight cliffs are much less affected by waves than irregular ones); (4) solubility of rocks and rock cement; (5) height of cliffs (high ones recede slower); and (6) nature of wave attack.

Along a regular stretch of beach, uniform transfer of material may leave the shoreline configuration unaltered. However, measurements normally show that the beach had suffered a loss of several feet of sediment. This emphasizes that visual (qualitative) comparison of the shores is not sufficient by itself in determining shoreline changes, and that only through accurate measurements of beach profiles in the field or of beach width on air photographs can rates of erosion or deposition be determined and interpreted correctly.

The paper by Lucke (Paper 8) is one of the earliest studies to utilize air photographs in the interpretation of coastal features including channels, shoals, and tidal deltas. The paper involves a detailed study of a tidal inlet on the New Jersey shoreline. Shoreline changes at the inlet were analyzed from old records as well as from sedimentary field data. Air photography of the region proved to be more revealing than any other method of observation. Photointerpretation of the shoals indicated active deposition by the flood tidal currents, which drop their sediment load on the sand flats, but continue to deposit sand until it spills over the edges of the shoal into the main channel.

The paper by Howard (Paper 9) and another paper by Nichols and Marston (1939) were the first to use air photographs in illustrating the effects of hurricanes on coastlines. Oblique photographs taken after the hurricane showed the new inlets that were opened across barrier beaches. Both papers dealt with the effects of the same hurricane, September 21, 1938, in different areas. Howard's study was concerned with the hurricane's modification of the offshore bar of Long Island, New York; whereas Nichols and Marston dealt with shoreline changes in Rhode Island that were produced by the hurricane.

Hurricane effects, in both areas, included the formation of new inlets, erosion of foredunes, erosion of the beaches, deposition of beach material in bays and marshes, and salt water flooding.

Under storm conditions, the destructive force of storm waves is enhanced by the rise of the tide. Such rise in water level results in bringing the zone of wave activity farther inland and allows larger waves to break against the foredunes or the shore cliff. Severe storms and hurricanes could result in changes, in a few hours, that are equivalent to hundred years of normal shore processes. The effects of one of these storms, and the flooding associated with it along the east coast of England are described in the paper by Professor J. A. Steers (Paper 10). The storm that hit the coast on the night of January 31, 1953, resulted in the flooding of 206,161 acres of land and caused great devastation of the forests in northeastern Scotland and profound damage to the low lands of Holland. Wide beaches and dune belts offered a certain amount of protection to the land behind them. Some walls were overtopped and others were breached. Some low areas were flooded to a depth of 1.5 m, and much valuable arable land was spoiled. Air photographs are used to illustrate the extent of damage and flooding by the storm. Two sets of before and after photographs are also included in the paper.

Similarly, Mosby (1939) described the effects of the Horsey Flood, on the night of February 12, 1938, on the east Norfolk coast in England. High waves resulted in erosion of 465 m of dunes, and salt water poured through this breach and flooded 7500 acres of low-lying hinterland. An air photograph shows the dimension of the breach and extent of flooding.

The effects of hurricane Audrey on the marginal deltaic plain of southwestern Louisiana were presented by Morgan, Nichols, and Wright (1958). Storm tides of the hurricane overwhelmed the beaches and resulted in reworking and spreading of their sediments. An immediate result of the storm was a zone of exhumed marsh that was left between the mean gulf level shoreline and the beach sediments. This zone of silty marsh clay has been suffering retreat by wave erosion during the months following hurricane Audrey.

The "Great Atlantic Coast Storm" of March 6–8, 1962, caused tremendous damage along the U.S. section of the Atlantic coast from New York to North Carolina. Although the winds associated with the storm did not reach hurricane force, it caused serious damage because it lasted through three successive high tides. Offshore bars, barrier beaches, sand dunes, and man-made defenses survived the first high-tide period in many cases, but the repeated hard action during a second and third high tide broke through in many places causing spectacular damage at Fire Island (Long Island, New York), along many kilometers of the New Jersey, Maryland, and Delaware shores, and at Cape May, New Jersey (Burton, Kates, and Snead, 1969).

Dietz (Paper 1) and Shepard (1950a) pointed out the importance of before and after air photography in studying shoreline changes. Zeigler and Ronne (1957) described a program of studying shoreline changes along the Atlantic coast of the United States by obtaining vertical air photographs of parts of the coast and oblique color photos of the entire coast, by a movie camera at a rate of two frames per second. The time-lapse air photography provided a means for comparing nearshore bars, giant ripple patterns, cuspate shores, and beach ridges in different locations. It showed that inlets change more often and more spectacularly than any other large features along coasts. Annually obtained time-lapse air photography would permit the observation of changes in land use patterns as well as progress of reclamation projects such as the filling of marsh areas. In all of these examples, only qualitative data can be obtained because of the difficulty of obtaining quantitative information from oblique photos.

Before and after air photographs were utilized by Tanner (1961) in studying shoreline changes in Florida due to hurricane Donna. By assuming a depth of 1.5 m for areas of erosion and deposition, he managed to convert areal changes to volume changes and concluded that the hurricane had accelerated existing active processes.

Immediately after the "Great Atlantic Coast Storm" of March, 1962, the U.S. Coast and Geodetic Survey initiated a storm damage survey for the purpose of updating nautical charts. Waugh (1962) reported that approximately 1620 km of the Atlantic coast of the United States were photographed with panchromatic film; and major inlets and harbor entrances as well as newly formed inlets were photographed with color film. The photographs showed that many of the nautical charts were rendered obsolete as a result of the storm and that color air photos were useful in storm damage investigations and in studies of shoreline changes.

Athearn and Ronne (1963) presented a study of shoreline changes at Cape Hatteras in which they utilized sequential air photography from 1945 to 1962 to obtain measurements of beach erosion. El-Ashry and Wanless (1963) also reported on the use of sequential air photography in studying hurricane effects on coastlines and in a later study (1965) reported on the rate of growth of a newly formed tidal delta in Pamlico Sound, North Carolina, after five days, two months, and ten months of the opening of an inlet, from air photographs.

The paper by Cameron (Paper 11) discusses the value of sequential air photographs, obtained at intervals ranging from two to twenty years, in studying long-term coastal changes. Two examples, Rose Bay and Advocate Harbor, Nova Scotia, Canada, were studied. Directions of littoral drift and growth of bars and spits were determined. The author concluded that a program of systematic study of existing photographs and supplementary re-photography would give results worth more than the cost involved. Cameron also published a study (1965) on shoreline changes on Sable Island, Canada, in which he supplemented sequential air photography with maps and charts dating back to 1776.

El-Ashry (1966) conducted an extensive study of storm effects and long-term coastal processes on shoreline features and shoreline changes in selected areas in South Carolina, the west coast of Florida, and Texas by utilizing sequential air photography and earlier surveys, maps, and coastal charts. Measurements of shore erosion or deposition were determined, and changes in beaches, beach ridges, barrier islands, tidal inlets, deltas, shoals, and other features were described. A method of dating old beach ridges from a comparison of air photographs and older maps and charts was also presented.

Paper 12 by El-Ashry and Wanless presents a number of examples from North Carolina, South Carolina, Florida, and Texas to illustrate the usefulness of sequential air photography in recording storm effects and long-term trends in shoreline changes as well as in studying shoreline features. The study points out that sequential air photography is an excellent tool for the study of coastal morphology and change. It provides an exact basis for measuring net changes in shape between photographic dates.

In the 1966 study by El-Ashry, up to ten dates of air photography were available for some areas. Not all photographs were of the same scale. Measurements from the photographs to determine amounts of erosion or deposition were scaled off directly from the photographs when the scales were known. Measurements were made from the high water line to fixed reference points or landmarks that appeared on all photographs. Where successive overlaps of the outline of the beaches at different dates were needed to illustrate visually the changes, the high water shorelines were traced from the individual photographs or charts and the reference points were plotted on them. All tracings were then reduced to the same scale using a Map-O-Graph. Similar projection systems could work as well.

El-Ashry and Wanless (1968) also presented a study of shoreline changes along a section of the barrier islands of the Outer Banks of North Carolina, between Capes Hatteras and Fear. Changes in coastal configuration and measurements of amounts of erosion and deposition were determined from sequential air photographs dating between 1945 and 1962, nautical charts and field surveys dating back to 1852, and topographic maps. Normal processes of coastal evolution are interrupted frequently by severe storms and hurricanes. Sixteen hurricanes struck the North Carolina coast between 1879 and 1943, and additional ones came later. The study reports on shoreline changes between dates of photography and maps, and on the effects of hurricanes and severe storms on barrier islands, inlets, capes, and shoals.

The effects of hurricanes and severe storms on that section of the North Carolina coast were found to include: erosion of the ocean side of the beaches, washover fans formed on the landward side of the barrier islands or in the lagoons behind them, new inlets opened and others closed by beach drifting, and hurricane effects are cumulative and most pronounced on low exposed beaches, cape points, and inlets. Material eroded from the beaches during storms may be moved in the direction of dominant longshore currents or offshore to form submerged bars or shoals, and finer material may get trapped in strong seaward-moving rip currents and move out on the continental shelf.

Wagener (1970) described a study of beach erosion in the Charleston Harbor area of South Carolina using photomosaics and photo index sheets that were taken in 1941, 1949, 1954, 1958, and 1964. Shoreline changes were interpreted from the mosaics, and current conditions of beach erosion, as well as the performance of erosion control measures, were evaluated from field observations.

Stafford and Langfelder (1971) evaluated a procedure to utilize existing air photography to conduct a survey of coastal erosion in North Carolina. The procedure consists of selecting stable reference points on air photographs taken at different dates and measuring the distance between these points and points on the transient beach. The measurements are then multiplied by the scale of the air photographs to produce ground distances. Differences in ground distances between the different dates represent the changes in location of the beach during that interval of time.

Maruyasu and Minatoku (1973) reported on a study of the erosion of Niigata Beach, Japan, from ERTS-1 imagery. One of the causes of coastal erosion in that area is the construction of flood-control dams, short cut channels, and breakwaters' which reduce the transportation of river sediments. They concluded that ERTS-1 imagery was helpful in showing the relations between conditions of erosion, effluent patterns affected by coastal currents, and engineering construction works. Similarly, Emplaincourt and Wielchowsky (1974) reported on studies of shoreline changes in Alabama from ERTS-1 data.

Many parts of the world are clearly suffering from beach and coastal erosion. The papers and comments already mentioned indicate the overwhelming interest in coastal studies and coastal changes along sandy beaches. Relatively little consideration, however, has been given to erosion of coastal cliffs. Perhaps one reason for the lack of such attention is the notion that beach erosion presents the more critical situation at the present time and that wave energy tends to be expended on the beaches rather than against the base of the cliff behind them. However, as Chieruzzi and Baker (1958) reported, in many areas where the toe of the bluff is protected from wave action by a beach, bluff failures still result and present serious problems. Bluff failures are similar in their characteristics to other types of natural slope failures. The only difference is that the additional forces caused by wave and ice action are present. In addition, the importance of ground water in the recession of coastal cliffs was also mentioned by Steers (1962). He stated: "The recession of cliffs in incoherent materials, such as boulder clay and comparable formations, is by no means wholly the work of the sea." He added that water percolating from above through heterogenous deposits may easily cause slips and slides, particularly if it reaches an impervious layer. The material that falls to the foot of the cliffs may later be removed by waves. The waves also cut into the foot of the cliffs and cause more falls.

The paper by Horikawa and Sunamura (Paper 13) presents a

study of the short-term erosion of coastal cliffs in two areas in Japan by using air photographs. The paper includes a general concept of the erosion of coastal cliffs, and instructions on the use of air photography in the study of erosion of coastal cliffs. The annual amounts of sand lost from coastal cliffs are estimated for one area, and the relationship between eroded distance of the cliff and the strength of rocks and the frequency of wave attack as well as prediction of future eroded distance is presented for the other area.

The problems of beach erosion, accretion in inlets and harbor entrances, coastal flooding, recession of coastal cliffs, high winds and waves as well as surges associated with hurricanes, are not the only ones facing all coastal areas of the world. Some coastal areas are located in tectonically active belts and encounter varying degrees of uplift or subsidence, as well as frequent earthquakes. Others are devastated by tsunamis (MacDonald, Shepard, and Cox, 1947). As an example of the effects of earthquakes on coastal areas, the paper by Kachadoorian (Paper 14) was chosen to illustrate the effects of the Alaska Earthquake of March 27, 1964, at Whittier, Alaska. Similar reports by the U.S. Geological Survey in other parts of Alaska include: "Effects at Homer" (Waller, 1966; with a section on beach changes on Homer Spit by K. U. Stanley), "Effects at Seward" (Lemke, 1967), and "Effects on Shore Processes and Beach Morphology" (Stanley, 1968). All of these studies include air photographs portraying damage to the coastal zone, and most of them include before and after air photographs.

Whittier was hard hit by the earthquake where thirteen persons out of the seventy people who were living at Whittier at the time were killed. A major part of the port facilities was destroyed and total damage was estimated in excess of five million dollars. Damage was caused by: (1) a 1.6-m subsidence of the landmass; (2) seismic shock; (3) submarine landslides; (4) at least two waves generated by landslides.

Stanley (1968) reported that some 18,000 km of shoreline in south-central Alaska was affected by subsidence or uplift associated with the earthquake. The changes in shoreline processes and beach morphology were similar to those caused by gradual changes in sea level acting over hundreds of years. In areas of subsidence, beaches tended to flatten in gradient and to recede landward. In uplifted areas, beaches were raised and stranded out of reach of the sea. New beaches are slowly and gradually developing to fit new sea levels.

8

Reprinted from pp. 44–51, 56–61, 92, 94 of *Shore and Beach* **2**(2):44–94 (1934), journal of the American Shore and Beach Preservation Association

A STUDY OF BARNEGAT INLET

BY JOHN B. LUCKE.

I. Abstract

A detailed study is made of Barnegat Inlet, a typical tidal inlet on a shoreline of emergence. The history of the inlet is deciphered by means of old records and sedimentary field data. Tidal conditions, including sea-level variations and current velocities are described. The sediments of the tidal delta are analyzed according to grain size, minerals and micro-fossils, and the field and laboratory technique is explained. All these data confirm the assumption that the tidal delta is composed of debris carried by longshore currents and deposited in the lagoon by flood tidal currents.

A new theory of an orderly evolution of lagoon deposits is presented. It is believed that tidal inlets may produce distinctive deposits in the lagoons, study of which may aid in deciphering shoreline history.

II. Introduction

THE Barnegat Inlet region (see *Chart*) is on the shore of a typical emerged coastal plain of the belted type. The relief of the area studied does not exceed 25 feet. The New Jersey coastal plain surface, developed on late tertiary to recent sands and marls, slopes seaward from four to six feet per mile. This gentle incline is bordered by marshes which vary from fresh to salt water flora, dependent upon the salinity of the water. In the Barnegat region the width of the mainland marsh fringe may be as great as two miles, but is usually much less.

The whole New Jersey shoreline south of Point Pleasant is a nearly continuous offshore bar, or barrier beach. This sandy beach ranges from a few yards to about a mile wide and attains an unbroken length of twenty miles between Bay Head and Barnegat Inlet. Tidal inlets, of which Barnegat is typical, cut the offshore bar and connect the ocean with a shallow tidal bay, called a lagoon in shoreline nomenclature.

The New Jersey lagoon is about five miles wide in the widest place, but averages about three miles in width. Its continuity is broken by marshy islands and peninsulas, which results in a chain of open-water bays separated by marsh, but connected by narrow tidal channels locally known as thoroughfares. The largest of such single bays on this shore is Barnegat Bay, which is nearly thirty miles long, and widest in the inlet region.

The landscape of the mainland as seen from the center of Barnegat Bay is that of a low, marshy or grassy shore, beyond which rises, very gently, a heavily wooded terrain. This dense growth of shrubs, small pines and red cedars, seldom represents virgin timber, because of the numerous forest fires in this sparsely populated area. The view east to the barrier beach also includes a shore fringe of marsh grass, but beyond this rises a hummocky ridge of sand dunes partially covered by grasses and shrubs. The dunes on the beach between Seaside Park and Barnegat Inlet, known as Island Beach, are perhaps the best developed dunes in New Jersey and may be as high as thirty feet. Just west of the dunes, the observer sees between the dune ridge and marsh shore many small groves of cedars, shrubs, and typical strand plants which never grow higher than the sand dunes because of the sand-laden winds from the beach. Some of the trees look as if they might have had their tops planed off at the level of the crest of the dunes. Of course, this did not happen, but when the trees reach this height they continue to grow only on a horizontal plane.

Between Barnegat Inlet and the mainland are many marshy islands rising but a few feet above the high-tide level of Barnegat Bay. Some are covered only by the typical high-tide salt grass, such as *Spartina patens*, as on Sea Dog Shoal (note *Chart*). Others have existed longer above water, like Clam Island, and support high meadows and shrubs, (*Figure 1*). Study of the

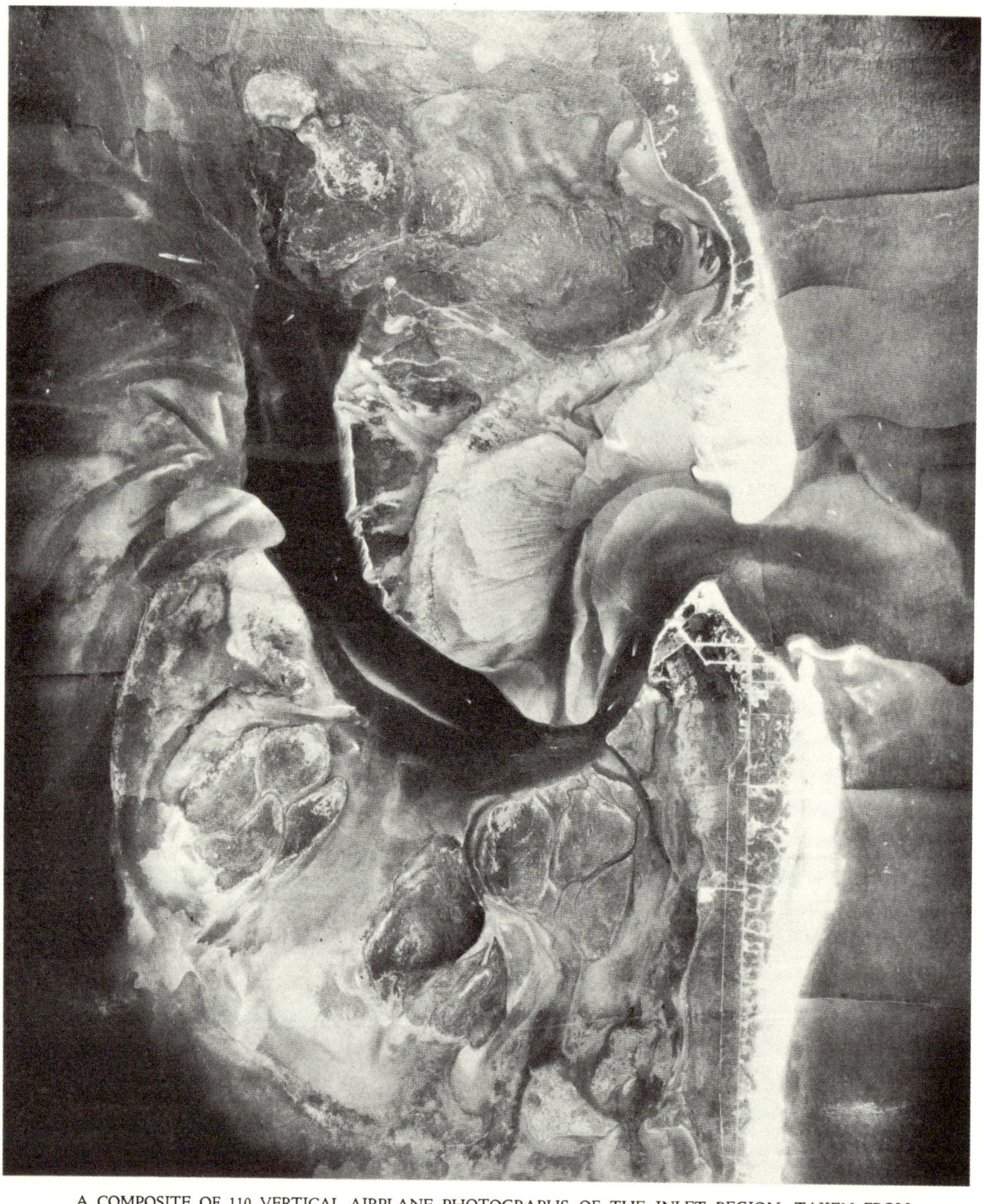

A COMPOSITE OF 110 VERTICAL AIRPLANE PHOTOGRAPHS OF THE INLET REGION, TAKEN FROM A HEIGHT OF 7000 FEET FOR THE STATE OF NEW JERSEY BOARD OF COMMERCE AND NAVIGATION. PHOTOS BY AIRVIEW INC., LONG BRANCH, N. J.

surrounding bottom surface, shown by submarine contours on the chart, reveals these marsh islands as but slight elevations on a submerged, semicircular platform with the inlet in the center. This great flat stretches almost to the mainland and is broken only by tidal, marshy, islands and the current channels which radiate roughly from the site of the inlet. Such an arcuate shoal, when associated with a tidal inlet and paired with a similar if smaller shoal seaward of the inlet, is known as a tidal delta. The main purpose of the present investigation is to establish the origin and precise sedimentary nature of this particular tidal delta. Any conclusions reached are then interpreted in relation to the known forces acting upon such a shoreline, to determine the significance of tidal deltas in the history of a shoreline of emergence. The *Frontispiece* gives a bird's-eye view of Barnegat Inlet and the tidal delta. This photograph is a composite of 110 vertical airplane photographs taken from an elevation of 7000 feet.

The area is reached only by boat or by automobile from Waretown, a small fishing and farming village on the New York to Atlantic City highway about 75 miles from New York City.

The locality is thinly populated and relatively inaccessible, considering its nearness to thickly settled areas. It is largely because of this that the region is a paradise for lovers of salt-water sports. Commercial exploitation of its natural advantages is so slight that the wild aspect of the land is undisturbed and gunners or fishermen may enjoy their sports free from the influences of civilization. The difficulty of access has probably prevented a rapid depletion of the extensive fish and game resources and has helped to maintain its great popularity with those who frequent these waters.

MAJOR CONTRIBUTIONS OF EARLIER WORKERS.

A brief classification of the principal types of work done heretofore may be of interest to other students of shoreline problems, although a full bibliography is provided at the end of this report and the material from other papers is individually cited in succeeding chapters. Most important contributions to shoreline studies of New Jersey or to related problems may be divided into six groups:

1. General Shoreline Treatises.

Comprehensive discussions of the broad field of shoreline geology are found in two noteworthy volumes: "Shore Processes and Shoreline Development," and "New England Acadian Shoreline," both by Douglas Johnson.[1] These books provide valuable shoreline data of a general and detailed nature and contain excellent classified bibliographies of nearly all important American and European works.

2. Studies of Shorelines of Emergence.

Shorelines of emergence are those which, like the New Jersey shoreline, have had their past histories and present forms determined by a lowering of the sea, a rise of the land, or both. A pioneer in this subject in America, also a leader in many lines of geologic thought, was N. S. Shaler, whose monograph on "Beaches and Tidal Marshes of the Atlantic Coast"[2] is frequently cited in coastal literature. Many papers were contributed by F. P. Gulliver, most notable among them, his "Shoreline Topography."[3] Most of these earlier works, as well as later reports, are ably summarized by Johnson in "Shore Processes."

3. Engineering Researches.

The New Jersey Geological Survey (now a bureau of the Department of Conservation and Development) has published many reports by engineers, primarily concerned with beach protection and improvement of inland waters. From the 1868 octavo report by George H. Cook to the time of the establishment of the Department of Conservation and Development (1916), Cook, C. C. Vermeule, H. B. Kummel, L. M. Haupt and others have written shoreline reports for this survey. W. B. Harris has long been a student of beach erosion and is credited with building the first protective jetties on the New Jersey coast. Recent investigations, partly unpublished, have been made by the U. S. Engineer Office of the War Department and the New Jersey State Board of Commerce and Navigation.[4] A later (1932) War Department survey is utilized in this report, but the official account is not yet published.

4. Studies of Coastal Subsidence.

The marshes which form so large a part of the Barnegat region are common coastal features from Nova Scotia to Florida. In the last century a group of scientists, led by George H. Cook, and others, gathered botanical and other supposed evidence to advance the hypothesis that the Atlantic seaboard was sinking at the rate of from 20 to 75 cm. per century.[5] This romantic hypothesis has been called in question by the investigations of Douglas Johnson.[6]

Section E of the American Association for the Advancement of Science held a symposium at Atlantic

[1] *Johnson, Douglas.* "Shore Processes and Shoreline Development," 584 pages, 1919. John Wiley & Sons.

Id. "New England Acadian Shoreline," 608 pages, 1925. John Wiley & Sons, Inc.

[2] *Shaler, N. S.* "Beaches and Tidal Marshes of the Atlantic Coast," Nat. Geog. Mon. I, No. 5, 137-168 (1895).

[3] *Gulliver, F. P.* "Shoreline Topography," Proc. Amer. Acad. Arts & Sci. 34 (1899).

[4] *Reports by Board of Commerce & Navigation, N. J.*, on "The Erosion and Protection of the New Jersey Beaches." 1922, 1924, 1930.

[5] *Johnson, Douglas.* "Botanical Phenomena and the Problem of Recent Coastal Subsidence," Bot. Gaz. vol. XVI No. 6, page 449, (1913).

[6] *Johnson, Douglas.* "The Supposed Recent Subsidence of the Massachusetts and New Jersey Coasts," Sic. n.s. 721-723, (1910).

Id. "Botanical Phenomena and the Problem of Recent Coastal Subsidence." Bot. Gaz. vol. XVI, No. 6, 449-467, (1913).

Id. "Is the Atlantic Coast Sinking?" Geog. Rev. 3, 135-139 (1917).

City, N. J., on December 28th, 1932, entitled: "Pleistocene and Recent Changes of Level along the North Atlantic Coast," showing the present high degree of interest in this problem. The papers read are abstracted in the Bulletin of the Geological Society of America for March 1933.

The study of salt marshes in New England dates from the early papers of Mudge[1] and Shaler[2] in the United States, and Dawson[3] in Acadia, to the detailed investigations of salt marshes by A. C. Davis[4] of the U. S. Bureau of Mines. Salt marsh studies, although primarily botanical, are closely related to the problem of coastal subsidence and to the evolution of certain types of shores. Johnson reviews the ideas on the origin of tidal marshes, classifying the different marsh types and discussing their importance in shoreline geology in the chapters on Tidal Marshes of the New England Type, and Tidal Marshes of the Fundy Type, in his "New England Acadian Shoreline."

5. *Map and Chart Studies.*

The oldest reliable charts of the Atlantic coast, and in particular of New Jersey, are those of the U. S. Coast and Geodetic Survey. The New Jersey Geological Survey prior to 1916 and the Department of Conservation and Development since 1916 are valuable sources of data on shore protection, artesian water, and topographic maps of the shoreline. The New Jersey State Board of Commerce & Navigation and the U. S. Engineer Office of the War Department both contribute reports on navigation, waterway improvement, maintenance, and tidal conditions.

6. *Tidal and Other Studies.*

The U. S. Coast & Geodetic Survey, Division of Tides and Currents, publishes most of the statistical data on tides in the United States. The action of the sea and tides in relation to shorelines is discussed by Johnson[5] and Marmer[6]. The U. S. Engineer Office of the War Department gathers tide and current information especially for waterway maintenance and improvement reports. Its nearest offices are at Wilmington, Del., and New York City.

ACKNOWLEDGEMENTS.

The author's debt of gratitude to those whose generous aid made this work possible is indeed great. Space will only permit mention of a few of those whose services are deeply appreciated.

Acknowledgement is made to the U. S. Engineer Office at Wilmington, Del., and especially to Col. E.

Fig. 1.—Clam Island from the east. *Spartina glabra* in the foreground, *Spartina patens* in the middle ground and bayberry bushes in the background. Clam Island is in the southwest corner of the *Frontispiece.*

I. Brown and Mr. H. M. Price, whose constant willingness to furnish valuable information made possible the use of unpublished tidal and sounding data. Captain R. S. Patton, director of the U. S. Coast and Geodetic Survey, generously loaned a current meter and took a friendly interest in the work. Professor J. K. Finch of the Department of Civil Engineering at Columbia University performed a like service.

The author is especially grateful to Mr. L. C. Widmer of the Atlantic, Gulf & Pacific Company, New York City, whose extremely friendly cooperation and loan of sampling apparatus made it possible to obtain cores of the bay bottom under the water.

Mr. Victor Gelineau, director of the New Jersey State Board of Commerce & Navigation has the thanks of the author for permission to use airplane photographs taken for the state, and for his constant willingness to extend the facilities of his office and the benefits of his broad experience to these researches.

Valuable counsel and encouragement throughout the study was given by members of the Geology Department of Princeton University, especially Professors A. H. Phillips, W. T. Thom Jr., Paul MacClintock and R. M. Field. Professor J. R. Sandige has generously contributed the material on the micro-fossils of the lagoon sediments.

Thanks are due many Columbia University friends, especially the author's fellows in the Seminar in Geomorphology whose criticisms have been extremely helpful. Professor R. J. Colony generously contributed his

[1] *Mudge, B. F.* "The Salt Marsh Formations of Lynn," Essex Inst. Proc. II, 117-119, 1858.

[2] *Shaler, N. S.* "Beaches and Tidal Marshes of the Atlantic Coast," Nat. Geog. Mon. I, No. 5, 1895.

[3] *Dawson, J. W.* "Acadian Geology," 2d Ed., 687 pages, London, 1868.

[4] *Davis, A. C.* "Salt Marsh Formation near Boston and its Geological Significance," Econ. Geo. V, 623-639, 1910.

Id. "Salt Marshes: A Study in Correlation," Assoc. Amer. Geog. I, 138-143, 1911.

Id. "Origin and Formation of Peat," U. S. Bureau of Mines Bulletin No. 38, 165-186, 1913.

[5] *Johnson, Douglas.* "Shore Processes and Shoreline Development," 584 pages, 1919. John Wiley & Sons.

Id. "Studies of Mean Sea Level," Nat. Research Council Bull. No. 70, 1929.

[6] *Marmer, H. A.* "Sea Level Along the Atlantic Coast and Its Fluctuations," Geog. Rev. XV, 1925.

Id. et al. "Oceanography," Bull. Nat. Res. Coun. No. 85, Chapter VII, "Tides and Tidal Currents." 229-310, 1932.

Id. "The Tide," 282 pages, 1926. D. Appleton.

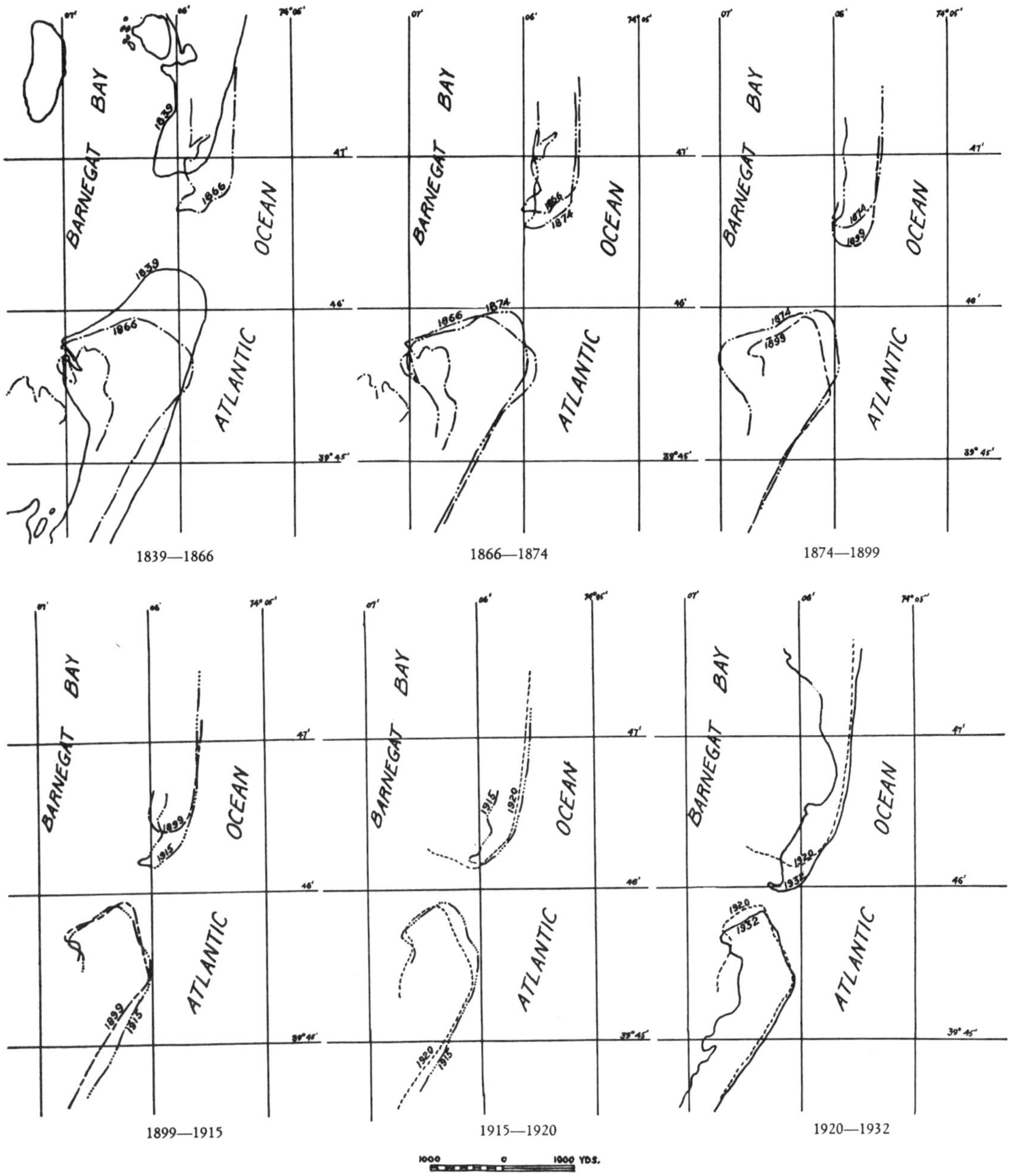

Fig. 2.—Shoreline changes at Barnegat Inlet, New Jersey. *Compiled by U. S. Engineer Office, Wilmington, Del.* Rearranged and traced by J. B. Lucke.

experience to the study of the heavy minerals and Dr. G. Marshal Kay offered valuable suggestions on the mechanical analysis of the sediments.

Professor Douglas Johnson directed the whole study as an interested adviser who gave constant encouragement and the inestimable advantages of his wide experience, without which the work would not have been possible. The great extent to which Professor Johnson's works have been drawn upon will be evident from the citations in the text.

III. History of Barnegat Inlet

Charts of mariners and early settlers indicate unbroken existence of Barnegat Inlet for at least 300 years. Accurate charts of the past century show steady southward migration at the rate of a mile per century. Sedimentary evidence indicates that Barnegat has not long exhibited this southward movement. The change from static to shifting conditions is probably due to the closing of the inlets between Barnegat and the headland[1] within historic times.

The inlet will probably attain equilibrium between deposition and erosion and continue to migrate southward.

A FEW explorers of the sixteenth century mapped North America, but so crudely that details of the shorelines were almost unrecognizable. The English and Dutch mariners of the seventeenth century produced the best maps and charts up to their time. One of the earliest accounts of a voyage along New Jersey was written in 1614 by Cornelius Mey. A countryman of his produced the Vanderdonck Map of the Atlantic coast in 1656, which was widely copied. Barnegat Inlet was present at that time, as were most of the inlets south of it, but each was labeled "Barnde-Gat." The name Barnegat is apparently a corruption of a Dutch word meaning a "narrow gap" or passage. The other inlets soon received English names such as Little Egg, Absecon, Hereford's, Corson's and Cold Spring.

More advanced charts were not drawn until the next century. The "American Atlas, with Views," by Henry Popple (1733) showed little change in the Jersey shoreline. Barnegat was then called "Burnigat or burning hole," and the highlands of Navesink were labeled "Neversink." A more famous map by Lewis Evans appeared in 1755, on which names are spelled as they are today. About halfway between Barnegat Inlet and the head of the bay was New Inlet, which, often known as Cranberry Inlet, was a point of entry for ships bound for Toms River, until it closed about 1812.[2]

The beach between Cranberry and Barnegat Inlets was named Island Beach, a name which has endured in spite of the change in conditions.

The Last 100 Years.

The era of modern, accurate charts of the coast of the United States began with the U. S. Coast & Geodetic Survey. The positions of Barnegat Inlet mapped in seven surveys from 1839 to 1932 are shown in *Figure 2*. Cross sections of the inlet mouth corresponding to four of these positions are drawn in *Figure 3*. *Figure 2* shows that the inlet has migrated southward about a mile in the past century. This confirms Haupt,[4] who in 1905 estimated this migration to be half a mile in 50 years.

The logical inference that this southward migration has been continuous for the life of the inlet is, however, not correct. Not only has the inlet shifted more rapidly in the past century than formerly, but it probably has never been far north of the 1839 position nor south of its present position in what may have been a long life. The reasons for this statement are discussed at length in *Chapter V* because the evidence found in the deposits examined by the author is entirely of a sedimentary nature.

If the sedimentary data constitute valid evidence that the inlet has not migrated continuously, then the recent and present migration of the inlet creates an erroneous impression about its past history. The old charts show Barnegat to be at least 300 years old, and, from its present position nearly at the center of a large arcuate delta, it would seem that the inlet was stationary, or nearly so, for a period of many centuries at least. However, this condition of stability, if true, changed abruptly in the early eighteen hundreds to the habit of steady, southward migration. The history of the past 100 years is represented in *Figs. 2 and 3*. The explanation of the striking change from stable conditions to progressive migration, accompanied by decrease in size, may be sought in old shoreline records.

Such a change may have been produced by either (1) an increase in the strength of littoral, or long-shore drift; (2) an increase in the quantity of material carried by the littoral drift, or both. These two conditions may be produced by changes in the forces acting upon the shore, or by changes in the shoreline itself. It seems unnecessary to postulate inexplicable variations of an astronomical or meteorological nature when the history of the shore affords an explanation worthy of consideration.

[1] "Headland" is a general term meaning any point of the mainland, projecting out towards the sea. In shorelines of emergence, such as New Jersey's, the term headland refers to that portion of the mainland which is severely attacked by waves and, therefore, serves as the source of material carried by longshore currents. "A headland which is bordered on either side by bay bars, or spits, is sometimes called a winged headland."* Typical examples are the Grassy Hollow headland of Long Island and the Long Branch headland of New Jersey. In both cases material is washed from the headland to form sandspits on both sides of it.

* *Johnson, Douglas.* "Shore Processes and Shoreline Development," 303, 1919, John Wiley & Sons, Inc.

[2] *Haupt, L. M.* "Changes Along the New Jersey Coast," N. J. G. S. Ann. Rep. 1905.

[4] *Haupt, L. M.* "Changes Along the New Jersey Coast," N. J. G. S. Ann. Rep. for 1905, page 46.

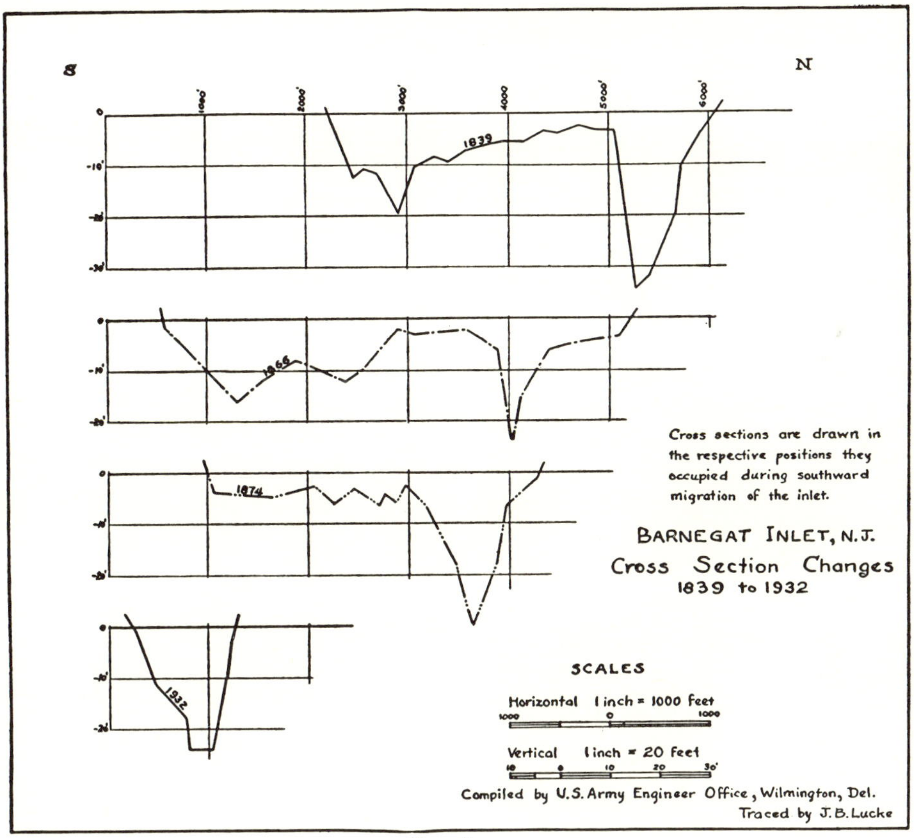

Figure 3.

The table on next page is reproduced from Haupt's report of "Changes along the New Jersey Coast."[1]

Erosion of the headland not only decreased the size of embayments north of Barnegat Bay, but the eroded debris kept filling the reduced bays. The small ponds between Seabright and Shark River were the first to be closed from the ocean, resulting in an increase of the debris being carried southward. Shark River and Squan (Manasquan) River receive most of their sediment from the attacks of northeast storm waves upon the headland. They would be permanently closed without the present extensive system of protective jetties.

If a constant amount of debris is eroded from the headland each year and carried southward by the longshore drift of violent northeast storms, this debris will eventually encounter an inlet as it moves southward. At the inlet nearest to the headland some of this debris will be caught in the tidal currents and deposited on the tidal delta in the lagoon. Much sediment may continue drifting away from the headland, but each succeeding inlet will reduce the amount of debris carried by the longshore drift.

When an inlet is affected by a large supply of longshore sediment two things may occur. The inlet may close or it may migrate away from the headland. Tidal currents maintain inlets only if the lagoon is large enough to produce tidal (hydraulic) currents of a sufficiently high velocity to erode one side of the inlet as fast as deposition occurs on the other side. If the tidal currents are not swift enough to produce this equilibrium between deposition and erosion the inlet may close. This has happened at Manasquan River and other small inlets to the north, as well as at Cranberry Inlet in 1812. Apparently Barnegat Bay could not maintain tidal currents swift enough to support two large inlets. When Cranberry Inlet closed, the supply of beach-drifted debris brought to Barnegat Inlet increased, narrowing the inlet by deposition on North Point and causing a corresponding increase in the velocity of the tidal currents.

Thus the recent situation at Barnegat Inlet, characterized by deposition on North Point and erosion on the south shore of the inlet, was created. If Barnegat should close, the increase of sediment carried southward would undoubtedly bring about a similar migration in the succeeding one or more inlets.

[1] *Haupt, L. M.* "Changes Along the New Jersey Coast," N. J. G. S. Ann. Rep. for 1905, page 38.

If, as Johnson[1] points out, waves tend to produce inlets, tidal currents to maintain them, and littoral currents to close them, Barnegat should continue to narrow until erosion on the south shore is equal to the deposition on North Point. When this state of equilibrium has been reached, the inlet will migrate steadily southward and probably maintain a deep, but shifting channel. The assumption that the inlet will attain equilibrium is based on the size of the tidal lagoon.

Barnegat Bay is probably too large to permit closure of the inlet, under existing conditions. Since erosion of the south shore of the inlet is not yet as rapid as deposition on North Point, and the velocity of the ebb tidal current approximates four miles per hour at the present time, it is believed that the increase in velocity of tidal currents which will be caused by further narrowing of the inlet will result in currents of more than sufficient velocity to maintain the inlet.

[1] *Johnson, Douglas.* "Shore Processes and Shoreline Development," page 368; John Wiley & Sons, Inc.

TABLE OF EXISTING AND EXTINCT INLETS NORTH OF BARNEGAT

Miles S. of Sandy Hook	Designation	Condition	Remarks
7	Navesink River	Closed	By R.R. in 1856
8.5	Shrewsbury River	Closed	By R.R. in 1856
14	Whale Pond	Closed	By Nature
15	Elberon Pond	Closed	By Nature
15.5	Brook (no name)	Closed	Seepage to beach
15.7	Brook (no name)	Closed	Seepage to beach
17	Deal Lake	Closed	Seepage to beach
17.7	Sunset Pond	Closed	Seepage to beach
18.2	Wesley (Long Pond)	Closed	Seepage to beach
19	Goose Pond	Closed	Seepage to beach
19.7	Duck Creek	Closed	Seepage to beach
20.2	Shark River	Open	*Artificially maintained*
20.7	Silver Lake (Perch Pond)	Closed	
21.7	Three Corner Pond (Como)	Closed	
23.2	Fresh Pond (Spring Lake)	Closed	Tributary to Wreck Pond
24	Wreck Pond (Sea Girt Inlet)	Closed	
25.5	Newberry's Pond	Closed	Once mouth of Squan River
26.2	Squan (Manasquan) River	Open	*Artificially maintained*
30	Metedeconck River Inlet	Closed 1755	
33	Kettle Creek Inlet	Closed	
38	Cranberry Inlet	Opened 1750, closed about 1812	Opposite Toms River

NOTE:—Remarks in italics are inserted by the author.

[*Editor's Note:* Material has been omitted at this point.]

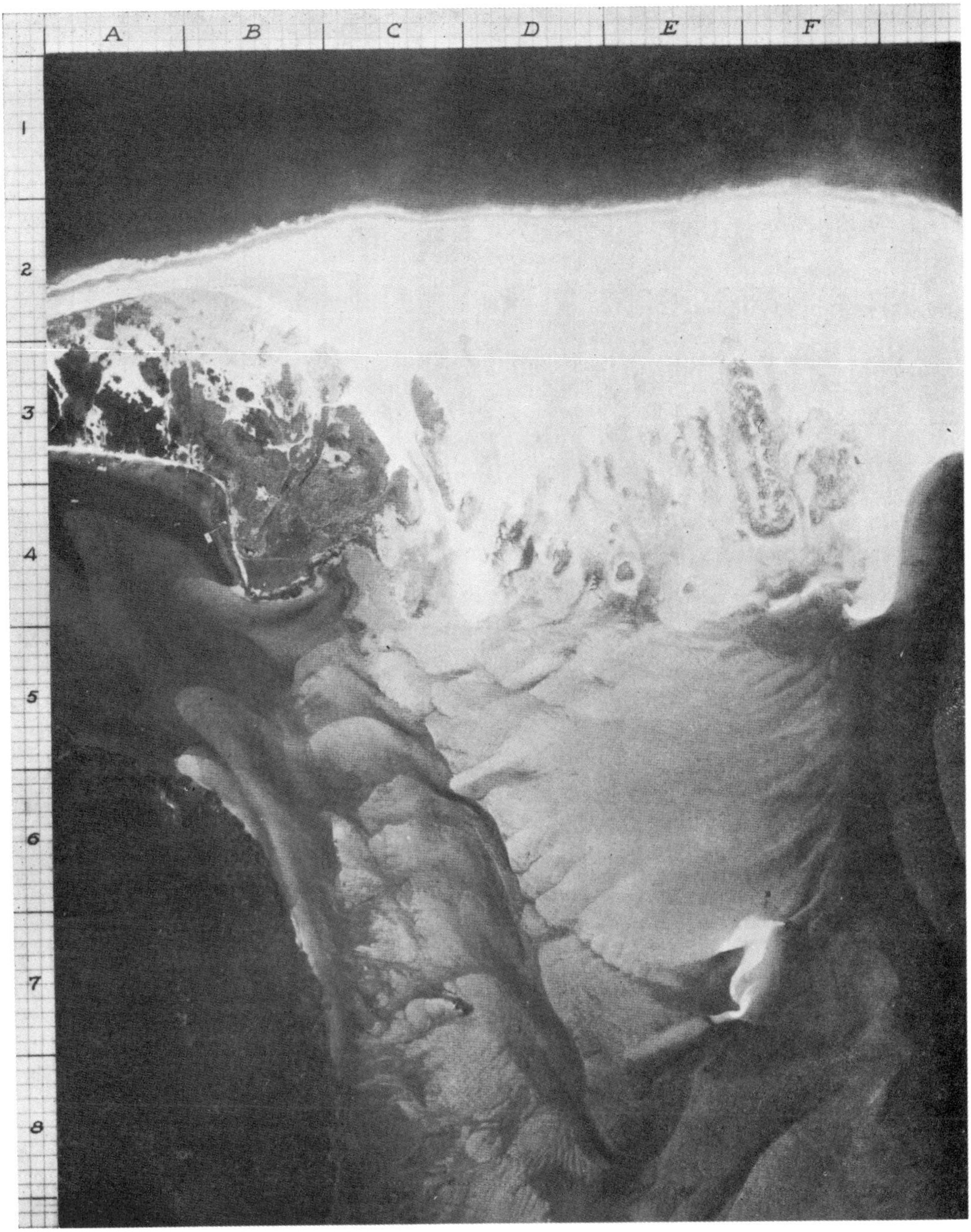

Fig. 9.—Airplane photograph of North Point Beach and the eastern part of Sea Dog Shoal. Reproduced by permission of the New Jersey State Board of Commerce & Navigation. Photograph by Airview Inc., Long Branch, N. J.

[*Editor's Note:* Material has been omitted at this point.]

3. Relation of Topography to Tidal Currents.

Until sedimentary data (to be discussed in the next chapter) afford proof of the origin of the shoals in the Barnegat area, it may be assumed that tidal currents have deposited them and cut the intervening delta channels. In that section of the area which has not yet attained the level of mean high water so as to support marsh growth, the details of the sand flats should provide valuable evidence of the activity and relative importance of both the ebb and flood currents. Such a section is found between North Point of Island Beach (the north side of the inlet) and the western edge of Sea Dog Shoal. This broad tidal flat is nearly free from salt marsh south of the Sedge Islands, except for the extreme western edge of the shoal (see *Chart*).

Airplane photography of this region is more interesting and revealing than any other method of observation. *Figures 9, 10 and 11* give the bird's-eye view of this critical section of the shoals. Refer to the *Chart* and to the *Frontispiece* for orientation.

The photograph in *Fig. 9* shows the ocean at the top and the marsh covered Sedge Islands at the lower left (in A-6, A-7, A-8 and part of B-6, B-7 and B-8). The narrow dark area in A-4 is the channel known as Winter Anchorage, and the small, white rectangular object in A-4 is a large houseboat. The large white area is the barrier beach north of the inlet, known as North Point. In C-3, 4, D-3, 4, and E-3, 4 are well-developed, concentric beach ridges on the barrier beach. All of the region shown, except for North Point (of the barrier beach) and Sedge Islands, is below water level.

In E-7 is a small sand island which ends in the upper right in a narrow projection barely discernible on the picture. This narrow point of the island is a half-buried wreck which has caused deposition of sand in its lee. Since the inlet is a little to the right of F-3 (just off the picture) the depositing agent was the flood tidal current, which also formed the small, curved spit in F-4.

The apron-shaped fans in A-4, 5, B-5 are also due to the deposition of sand by the flood water. They encroach upon the quiet waters of Winter Anchorage in symmetrical fans which indicate a current moving in only one direction, that is, from the inlet to the bay.

Fig. 10 overlaps the western (lower) end of *Fig. 9* and extends west to the edge of Sea Dog Shoal, marked by the white beach across the bottom of the picture. The dark, curved area in E-1, F-1, is the channel from the inlet which swings off the picture (see *Fig. 11*) at F-2, curves around the southern end of the Sea Dog Shoal and runs across the bottom of the picture below

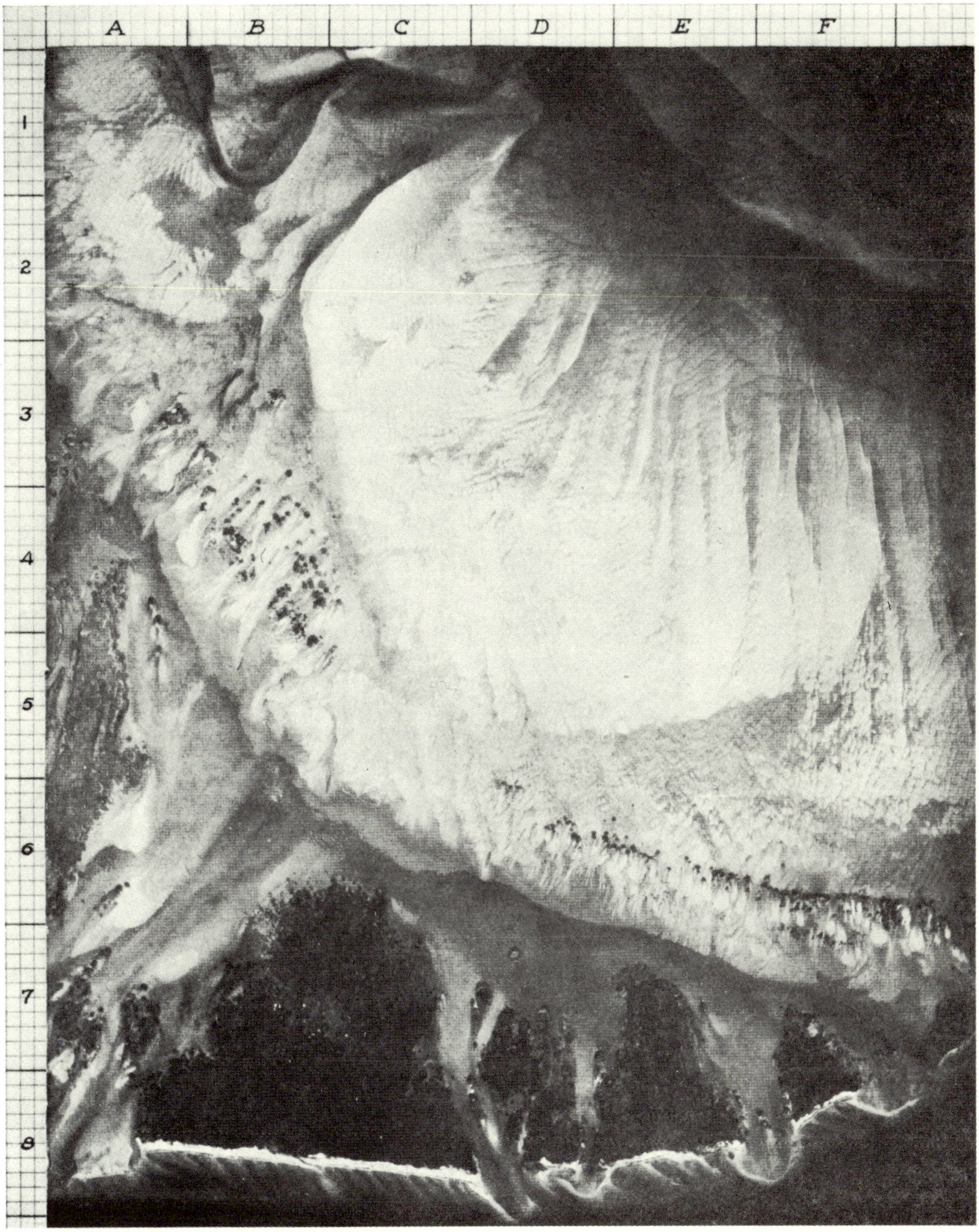

Fig. 10.—Airplane photograph of central part of Sea Dog Shoal. Reproduced by permission of the New Jersey State Board of Commerce & Navigation. Photograph by Airview Inc., Long Branch, N. J.

the narrow beach just mentioned. The white area in C-1 is the same island described in *Fig. 9*.

The ridges and swales of sand in D-3, 4, E-3, 4, and F-3, 4 radiate from the inlet (above this picture) and merge into a great fan of sand in C-3, C-4, D-5 and E-5 which is highest (nearest to the water surface) at its western (lower) edge. Farther from the inlet (farther down on the picture) the shoals occasionally attain high-water level and support tufts of salt grass, visible as dark dots in B-4, D-6 and E-6. Between each two bunches of marsh small fans are building westward, away from the inlet. These tiny fans are best shown in F-7 where there is aboslutely no deposition on the inlet (upper) side of the marsh tufts. This affords a clear indication of the activity of flood waters, but apparent absence of ebb currents on the Sea Dog Shoal. Other effects of the flood currents are visible in A-8, D-8, E-8, and F-8, where the sand on the shoal has been carried westward by flood waters to form deltas on the steep erosion scarp of Sea Dog Shoal (see *Chart*).

Fig. 11, with the two preceding pictures, completes the bird's-eye view of Sea Dog Shoal. The main delta channel enters the picture at A-1, flows southwestward to E-3 and turns northwestward to flow up the bay,

Fig. 12.—The Northwest shore of Barnegat City; showing undercutting of the bank by the ebb current.

skirting the westward edge of the great shoal. The white streaks in C-1, D-2 and C-7 are the foaming wakes of fishing boats returning home from the sea. The point of land in C-1 is the western edge of Barnegat City.

Nearly all of the features on the shoals described in the three airplane photographs indicate active deposition by the flood tidal current, which drops its burden on the sand flats, but continues to deposit sand until it spills over the western edge of the shoal into the main channel.

The shoals tend to grow outward from the inlet as deposition continues, but in moving westward the sand encounters the main delta channel. The swift ebb current in this channel is able to decrease, if not entirely counteract, the westward migration by erosion, producing the steep scarp at the western edge of Sea Dog Shoal. It may be possible for such a steep slope to be produced by deposition on a delta front, but in this case the scarp is neither lobate nor arcuate, but nearly

Fig. 13.—Looking northeast along the same shore. Old salt peat outcropping from under a dune cliff.

straight, and is believed to be the product of erosion by the ebb current.

The western edge of the shoal (see *Fig. 10*) is slightly above the high-water level, therefore the ebb waters must flow around the shoal and cannot alter the deposits formed by the flood tidal current. The ebb and flood currents, therefore, are far from being equal and opposite forces. The flood waters enter the bay in two ways, one part flowing in through the main channel, but a large part flowing over the shoals north and west of the inlet. The ebb current, on the other hand, is almost entirely confined to the main channel. It is evident from the topography on the *Chart* that the ebb current is rapidly eroding the concave bank of the channel south of Sea Dog Shoal. Deposition on the southern end of the shoal tends to narrow the channel and increase the erosion of the south bank.

Two important factors which may contribute to the migration of the point of Sea Dog Shoal with corresponding erosion on the Barnegat City shore are: (1) Most of Barnegat Bay is north of the inlet. It is probable that the stronger tidal currents and storm waves from the north side of the delta have produced a larger distributary channel than any other, which eventually became the main channel by erosion on the south bank and deposition on the north bank of its curve. (2) The gradual migration of the inlet in the past century, due to strong southward longshore drift, caused the addi-

Fig. 14.—View from the top of the lighthouse, looking southwest along the same shore. Notice the edge of Sea Dog Shoal in the right background.

Fig. 11.—Airplane photograph of the southern end of Sea Dog Shoal, showing part of the main channel. Reproduced by permission of the New Jersey State Board of Commerce & Navigation. Photograph by Airview Inc., Long Branch, N. J.

tion of material to North Point and Sea Dog Shoal, which in turn deflected the main channel westward and southward to its present position.

Both factors are undoubtedly at work, but the writer believes the migration of the inlet to be of primary importance. Another view[1] is that although the longshore drift caused growth on North Point, the southward advance of Sea Dog Shoal is primarily due to the effects of northerly storms in Barnegat Bay, resulting in southward movement of sediment along the shallow eastern side of the bay.

A more vivid picture of the erosion on the south shore of the inlet channel is afforded by three photographs taken at close range, *Figures 12, 13, 14.* That the situation has recently become alarming is evident from the hue and cry raised by the local residents. This resulted in the tidal survey by the Corps of Engineers (1932) and probably will be followed by further federal action. It is doubtful that Barnegat Inlet could be permanently closed, but even temporary closure would be a misfortune.

[1] *Report* by the Board of Commerce and Navigation, New Jersey, on Erosion and Protection of New Jersey Beaches, 23, 1922.

[*Editor's Note:* Material has been omitted at this point.]

VIII. Summary

BARNEGAT Inlet has been shown to have had a life of at least a few hundred years and probably much longer. Sedimentary data indicate moderate southward migration during the past century, but no extensive movement prior to this period. It is believed that the lagoon deposits may be used to decipher inlet history if the many variable shore processes are given due consideration.

The tidal and sedimentary studies afford a clear picture of the processes at work at such a tidal inlet. The tidal survey confirms Johnson's deductions about the sea level variations caused by tidal inlets and demonstrates the manner in which such variations may take place. The sedimentary data enables the tidal delta to be characterized as follows:

1. The largest particles are beach sands or gravels and average size decreases landward from the inlet, that is, north and south as well as west of the inlet, if the inlet has always been stationary. 2. Organic material increases landward from the inlet. 3. The delta consists largely of medium to fine quartz sand, well rounded. Heavy minerals are those common to the beach sands north of Barnegat Inlet. 4. The delta contains very little silt or clay. 5. The delta beds near the surface show little or no stratification. Therefore, the assumption that such tidal deltas are formed of longshore debris deposited by flood tidal currents is fully confirmed.

A theory of the evolution of lagoon deposits on young shorelines of emergence has been evolved. The shorelines of both New Jersey and Long Island seem to confirm the theory that inlets may produce distinctive shoals and marsh deposits in the lagoons; stationary inlets of one type, and rapidly migrating inlets of a very different type. Study of coast charts may serve to decipher inlet history by means of the deduced evolutionary stages of lagoon deposition. In New Jersey the charts are shown to indicate great stability of the inlets, followed by a recent, short period of migration. This interpretation is confirmed by old charts and by the sedimentary data at Barnegat Inlet. The change from static to shifting conditions at the New Jersey inlets is attributed to the closure of many inlets between Barnegat and the headland within historic times.

The field methods, especially the problems of transportation and sediment sampling are explained. The instruments used are described and the author's opinion of the utility of each instrument under varying field conditions is stated.

Bibliography

Colony, R. J. Source of the sands on the south shore of Long Island and the coast of New Jersey. Jour. sed. petrography, Vol. 2, No. 3, 150-159, 1932.

Davis, A. C. Salt marsh formation near Boston and its geological significance. Econ. geol. V, 623-639, 1910.

Id. Salt marshes; a study in correlation. Assoc. Amer. geog. I, 138-143, 1911.

Id. Origin and formation of peat. U. S. bureau of mines, bulletin No. 38, 165-186, 1913.

Goldman, M. I. The petrography and genesis of the sediments of the upper cretaceous of Maryland. Md. geol. sur. bulletin 1916, 112-123.

Gulliver, F. P. Shoreline topography, Proc. Amer. acad. arts and sciences 34, 1899.

Haupt, L. M. Changes along the New Jersey coast. N. J. G. S. ann. rep. for 1905.

Johnson, Douglas. Shore processes and shoreline development. 584 pages, 1919, John Wiley & sons, inc.

Id. New England Acadian shoreline. 608 pp, 1925. John Wiley & sons, inc.

Id. Botanical phenomena and the problem of recent coastal subsidence. Bot. gaz. XVI, No. 6, 1913.

Id. The supposed recent subsidence of the Massachusetts and New Jersey coasts. Sci. n. s. 721-723, 1910.

Id. Is the Atlantic coast sinking? Geog. rev. 3, 1917.

Id. Studies of mean sea level. Nat. res. council. bull. No. 70, 1929.

Krumbein, C. W. History of mechanical analysis. Jour. sed. petrography, Vol. 2, No. 2, 1932.

Lahee, F. H. Field Geology. 3d ed. 785 pages, 1931, McGraw-Hill.

Marmer, H. A. Sea level along the Atlantic coast and its fluctuations. Geog. rev. XV, 1925.

Id. et al. Oceanography. Bull. nat. coun. No. 85, chapter VII "Tides and tidal currents," 229-310, 1932.

Id. The tide. 282 pages, 1926, Appleton.

Id. Mean sea level and its variations. Assoc. Amer. Geog. annals XV, 106-118, 1925.

Id. Tides and currents in New York harbor. U. S. coast and geodetic sur. spec. pub. 111, 1925.

Milner, H. B. sedimentary petrography. revised, 635 pages, 1929, Murby, London.

Mudge, B. F. The salt marsh formations of Lynn. Essex. inst. proc. II, 117-119, 1858.

Patton, R. S. Personal communication to the author.

Rubey, W. W. Lithologic studies of sedimentary rocks of the Black Hills region. U. S. G. S. prof. paper 165-A, 1-54, 1930.

Sandidge, J. R. Personal communication to the author.

Shaler, N. S. Beaches and tidal marshes of the Atlantic coast. Nat. geog. mon. I, No. 5, 137-168, 1895.

Trask, P. D. Origin and environment of source sediments of petroleum. 1932, Gulf pub. co.

Twenhofel, W. H. Treatise on sedimentation. 2d ed., 926 pages, 1932, Williams and Wilkins.

Udden, J. A. Mechanical composition of clastic sediments. Univ. Iowa studies v. 11, No. 11, 1926.

Wentworth, C. K. Methods of mechanical analysis of sediments. Univ. Iowa studies v. 11, No. 11, 1926.

Id. A scale of grade and class terms for clastic sediments. Jour. geol. XXX, No. 5, 377-392, 1932.

Widmer, L. C. Personal communication to the author.

9

Reprinted by permission from *Geog. Rev.* **29**(3):400–415 (1939)

HURRICANE MODIFICATION OF THE OFFSHORE BAR OF LONG ISLAND, NEW YORK*

Arthur David Howard
New York University

AN account of the New England hurricane of September 21, 1938, and some of its effects has already been given in the *Geographical Review.*[1] It is the purpose of the present paper to report on a further aspect of the storm—the havoc wrought on the slender thread of sand constituting the offshore bar of southern Long Island.

The damage to the bar was caused by huge storm waves striking from the crest of a swollen tide. Great gaps were opened in the bar, the high-water shore line was driven scores of feet inland, and acres of protective dunes were washed away. Many thousands of dollars have already been spent in reclamation work, and far larger sums will be involved if the comprehensive program of reconstruction now under consideration is adopted. The essential features of the proposed scheme are summarized on a later page.

The present report not only describes the damage but attempts to analyze the processes involved. The post-hurricane modification of the bar, both natural and artificial, is considered, and the efficacy of the methods of artificial reclamation is examined.

The changes effected in the bar may be summarized as follows: (1) formation of new inlets; (2) destruction of sand dunes; (3) development of shore-line irregularities; and (4) aggradation of the bar behind the dune line. In addition, the modification of the bar has so disturbed tidal conditions in the bays behind it that some areas formerly beyond the reach of even the highest tides are now periodically flooded.

New Inlets

For some time before 1931 the offshore bar of Long Island was unbroken for 50 miles, from Southampton to Fire Island Inlet. In February of that year storm waves pierced the bar at Moriches Bay and Moriches Inlet was formed; it has been kept open by vigorous tidal currents ever since.[2]

---

*The Works Progress Administration of New York State coöperated throughout the investigation. The writer expresses his deep appreciation to Lester W. Herzog, state administrator, and to his capable aids engaged in the reclamation work on the bar. The writer is also much indebted to the personnel of the Long Island State Park Commission for their lively interest in the problem and for their helpful suggestions. The investigation was undertaken at the request of the American Geographical Society of New York, whose generous aid the writer gratefully acknowledges.

[1] C. F. Brooks: Hurricanes into New England: Meteorology of the Storm of September 21, 1938, *Geogr. Rev.*, Vol. 29, 1939, pp. 119–127; H. H. Bennett: A Permanent Loss to New England: Soil Erosion Resulting from the Hurricane, *ibid.*, pp. 196–204.

[2] R. S. Patton: Moriches Inlet: A Problem in Beach Evolution, *Geogr. Rev.*, Vol. 21, 1931, pp. 627–632.

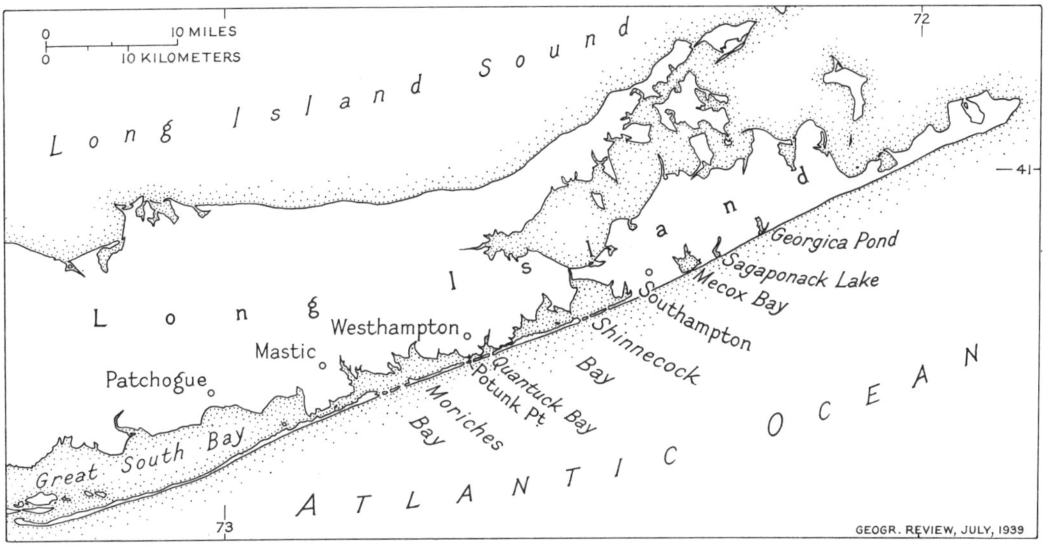

FIG. 1—Map showing position of inlets formed during the hurricane of September 21, 1938. Of the nine inlets between Mastic and Southampton all but Moriches Inlet, the easternmost of the Moriches Bay group, were formed at that time. Inlets were also formed through the bars at the mouths of the three bays east of Southampton.

FIG. 2—Offshore bar at Moriches Bay a few days after the hurricane. Note the irregularities formed in the ocean shore line by erosion, and in the lagoon shore line by deposition. The irregularities on the seaward side are rapidly being obliterated by natural processes, whereas those on the lagoon side remain essentially unchanged. (Photograph by John D. Kay.)

FIG. 3—The western inlet, Moriches Bay. (Photograph by John D. Kay.)

The inlets formed during the recent hurricane were so spaced that, together with Moriches Inlet, they occurred in groups of three (Fig. 1). All of them were formed in the eastern half of the bar. Inlets were also formed through bars at the mouths of small bays east of Southampton—one at Mecox Bay, another at Sagaponack Lake, and a third at Georgica Pond. The present study is confined to the bar west of Southampton.

TABLE I—APPROXIMATE DIMENSIONS OF NEW INLETS SHORTLY AFTER FORMATION

INLET	BREADTH	DEPTH
Moriches Bay Group		
Western	300′	Very deep. Estimated by boatmen at about 25′
Central	750′	Shallow
Westhampton Group		
Potunk (Western)	300′	No figure given, but estimated by men engaged in reclamation work at 15–20′
West Bay (Central)	675′, but interrupted by three small islets	Shallow
Quantuck (Eastern)	500′, but interrupted by island	4′?
Shinnecock Bay Group		
Western	90′	Shallow
Central	300′	18½′ (from *Engineering News-Record*)
Eastern	150′	Shallow

The inlets are concentrated in the eastern half of the bar because the winds on the east side of hurricanes are exceptionally strong. The center of the recent hurricane passed over the bar at a point about 15 miles from its western terminus. The diameter of the calm eye of the storm was about 40 miles, so that the strongest winds spent their fury on the bar from a point some distance east of its mid-point. The pressure of these winds raised the water level as much as 15 feet above mean high water and permitted huge waves to break inland beyond the normal high-water shore line.

Table I gives the dimensions of the newly formed inlets as determined by A. L. Shalowitz, cartographical engineer of the U. S. Coast and Geodetic Survey, from Army Air Corps photographs taken three days after the hurricane, but without actual ground inspection. The widths and configuration of inlets change so rapidly, however, that precise dimensions at any one time are of only academic interest. The western inlet of the Moriches group was widened appreciably during the storm of December 25 and December 26, and Potunk

Inlet, the westernmost of the Westhampton group, tripled in width within four months of its opening.

The inlets owe their locations to different factors. Considerable parts of the bar opposite Moriches Bay were very low even before the hurricane, and in heavy weather the seas swept across them. The new breaks at Moriches Bay occupy the sites of two old inlets, Western and Old Franklin. Before the hurricane the site of Old Franklin Inlet was a trouble spot, flooded even in moderate weather. Half-hearted attempts had been made to bolster this area for some time before the storm. At the time of this writing (April 22) the western inlet was being closed artificially. The task was within a few weeks of completion. The center inlet was to be closed next, restoring the bar to its pre-hurricane form, with only Moriches Inlet maintaining passage to the ocean.

Of the remaining six inlets, only one was still open at the time of this writing (April 22). The inlets of the Westhampton group had been closed artificially; two of the Shinnecock Bay inlets had closed naturally. The Westhampton inlets presumably owed their locations to the presence of low places between sand dunes. Many of the dunes in this part of the bar are 30 feet high. The unusually high tide that prevailed during the hurricane permitted the ocean waters to pass directly across the bar through the low places. The center inlet of the Shinnecock Bay group is still open and, judging from present indications, will be maintained as an aid to navigation. According to the men of the near-by Coast Guard Station, the other inlets of the Shinnecock Bay group were very shallow, barely submerged at low water. Within a few days their floors were built above low-water level by natural processes, and after a few weeks they were completely closed.

It is not readily apparent why the Shinnecock Bay inlets formed where they did. A reasonable explanation is that here, too, there were low places in the line of dunes. Another possibility suggests itself, however. Brooks, in his description of the hurricane, mentions[3] occasional gusts of wind of extraordinary velocity, which he attributes to "the penetration of projectiles of freely moving air from a height of a few thousand feet through the friction-hindered surface layer." Along the coasts where the wind averaged 80 miles an hour these occasional blasts had velocities of more than 100 miles an hour. Locally these gusts may have increased the height of the waves and raised the water level sufficiently to permit the ocean waters to spill across comparatively low places in the bar, though such points may actually have been higher than places elsewhere along the bar that were not affected by the gusts.

In contrast with the varying height of the dune ridge near South-

[3] *Op. cit.*, p. 122.

ampton and Westhampton and in the stretch opposite Mastic, a long stretch of the bar on the west side of Shinnecock Bay is characterized by an even, unbroken crest, the lagoon side of which slopes gently inland at an angle of 2° to 4°. At present the average elevation of this part of the bar is 10 to 15 feet above high water; how much higher it stood before the hurricane, when its crest was farther seaward, has not been determined, but presumably it was at least several feet higher. The absence of inlets through this stretch seems rea-

FIGS. 4 and 5—Miniature canyons 4 feet deep on lagoon side of dune ridge about a third of a mile west of the Quogue Coast Guard Station. Eroded by storm waters flooding over a sag in the dune crest.

sonably explained by the uniform height of the ridge crest, which served as an unbroken barrier against the storm waves.

The opening up of the inlets may have been aided locally by waters flooding across the dune ridge and flowing down the far side as simple streams. Where the lagoon side of the dune ridge is gentle, as at Shinnecock Bay, the effects of erosion by these streams are negligible. Where the lagoon-side slope is steep, however, as at Westhampton, miniature canyons with vertical headwalls occur in many places (Figs. 4 and 5). Headward erosion of such "canyons" may have aided in breaching the dune ridge.

Destruction of Sand Dunes

The second direct result of the hurricane on the offshore bar was the destruction of great stretches of sand dunes, which previously had served as protective barriers screening the dwellings behind them from the fury of Atlantic storms. According to Potter,[4] some of the Long Island dunes that were washed away in a few hours had taken a hundred years to build.

In some stretches the dunes were eroded only by frontal attack of the waves. Elsewhere they were reduced from the sides as well, by wave and current action in the low places between dunes that were flooded during the storm but were not deepened sufficiently to become true inlets. The shape of these storm-water passages is characteristically that of an angular U (Fig. 6), the base ranging in width from

[4] R. D. Potter: The New England Hurricane, *Science*, Jan. 20, 1939, Supplement p. 10.

a few feet to several hundred feet and the depth sometimes reaching ten feet. For the sake of convenience, these passages will be referred to here as "storm sluices."

The floors of the larger sluices are at about the same level as the seaward base of the dune cliffs; i.e. the rising slope of the beach passes without interruption through the gaps in the dunes. Inasmuch as the elevation at the base of the dune cliffs marks the level of effective

FIG. 6—Storm sluice on the offshore bar at Great South Bay. Photograph taken a few days after the hurricane by Arthur W. Murray of Patchogue, Long Island.

wave attack at that position, it seems as though the level of the floors of the larger sluices was likewise determined by the limit of wave attack. The smaller sluices stand at various heights above this level. As they are deeper in comparison with their width than the larger sluices are, it is probable that the sluices start as small features, which at first deepen more rapidly than they widen. When the floor has been lowered to the level of effective wave attack, however, downcutting as a result of wave activity alone ceases and widening becomes the dominant process. Thus the height of the limiting walls depends largely on the height of the dunes that are being cut into.

According to one inhabitant of the bar, the water flowed through the highest sluices in spurts, impelled by the succession of breaking waves. Many of these highest sluices were occupied by water only at the very height of the hurricane tide. The sag in the dune line at the head of the canyons shown in Figures 4 and 5 shows practically no erosion; only the flattened dune grass converging downward toward the canyon heads proves that floodwaters passed through here. The shortness of the period during which the hurricane waters stood at their highest level makes it seem unlikely that many of the larger, deeper sluices could have been fashioned from these highest spillways. It seems likelier that the larger sluices formed at the sites of somewhat lower gaps in the dune ridge, where erosion started earlier and continued during the rising hurricane tide. Through such sluices, according to a coast guard at the Tiana Coast Guard Station, the flow of water at the height of the hurricane was steady.

The inlets may or may not have begun as sluices. Some of them, such as those near Moriches Inlet, formed where the bar was so low that only a slight rise in water level was required to submerge it completely. The lowest places were naturally flooded first, and the depth of the currents passing through them increased with the hurricane tide. Like all other vigorous streams with appreciable velocity, these floodwaters eroded their channels, and the depths reached were sufficient to keep the channels flooded even after the storm ceased and the rest of the bar emerged.

Some of the remaining inlets, however, probably started as storm sluices. The remaining inlet into Shinnecock Bay is trenched into the floor of a broad sluice more than a thousand feet wide; i.e. the dunes have been sheared away for hundreds of feet on each side of the inlet. The writer believes that the inlets formed in the floors of sluices may be reasonably explained in either of two ways. The local deepening may have been due to concentrated storm activity, caused either by the high-velocity wind gusts mentioned by Brooks or by chance combinations of smaller waves into exceptionally high ones. The concentrated storm activity may have created low places in the sluices by locally pounding the floor and sweeping the disturbed sediment into the lagoon. These low places would then have determined the position of the strongest current. It is also possible that the sluice currents, by concentrating in the shallow valleys leading lagoonward from the sluice floors, may have caused deepening and headward erosion of these valleys until they were projected headward across the sluice floor. Other inlets through sluices, however, may have been formed by storm waters flowing through low places between dunes, and the associated sluice forms may have been developed later by the planing away of the dunes for some distance on each side by wave attack.

The widening of the sluices is accomplished at the expense of the limiting dunes. That destruction of the dunes by erosion of their sides during widening of the sluices is more rapid than destruction from the seaward side is indicated by the great breadth of some of the sluices as compared with the small amount of retreat of the seaward face of the dunes. Although erosion on their flanks proceeds at a faster rate, the greater frontage means that a larger area of dunes was obliterated by frontal attack than by lateral attack. Some information is available on the extent of dune destruction. As part of the reclamation program miles of jetty were constructed parallel to the shore. This jetty has been located at the site of the former crest of the dune ridge as determined from maps showing its position with respect to the auto road. For instance, for nearly two miles along the bar west of the Tiana Coast Guard Station the jetty is nearly 100 feet away from the present dune crest. Surveys by the Long Island State

Park Commission show that the high-water shore line in front of the main building at Jones Beach was pushed back 300 to 400 feet. However, much of the loss at this place has been compensated by accretion since the hurricane. It may be stated as a general truth that the hurricane caused a recession of almost the entire shore line; but the amount of recession was not everywhere the same, and there are some places where little or none seems to have occurred. The greater part of the recession was at the expense of the sand dunes, which in some places, such as west of Westhampton, have been beaten back almost to their lagoon sides and offer only a thin barrier against future storms.

Irregularity of the Shore Line

In many places the former straight shore line was made irregular, even crenulate, in pattern. Many of the indentations (Fig. 2) are associated with storm sluices; this association suggests that some of them may be causally related to the sluices. It has already been pointed out that lagoonward of each sluice is an apron of debris spread out by the waters that passed through the sluice. Much of the sediment undoubtedly was furnished by the destruction of the dunes themselves, but part of it may well have been contributed by the beach. It seems reasonable to suppose that at least a part of the sediment that was stirred up by the waves opposite the mouths of the sluices would be swept through them and cause a retrogression of the shore line at these places.

Other indentations, however, seem unrelated to sluices and probably reflect local concentrations of storm activity. Again the exceptional wave attack may have been due to high-velocity wind gusts or to chance combinations of smaller waves into a series of exceptionally large ones. In these cases the material eroded from the beach may have been either contributed to the longshore current or returned seaward by backwash and undertow. Profiles of the beach made at many places on the offshore bar during a series of years by the members of the Long Island State Park Commission show that after storms which have been destructive to the beach the water offshore is shallower than before, and this suggests that the bulk of the material eroded from the shore is returned seaward and deposited some distance offshore. In the intervals between destructive storms the material is restored to the beach. At Jones Beach, as was said above, the high-water shore line locally receded 300 to 400 feet; yet at present this shore line is approximately where it was before the hurricane.

The indentations of the shore line are healed in an interesting manner. The longshore current builds a sandspit part way across the reëntrant, and the waves drive the sandspit on the beach during heavy weather. A new spit is then built forward and in turn is driven

FIG. 7—Formation and migration of sandspit at site of West Bay Inlet near Westhampton. The jetty was built across the reëntrant left in the shore line when the neck of the inlet was closed. Upper view taken October 28, 1938, shows sandspit building westward (toward observer) across reëntrant. Center view, taken November 15, 1938, shows spit driven toward jetty. Sand between oblique groins is from two to three feet deep. Lower view, taken December 6, 1938, shows the spit driven completely into jetty and the sand almost five feet deep between the groins. (Photographs by courtesy of Works Progress Administration of New York State.)

on the beach. In this way the reëntrant is filled from the upcurrent side. In no one case has the writer himself been able to follow the process through to completion; but the photographs of Figure 7 show various stages in this process as observed by one of the men in charge of the reclamation work on the bar. It is of course possible that during long periods of comparative calm the longshore current might be able to build the sandspit completely across the indentation, leaving a narrow swale between it and the beach proper. In such a case, were the bar to be driven into the reëntrant, the reëntrant would prograde directly seaward rather than in a downcurrent direction.

Still another process may hasten the filling of the indentations. Surveys by the Long Island State Park Commission show that a shallow submarine bar is formed at distances of as much as 1000 feet from the shore during the milder storms of summer but that this bar disappears during the severer weather of winter. Inasmuch as the profiles are surveyed at only two widely spaced intervals during the year, the possibility exists that unusually violent summer storms may also cause the bar to disappear. Hence a second summer bar might form. No information is available concerning the ultimate resting place of the material of these bars, but it is reasonable to suppose that they are driven on the shore like other bars. This supposition seems to be verified by the history of the spit shown in Figure 7. The part of the bar thrown up within the reëntrant would be protected from later erosion by the growth of a sandspit from the upcurrent side, whereas elsewhere the bar would be subject to erosion.

The explanation for the upbuilding of one or more new submarine bars each year may not be immediately apparent. It is generally recognized that the formation of a bar by storm waves causes a deepening of the water seaward. It might be expected that the great waves of the winter storms which drive the bar on the land would so deepen the water seaward that no new bar could be created. That new bars are formed suggests that the bottom is later shallowed. The shallowing is accomplished by the importation of sand by the longshore current and by seaward movement of beach sand during storms destructive to the beach.

The writer found no beach terraces, or berms,[5] formed by the hurricane. In an article published by the U. S. Beach Erosion Board[6] it is stated that "a narrow berm remained at about elevation +10 feet" at Westhampton Beach on September 24. This berm may have been obliterated by the time the writer visited the area. It is probable, however, that the apparent absence of the berm in other areas is due simply to the recession of the front scarp of the berm under wave attack until the scarp coincided with the dune cliff itself. As the scarp was beaten back, the upper surface of the berm, extending from the scarp to the dune cliff, would decrease in width and finally disappear.

Aggradation of the Bar

Much of the sand derived from the destruction of the dunes and of the beach itself was carried through the storm sluices and spread

[5] A terrace formed on a beach as a result of the decrease in the gradient of the beach profile seaward of the line of wave attack is referred to by engineers as a "berm," a term originally employed in geology to designate an erosion surface of fairly great extent. The gentler beach profile seaward of the front scarp of the berm is probably due to the greater strength of the undertow and backwash.

[6] "Inspection of Beaches in Path of the Hurricane of September 21, 1938," *Shore and Beach*, Vol. 7, 1939, pp. 43–47; reference on p. 44.

out as alluvial trains or fans on the lee side of the dunes. The sediment in many places is finely bedded, the layers averaging about an inch in thickness. The thickness of the deposits as a whole varies, but locally it is more than three feet.

The terminal parts of many of these alluvial deposits, although now at some distance from the lagoon shore line, were probably formed as true deltas at the high-water stage of the lagoon. This is suggested by their elevation, which is below that reached by the waters of the lagoon during the height of the storm, and by their relatively steep fronts. Other deposits continue into the lagoon as deltas at the present time. These deltas, and the deltas at the inlets, add to the irregularity of the lagoon shore line of the bar and suggest that many of the other projections into the lagoon are deltas formed in association with former storm sluices or inlets.

Lagoon Tides

The new breaches in the bar permit a greater influx of ocean water into the bays during the incoming tide and a readier exit of water during the outgoing tide. Consequently, the tidal range within the lagoon has increased. No precise measurements are available; but according to information supplied by fishermen and boatmen at Moriches and Shinnecock Bays the mean high-tide level is now about a foot higher, and low tide about a foot lower, than before the storm. These figures are only approximations, and the tidal range undoubtedly varies according to the configuration of the mainland shore.

The bimonthly spring-tide range seems to have increased by an even greater amount. This is probably to be explained by the fact that the cross profiles of inlets widen upward, so that the higher the tide, the greater the amount of inlet space available. Thus a series of new inlets would furnish more high-tide passage space than low-tide space and more spring-tide space than mean-high-tide space. At Mastic Beach a boathouse resting on piles formerly stood 18 inches above ground level. The water rarely stood more than a few inches deep under it, and then only at certain combinations of wind and tide. The entire area is now periodically submerged by water which, near the shore, stands more than two feet deep and which floods the surrounding flats for several hundred feet inland. As a result of the higher tides the boathouse has had to be raised several feet.

The tidal changes caused by the opening of the new inlets are of interest in connection with the problem of sea-level changes. Local increases in high-tide level have produced results that more than once have been interpreted as indicating subsidence of the land or a general rise in sea level. Among the results of such changes in water level may be mentioned the landward expansion of tidal marshes and the killing of trees along the shore. Because of the expansion of the marsh, tidal grasses may cover fresh-water grasses along its borders, and tidal grass characteristic of a lower depth zone may overlie grass of a higher zone offshore. These conditions, and also the killing of marginal forests by contamination with salt water, may result from a rise in water level, regardless of cause or extent, and hence are not in themselves evidence of a subsidence of the land or of a world-wide change in sea level.[7]

The increased tidal range has created a property loss of a type not generally recognized when taking stock of the hurricane damage, namely the periodic flooding of property that was formerly beyond the reach of the highest tides. Many summer dwellings in the low areas on the mainland are now inaccessible during these tides except by boat.

Reclamation

Reclamation work on the bar was begun a few days after the hurricane. The work was started by individual communities but was shortly taken over in large part by the Works Progress Administration. The plan was to close the new inlets, which now divide the bar into a number of sections, and to restore the beach as nearly as possible to its original dimensions. In the conviction that restoration alone is insufficient, an elaborate program of construction has been proposed to ensure the protection of the bar from future damage comparable with that of the last hurricane.[8] Some such program is highly desirable.

The artificial closing of the inlets was accomplished with a varying degree of difficulty and furnished interesting information on the probable natural history of inlets. Consider the case of Quantuck Inlet. According to Shalowitz, this inlet was originally about five hundred feet wide, and its waters were about four feet deep. At first an attempt was made to fill the inlet by dredging sand from the bay, depositing it on the bar, and pushing it with the aid of a bulldozer into the inlet from the east. Tree stumps were also pushed into the inlet in the hope that their spreading roots would hold the sand in place. The inlet was narrowed appreciably by this method; but as the narrowing progressed, the velocity of the tidal current increased, until a point was reached

[7] For a fuller discussion of this problem the reader is referred to the works of Douglas Johnson: see, for example, among other papers, "The Supposed Recent Subsidence of the Massachusetts and New Jersey Coasts," *Science*, Vol. 32 (N.S.), 1910, pp. 721–723; "Botanical Evidence of Coastal Subsidence," *ibid.*, Vol. 33 (N.S.), 1911, pp. 300–302; "Botanical Phenomena and the Problem of Recent Coastal Subsidence," *Botan. Gazette*, Vol. 56, 1913, pp. 449–468; and "Mean Sea-Level as Affected by Shoreline Changes," *Science*, Vol. 66 (N.S.), 1927, pp. 60–61; and Douglas Johnson and Elizabeth Winter: Sea-Level Surfaces and the Problem of Coastal Subsidence, *Proc. Amer. Philos. Soc.*, Vol. 66, 1927, pp. 465–496.

[8] See report by W. E. Andrews on "Restoration and Protection of Fire Island," 1938. Prepared at the suggestion of Robert Moses, chairman of the State Council of Parks, and distributed by the Long Island State Park Commission, Babylon, Long Island.

FIG. 8—Late stage in the filling of Potunk Inlet. The far (west) bank was being cut back by the current as rapidly as the east side was built forward.

at which the dredges merely held their own, the current eroding the west bank of the inlet as fast as the east bank was pushed forward. The closure was finally effected by piling up a large store of sand on the west bank and pushing it into the inlet with the aid of two bulldozers while a bulldozer assisted on the east bank. The operation was aided by sandbags. At the time of closure, at low tide, there was ten feet of water in the inlet, as compared with the original depth of about four feet.

A similar experience was encountered at Potunk Inlet, which at

FIG. 9—Potunk Inlet shortly after opening. Note the double funnel or hourglass pattern. Streamers of sediment may be seen leeward of storm sluices. (Photograph by John D. Kay.)

the time the work started was probably more than a thousand feet wide. Figure 8 shows a late stage in the closing of this inlet. The effects of lateral corrasion on the far (west) side are reflected in the steep, undercut slope. A storm a few days earlier had destroyed a sandbag barricade on that side. At about the date this photograph was taken the water stood 17 feet deep in the inlet at low water. This inlet was also closed at slack water by pushing sand in from each side behind a sandbag barricade.

The marked increase in the velocity of the inlet currents as their

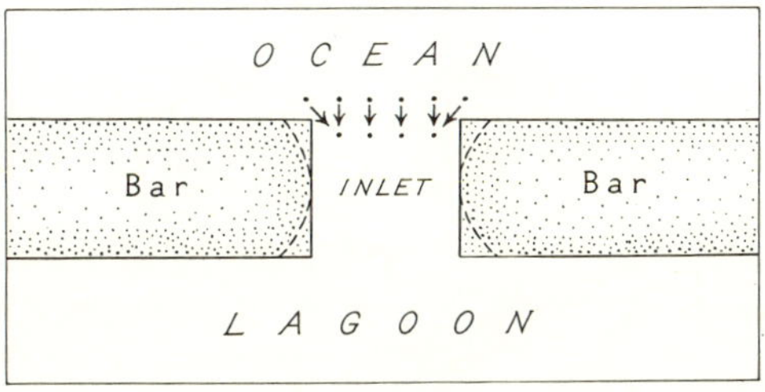

FIG. 10—Role of oblique currents in the formation of hourglass inlets. During the rising tide the water particles on either side of the mouth of the inlet, as well as those directly opposite the mouth, stand higher than those within the inlet. The marginal particles move down the hydraulic gradient as oblique currents which erode the sharp corners of the inlet to create the funnel shape. Similar currents moving out of the lagoon during the outgoing tide create a funnel shape on the lagoon side. The funnels together form an hourglass pattern indicated by broken lines.

width decreased is of interest. The inlets are characteristically funnel-shaped; or, as funneling occurs on both the ocean and the lagoon sides, they may often show an hourglass pattern (Fig. 9). An explanation of this pattern is given herewith.

Let us assume a newly formed inlet whose sides are parallel from ocean to bay; no funnels are present (Fig. 10). During the incoming tide the surface of the water slopes downward from the higher level of the ocean to the lagoon. Because the water particles inside the inlet are at lower levels than the water particles just outside, there is flow into the inlet. The water particles on the outside and just beyond the confines of the inlet move down the gradient into the inlet as oblique currents. These oblique currents are probably the cause of the erosion of the corners of the inlet, which produces the funnel shape. Actually, in nature, other processes also probably play an important part in developing the funnel shape; namely wave attack on the projecting corners, the deflection of longshore currents into the inlet by the tidal (hydraulic) current moving in, and erosion of the projecting corners by the shifting currents on the deltas at either end of the inlet. The same processes operating during the outgoing tide account for the funnel shape on the lagoon side.

Because of this funnel shape the incoming waters are crowded into the inlet, with a resulting increase in velocity. Narrowing the neck of the inlet accentuates the funnel shape and causes further crowding. The additional increase in velocity leads to more rapid erosion and deepening of the channel.

FIG. 11 (above)—Jetty and groin system used in reclamation work on the bar. For economic reasons the brush filling (background) of the jetty and groins has been locally supplanted by zigzag snow fencing. (Photograph by courtesy of Works Progress Administration of New York State.)
FIG. 12 (below)—Jetty system opposite Quantuck Bay near Westhampton. Five feet of sand accumulated at the jetty line in about a month.

The above considerations seem to indicate that a better method of closing inlets would be either to fill the inlet up to the surface at a uniform rate throughout its breadth without narrowing the neck or to eliminate the funnel shape through deposition of fill at the outer margins of the inlet and then to fill in the neck.[9]

In view of the prevalence of the funnel shape it is surprising that nature itself closes so many inlets. In the writer's opinion, many inlets are probably closed by the sea, not by longshore deposition of

[9] The writer visited the offshore bar after this article was in proof and was gratified to find that the engineers in charge of reclamation work had independently arrived at the same conclusions. The western inlet at Moriches Bay was being closed by filling it up to the surface. Sandbags weighing several tons and enclosed in wire netting are lowered into the channel by crane. Nature is aiding the work by building a sandspit across the mouth of the funnel, thus cutting down the volume of water entering the inlet.

debris on the upcurrent side of the inlet, but by the driving of a submarine bar into the inlet at stages of the tide when the current is too weak to prevent such activity. By such shallowing the mass of water funneled into the inlet would become less and the velocity would decrease, with the result that conditions in the inlet would become increasingly more suitable for deposition and ultimate filling.

The beach itself is being restored by the Works Progress Administration through the construction of a jetty and groin system, shown in Figure 11. The rear line of posts is located on the site of the former crest line of the dunes. Sometimes in a single storm the reëntrants between groins are filled with sand to the top of the five-foot cyclone-wire fence. After West Bay Inlet was closed, a jetty was built across the reëntrant in the shore representing the funnel of the inlet (Fig. 12). Aggradation at the jetty line resulted in the formation of a lake between the jetty and the bar proper. The lake basin was gradually filled with sand washed over the bar by the waves and blown into the depression by the wind. As a result of such work the beach has in many places been restored to its original dimensions, only the dunes being missing.

A plan has been sponsored by the chairman of the State Council of Parks to develop the seaward side of the bar as a state park.[10] It is proposed to dredge in sand to make a broad, shallow ridge parallel to the sea. On the summit of this ridge, well above the level reached by the sea during the recent hurricane, a concrete parkway is to be built. This ridge, with slopes carefully graded and matted with vegetation, has been designed to furnish protection to the dwellings on the lagoon side.

[10] Andrews, *op. cit.*

10

Reprinted from pp. 280–282, 286–295 of *Geog. Jour.* **119**(3):280–297 (1953)

THE EAST COAST FLOODS

JANUARY 31 – FEBRUARY 1 1953

J. A. STEERS

THE FLOODING which took place at so many places on the east coast between the Tees and Dover on the night of 31 January 1953 was serious both to life as well as to property, whether houses or land (Fig. 1). It is estimated that 206,161 acres [1] were definitely flooded (and possibly another 11,732), and that 307 people lost their lives. It is impossible to minimize this disaster, but nevertheless it is right to realize that the results might have been far more serious. We are more than fortunate in that the rivers, particularly those flowing through the fenlands were not in flood; we were fortunate too in that the surge did not occur at the top of the tide and, what is of still greater significance, did not occur on a high spring tide.[2] The predicted tides for January 31 were 1–3 feet less—according to locality—than can occur at other times of the year. There is little doubt that if the surge had occurred under conditions of a high spring tide and flooded rivers, the whole of the Fens, and other similar places, would have been submerged, mainly by fresh water.

Surges are not uncommon, but it is only of comparatively recent years that they have been investigated. It is a matter of great regret that Dr. R. H. Corkan, who had studied the subject more fully than anyone else in this country, died a young man, only a few years ago. The account he gave of the surge on 8 January 1949, and of the conditions under which it was generated, are applicable in a remarkable way to the surge of 31 January 1953. On the other hand, the violence in 1953 of the storm in the northern part of the North Sea was exceptional. The devastation of the forests in north-eastern Scotland moreover and the abnormally high northerly winds in the Shetland–Orkney area meant that, despite the far less disastrous wind action in the southern part of the North Sea, there was very rough water—far more so than is usually associated with such north-westerly, or north-north-westerly winds, as prevailed.

In this paper no account is given of the profound damage that occurred in Holland. The surge which travelled southwards along our east coast later turned and moved northwards along the coast of the Low Countries, where it was accompanied by an on-shore wind in which occasional gusts exceeded 90 miles per hour.[3] It needs but little imagination to understand what

[1] I am particularly indebted to Dr. E. C. Willatts and his colleagues in the Ministry of Housing and Local Government who gave great help not only with the maps but also with the photographs The figures are from measurements made in Dr. Willatts' office.

[2] "The peak of a surge of short duration may occur at any state of the tide." K. F. Bowden, *Weather*, March, 1953, 82.

[3] "The on-shore direction of the winds on the coast of Holland averaged 50 to 60 knots for 6 to 9 hours before high water." H. A. Q. v. Ufford, *Weather*, April 1953, 116.

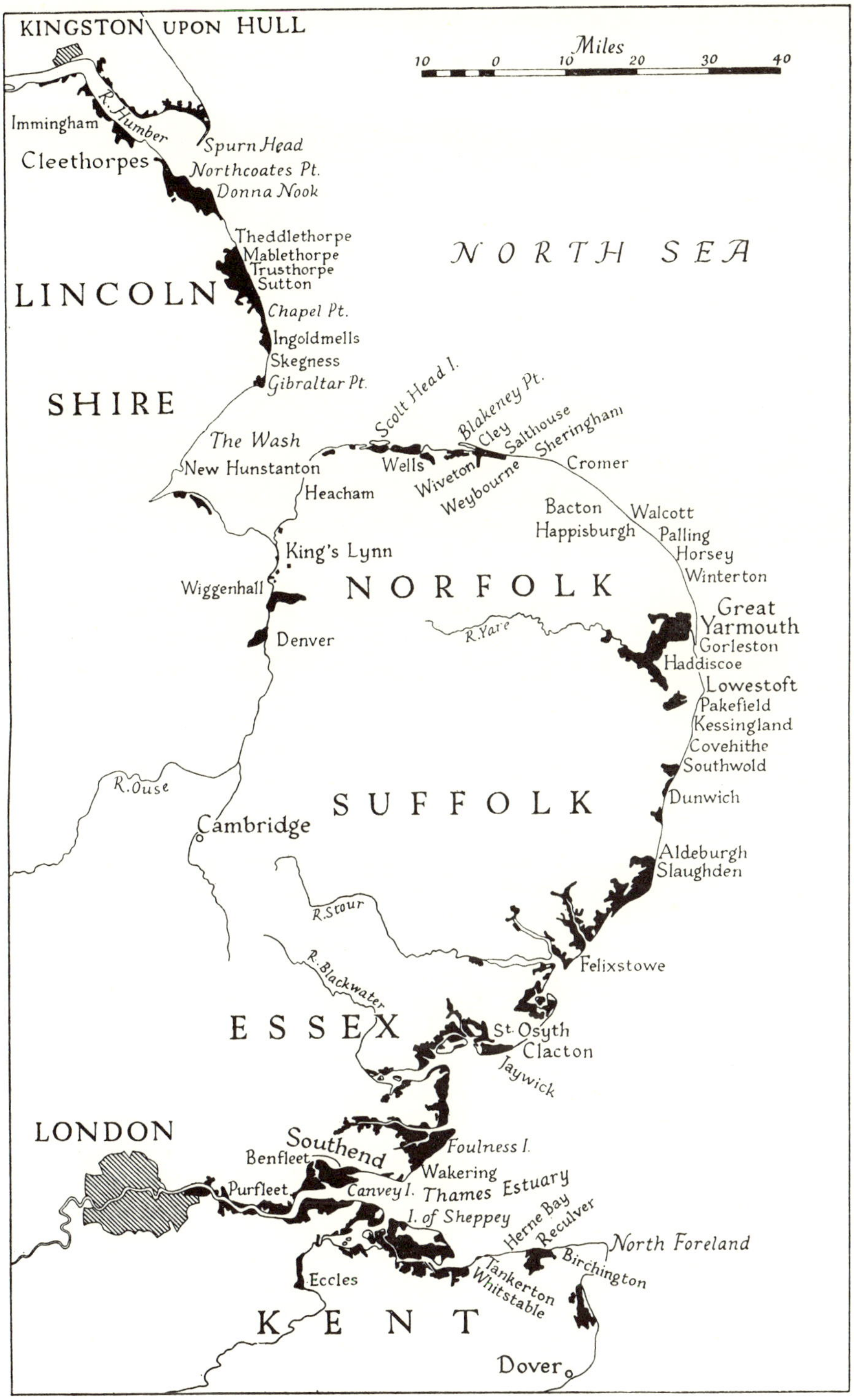

Fig. 1. Flooded areas on east coast, 1953

happened under such circumstances to a country almost entirely protected by dykes or dunes.[1]

[*Editor's Note:* Material has been omitted at this point.]

The effects of the flooding on the coastal areas

Some damage occurred in Northumberland, Durham, and on the coasts of north-east Yorkshire, but in Holderness, near Easington, the sea broke over the dunes and the flood water made exits to the Humber near Skeffling. Spurn Head does not appear to have suffered much change.

The Lincolnshire coast is nearly all low and flat. There is a small cliff at Cleethorpes, now protected by the promenade. From there right round to the Wash no solid ground reaches the sea. Nearly all the area—known as Lincoln Marsh—is reclaimed ground protected by walls and dunes. Between Cleethorpes and Donna Nook there was some flooding, and breaches were made in walls. The airfield at Northcoates Point was submerged. Southwards from Donna Nook, even as far as the northern end of Mablethorpe, the amount of damage was relatively small. Where the dune belt and the beach are both wide the sea did little harm. Part of the energy of the waves was expended before they reached the dunes. These were certainly cut away, often deeply, but they made a real barrier. The most striking effect of dune protection was to be seen at Mablethorpe. About a mile north of the Coastguard Station the dunes held, even though they are much narrower than at Theddlethorpe. But in Mablethorpe itself there were serious flooding and damage. This was continuous with the damage at Trusthorpe and Sutton—and represented the most impressive on the whole of the east coast. Between Mablethorpe and Sutton the dunes were narrow, for the most part protected by sea walls. The sea level rose higher than the pull-overs to the beach and flooded the land within. The narrow beach allowed access of bigger waves, and the abnormal height of the water washed away dunes and walls and invaded the towns. The damage in this area was considerable, and once the dunes or wall had disappeared the sea in some places could directly attack the houses. This was so at the southern end of Sutton where the Beach Hotel suffered. It must be remembered, too, that in Mablethorpe and Sutton, and to a lesser extent in Trusthorpe, the main coast road runs immediately inside the dunes or wall and is particularly vulnerable.

South of Sutton the "Roman" Bank acted as a very sound inner bank and was not breached. At Anderby the dunes are somewhat wider, and although seriously eroded, the line remained fairly intact. Several houses on the dunes were partially undermined. Much the same occurred at Chapel Point, where the sea swept over the wall. The dunes to the north, although not very wide, stood up well. Serious breaches were made at Ingoldmells, and there was a good deal of flooding—including the large Butlin's Camp—south of the point. With the widening of the beach towards Skegness the damage decreased, and no marked alteration took place at that town or at Gibraltar Point.

Miss King and Mr. Barnes, both lecturers in geography at the University of Nottingham, had studied Gibraltar Point in some detail, and after the flood examined not only it, but also took profiles through the beach and dunes along nearly all the Lincolnshire coast.[1] Their sections are shown in Fig. 7 and illustrate remarkably clearly the effects of steeper and narrower foreshores in leading to greater destruction of dunes and sea walls. One effect of the storm was to scour away the beach and leave extensive areas of mud, and in some places of peat, exposed. Much of this beach material was thrown inland.

Apart from occasional flooding, due to high levels, little happened in most parts of the Wash. The surge at King's Lynn was estimated at about 8·1 feet. Most fortunately the Fen rivers were not in flood. All the sea banks in the territory of the Great Ouse River Board have been raised and strengthened in the last half century. The reason for the breach in the Ouse walls at Wiggenhall was primarily the unusual height of the surge, which overtopped the wall. The river is tidal up to Brownshill Staunch. The high water level at Lynn was 18–21 inches above the level of the sea at Hunstanton, and the level at Denver was somewhat below that at Lynn. It was too much for some of the banks. The cause of the bank failures was erosion of the *back* of the bank, except possibly in one case where it may have been due to a rat hole.[2]

On the east side of the Wash at Heacham and Hunstanton there was much damage and loss of life. The bungalows and other light buildings faced west, and even the narrow width of the Wash seems to have had some effect, since the wind and waves were directly on shore. Moreover, behind most of the bungalows the ground level falls for a short distance, so that it was impossible for people to make quickly for higher ground. The damage to the promenade at Hunstanton was noteworthy, especially in view of its relatively sheltered position. The cliffs were also cut away here and there.

Along the coast of north Norfolk, from Old Hunstanton to Weybourne, the most obvious effect of the storm was the flooding of all the reclaimed marshes. The sea walls were overtopped, and breaches formed in most cases from the rear. The great bank running from Wells Quay to the harbour mouth was partly breached in this way, and walls and outhouses facing the marshes were knocked down by the waves inside this wall. The water ran up the old Warham Slade and for a time was 3 or 4 feet deep across the main road at the head of the Slade. The water from the entrance to the Slade also swept into the town and for a time submerged the Station.

On the other hand, natural salt marshes suffered not at all. Dunes were often eroded, but they too often stood remarkably well. On Scolt Head Island the seaward foot of the high dunes near the Hut was cut back 8–10 yards, and changes of that order, or even more, occurred in lines of dunes facing west or north-west. Low dunes were overrun and often washed away. Shingle spits were driven in, but apart from the washing away of the low dunes which face much of the foreshore of the island, the waves did little else than overrun the beach, somewhat flatten it and wash shingle and sand on to the marshes. Only in one place, already partially breached, does it seem likely that a

[1] In *Survey*, Nottingham University, 3, 29, 1953 (March).

[2] Information in letter from Mr. Doran (the italics are mine, J.A.S.).

long-lived gap will endure. Blakeney Point suffered more, partly because along much of it there is deeper water. The Headland and Far Point, in many ways like the corresponding features at Scolt Head Island, were cut into and partly washed away. The main shingle beach was overswept and pushed inland throughout its length. Opposite Salthouse a definite breach was made. This village suffered greatly, and not for the first time in its long history. Since there is high ground immediately behind, refuge was easy. At Cley and Wiveton flooding was severe, and conditions in the Glaven Valley must have resembled those in the great flood of 1897.

Along the cliffed coast of Norfolk there was some loss, particularly where the cliffs are low as at Bacton and Walcott. The main street of Walcott was destroyed, and many buildings wrecked. At Sheringham the cliffs behind the promenade were severely attacked; Cromer, too, suffered somewhat in this way. Erosion has long been serious hereabouts, and throughout their length the cliffs offer very little resistance. Moreover, the beach tends to steepen, especially beyond Happisburgh. A few miles farther east the cliffs cease and give place to a dune coast. The dunes have been "trained" and blowouts mended, the summit of the sandhills is kept fairly level, and marram grass planted. Nevertheless this is a very vulnerable part of the coast, and in 1938 there was a serious break at Horsey which led to the flooding of 7,469 acres. In January 1953, the only break occurred at Palling, but the dunes have been seriously pared away, and it is more than fortunate that storm and surge conditions were relatively shortlived, otherwise severe flooding in Broadland would have taken place.[1] The wall built across the 1938 gap at Horsey helped greatly in maintaining this stretch. At Winterton the dunes widen again, but have suffered loss all along their front. Here too the beach is fairly steep.

In general, eastwards of Cromer the winds were more and more offshore, and beyond Winterton Ness and as far as the Thames Estuary this was especially noticeable. In terms of direct erosion the sea did comparatively little damage except locally. At Kessingland, for example, at the northern and unfinished end of the new sea wall the cliffs were attacked at an unusually high level and were cut back. The same is true of the soft cliffs at Covehithe. At Dunwich the cliffs, equally soft, scarcely suffered at all at their northern end. On the other hand, flooding was often serious, and large areas of Lowestoft were covered with water which came in both behind and over the sea walls which themselves stood rigid. Even as far inland as Haddiscoe the water coming in through Lowestoft, and to some extent through Yarmouth, caused no little damage to a railway embankment. All the valleys, many of them blocked by shingle bars, along the Suffolk coast were flooded. At the south ends of Southwold and Aldeburgh the low lying ground was inundated and damage done to property. The bungalows that suffered at Southwold were situated behind a seawall on and in front of which small dunes had formed, and were in much the same position as were the houses at Mablethorpe and Sutton. But the barrier between them and the sea was much weaker.

The Alde did not break through at Slaughden. The walls on its north bank broke, and the shingle bank between Aldeburgh and Slaughden was flattened.

[1] The width of the dune belt near Waxham is about half what it was.

The breach at Palling, Norfolk. Compare damage to dunes on right of gap (east) with photograph of Waxham, facing p. 297

For a time there was an overflowing of the sea at high water, but the river maintained its course. Even if matters had been left to nature I think this would have been so, but the flattened shingle was bull-dozed up from the inside, and a new, even if thinner, ridge was soon made. However, the original ridge all the way from Aldeburgh to and beyond the Martello Tower was displaced perhaps 50 yards landwards.

Apart from flooding, which occurred all along the coast, the breaking of some dykes and local erosion of cliffs, there was nothing of major importance until, just to the south of Clacton, the bungalow town of Jaywick was severely hit. The sea once again overtopped the wall at the south end of the settlement and soon made a breach. The water also came in from behind. The whole place is below high water mark, and flooding to a depth of 5 feet or so occurred over much of it. The caravans, as at Ingoldmells and other places in Lincolnshire, were swept far from their original sites and usually much broken. Even some permanent bungalows floated away and overturned. But the masonry sea wall along the main front was undamaged (Fig. 8).

The same kind of calamity affected Canvey Island. The major walls along the Thames held, and the flooding took place mainly over the lower and weaker walls along Benfleet and other creeks. Here, too, once the walls were overtopped, these were eroded from the inside. Canvey has far more permanent inhabitants than Jaywick, and so material damage and loss of life were very serious. In both cases the settlements are below high water mark. At Foulness Island the walls at the north and north east were breached, and another big gap was made opposite Wakering. The whole island, together with its smaller neighbours, was inundated, and much valuable arable land spoiled (Fig. 9).

On the south side of the Thames Estuary the winds were more directly on shore. There was considerable erosion in Sheppey and also in the soft cliffs at Beltinge just east of Herne Bay. The ruined church of St. Mary at Reculver had been protected, but there was some erosion there. The low ground west of Whitstable was flooded, and huts and houses at Tankerton damaged. The low ground between Reculver and Birchington was flooded, and the main north Kent railway put out of action. Beyond the North Foreland there was far less damage, and although flooding occurred, the effects of the storm were of minor significance.

In the Thames and other rivers as far north as the Tees, inundations due to overtopping of walls and the formation of breaches took place. There is no need to enumerate all these. The same series of events occurred in each place, and Fig. 1 shows far better than a long description how much of the coastal district was flooded.

Effect on industry

The damage to industry is perhaps not so fully realized as is that to domestic housing. The flooding in rivers and estuaries was extremely serious in this respect, and damage including that to the West Norfolk Farmers Manure Co. at King's Lynn and to Birds Eye Foods Ltd. and other firms at Yarmouth was great. In the Orwell estuary there was also severe flooding; twenty-two firms, as well as a Gas Works and a Power Station, were seriously affected.

19

Canvey Island, after the storm: main Thames wall on right

Much the same happened in the River Colne, and some works here and elsewhere had to cease production entirely for periods of one to four weeks. The same conditions, often intensified, occurred in the Thames, and factories at Purfleet and other places, including those of Messrs. Van den Bergh and Jurgens, and the Thames Board Mills Ltd., are examples of many that suffered seriously.

In estimating the effects of the storm, full allowance must be made for the loss suffered by industry both as a direct result of the flooding and also because of the fall in production as a consequence of it. Many industrial plants are situated on low ground and so preserve land which is often better used for other purposes. Moreover, proximity to navigable water is essential, and adequate protection of works and factories in these situations is of vital importance to the country as a whole.

Transport was locally much disorganized. Two hundred and twenty miles of track and eighty stations were temporarily out of service. Many hundreds of railway carriages were flooded. Several premises of London Transport, British Road Services, and the Tilling Group were also flooded. Docks at King's Lynn, the Humber ports, and the Hartlepools suffered, as also did stocks in the warehouses.[1]

The effects on agricultural land

Land flooded by salt water presents several problems. Flooding, whether by salt or fresh water, temporarily destroys the tilth and makes the soil pasty and largely impervious to air. Salt has the further disadvantage of causing damage to crops, the extent of which will vary with the salt tolerance of the crop—barley, for example, can tolerate up to 0·25 per cent. and some grass up to as much as 0·3 or 0·5 per cent. Evaporation in summer may raise the salt content to lethal proportions.

But the more serious long-term damage arises as a result of the substitution of calcium by sodium in the clay fraction of soils. This is a factor which is of the greatest significance in heavy soils, which may become very sticky and dry into hard lumps which will not crumble when subjected to frost or wetting.

It is essential to get rid of the flood water as soon as possible by whatever form of draining is most appropriate. Sometimes flooding with fresh water may be proper, and the water should be applied as fast as the soil can absorb it. This process is unlikely to meet with much success on heavy land.

There is no simple method of estimating how long land will be out of production, because there are too many variable factors such as the permeability of the soil, the efficiency of the drainage system, the local *winter* rainfall and the amount of salt in the soil and its distribution therein. No field flooded by ordinary sea water this winter is at all likely to bear a crop this year, but if the water is only somewhat brackish, a summer crop is not an impossibility, especially if other factors are favourable. Many cases are known where land flooded by sea water carried quite good crops after one whole winter's rain. On the other hand, heavy and poorly drained soils may

[1] *British Transport Review* 11 (1953) 338, and *The Railway Magazine*, May (1953) 301.

need five or more years before they can bear crops. Past experience in the coastal parts of Lindsey shows that strong clay land is best left quite untouched for a year and, when work starts, as little weight as possible should be used. Cultivation should be light and begin from the surface. But on the light and well-drained silt lands on which the flooding was of short duration, it may be possible to grow crops this year in Lindsey—and there is little doubt that barley is the most successful crop following flooding by salt water. But it is important to bear in mind that just because a first crop is successful the land has not necessarily recovered. In Holland a second crop has often failed largely because damage to soil structure does not reach the maximum until after the bulk of the salt has been washed out. On some land gypsum treatment may be effective. The gypsum is spread on the surface and is gradually dissolved and washed into the soil. In an average winter little more than a ton of gypsum per acre is likely to be dissolved, but 2–2½ tons per acre should be spread in the dry weather of summer or autumn this year.[1]

These few remarks are sufficient to emphasize that the problem of recovering land from flood water is complicated. Quite a different group of problems arises in considering the feeding of livestock on flooded areas. There is no space to develop this subject, and it must suffice to note that some forage crops may be killed, but that some salt-tolerant grasses may survive. Moreover the salty deposits left in the land may be toxic to animals, and some food crops may absorb salt. After effects, such as moulds forming on flooded hay, may all too easily develop. Furthermore the salt content of drinking water used for animals may easily become too high and bring further trouble.[2]

The subsidence of eastern England

A subject which has given rise to considerable interest is the downward movement which is known to be taking place in south-eastern England. There is little need for me to deal with this matter in detail, since Dr. Valentin, who has been working in the Department of Geography at Cambridge, discusses it at some length in the paper immediately following this. It is however important to emphasize that the movement at the present time can be measured with reasonable accuracy from tide gauge observations. In the region of the Thames approaches it is of the order of 1–2 millimetres a year, which is equivalent to a foot in 150–300 years. This is a significant amount, and throughout a long period of time has made itself felt in various ways. The Thames is now tidal to Teddington; in Roman times it was probably unaffected by tidal influences above London Bridge. The change of level since Roman times may amount to 15 feet, but the subsidence dates back much earlier, and the alternate silt and peat bands that have been so often demonstrated in dock sections imply a sinking which possibly occurred in stages from

[1] The physical properties of clays depend on their chemical composition. In particular, the replacement of calcium, present in normal clays, by sodium from the salt left by flooding with sea water causes the clay to assume intractable properties. The addition of gypsum (calcium sulphate) provides once again an excess of calcium which restores the clay to its original chemical composition and physical properties.

[2] See particularly: Min. of Agriculture and Fisheries Advisory Leaflet on "Treatment of land flooded by sea water," March 1953, and Supplement to the Lindsey A.E.C. "Cropping and treatment of land which has been flooded by salt water," 1953.

the time when the North Sea floor was exposed. The earlier stages are the concluding episodes of a long and tangled story dating back to Pliocene times; but the superposition of Roman on Iron and Bronze Age sites at Brentwood gives us a good measure of the prehistoric and historic part of this movement which is still in progress. That the movement may be partly isostatic—the result of subsurface changes resulting from the relief of the weight of the great ice sheets of Pleistocene times—and partly eustatic, in the sense that it is associated with the melting back of the Arctic glaciers today, is discussed in Dr. Valentin's account which also gives references to relevant papers.

Future problems

Any long-term planning must, of course, take these isostatic and eustatic movements into consideration. The great flood of last January would have occurred even if this movement had ceased a century ago, but we must realize that two or three centuries ahead the general level of our east coast may well be a foot or more lower than now. This in its turn implies not merely a raising of the level of walls, but also strengthening them so that at their increased height they are still able to hold out the sea. The lessons of the flood also emphasize everywhere that the landward side of the banks needs protection. This is a serious matter, and one that requires much money if it is put into effect. Everywhere on January 31 we heard of the sea overtopping the walls and then scouring out behind them, and in this way producing most of the sea breaches. It may well be that in very vulnerable areas the walls should be duplicated, not so much to prevent entirely an overspill from a very high surge, but in order to have a second line of defence if the outer wall is breached. Whilst it is true that much damage would have been done if the walls had not been breached, it is undoubtedly the case that far more serious flooding resulted from the breaches through which the water rushed.

A study of natural coasts like Scolt Head Island and Blakeney Point demonstrates several points. First, that where beaches and dunes are wide damage may be small. Next, that where the beach is steep big waves break directly on it and override it and, as opposite Salthouse, may form breaches. Where, too, there is a wide area of sand and salt marsh fronting the coast, little trouble occurred, since the sea spent some of its force well away from the coast. Nevertheless, the unusual height of the surge meant that cliffs and slopes were attacked in unexpected places. The low cliffs near Bacton and Walcott illustrate this point very well. At Winterton, another natural district, the much steeper (*i.e.* as compared with, say, Scolt Head Island) beach led to a more severe attack on the dunes behind.

All this means that every effort should be made, when it is at all possible, to help the natural accumulation of beach, dunes, shingle ridges, and salt marsh. The narrow dunes between Happisburgh and Winterton stood up, but they have been sadly weakened. Another storm like that of last January would bring devastation to this stretch of coast and hinterland. It is not easy to alter the beach profiles in this part of Norfolk, nor is there any obvious means of bringing about faster dune growth. The sea wall now being built between Eccles, Horsey, and Winterton is the first necessity.

A very similar, but even more urgent problem, faces Lincolnshire. At Sutton and other places the dunes and walls disappeared almost *in toto*. A remarkably good job of new walling and defence has already been done. But if Sutton, Trusthorpe, and Mablethorpe are to remain where they are they will need every protection that man and nature can devise. Doubtless the beach will gather again, but can dune growth be engendered? Theddlethorpe and south of Skegness are well protected in this way, because the beach is wider and blowing sand can easily build the dunes. A further trouble in Lincolnshire is the paucity, or even absence, of shingle. There is little or nothing to form a ridge on which dunes might grow. For many years it seems more than probable that the safety of this exposed part of the Lincolnshire coast must depend wholly on man.

Since often the best natural protection is that of a wide beach on which the waves can expend their energy, it follows that it is essential to take an overall picture of the coast. It should be too well known to need emphasis that if the occupiers of a piece of foreshore build groynes to catch the sand and shingle, their neighbours may possibly suffer. Great Yarmouth has undoubtedly gained at the expense of the coast between Gorleston and Lowestoft; Pakefield has lost as a result of Lowestoft harbour works; Southwold harbour holds up material on its north side. These are but examples; many others could be found on our east and south coasts. There may be very good reasons why a particular place should cause beach to accumulate, but are there equally good reasons why, if it so does, another place should suffer? That is the question that has never been satisfactorily answered, and the answer can only be attempted by a central authority trying to see the coast as a unit. Obviously all sorts of problems arise, economic, administrative and others, and to say that what Yarmouth gains is another place's loss is not in itself very helpful since it is difficult to assume that a thriving town could be made to make fundamental alterations in its harbour and promenades.

The loss of life and property at places like Canvey, Jaywick and the central parts of the Lincolnshire coasts is to be regretted in every way in so far as the immediate sufferers are concerned. It is however right to say that the nation acting through its central and local government authorities is in a sense responsible, because if houses and towns are built close behind sea walls and at levels which mean that they could be flooded at any high, or even ordinary, spring tide, those houses are in a potentially dangerous position. Conditions like those existing on January 31 are fortunately rare, but they can and doubtless will recur. They have also occurred in the past. The reasons why the storm of last January will be long remembered are because of the great loss of life and property resulting almost entirely from recent bungalow development and because of its having been so carefully recorded by photography from the air and so fully discussed in the Press. The storm of 1897 and those of earlier times were just as severe, but the number of houses that then existed on reclaimed salt marshes or tidal flats was negligible. Even if this flood had occurred in 1939 the recovery would have been far slower and less effective than it is today. The bulldozer and other war-time and American machines have made a vast difference. Until however all houses are out of reach of tidal water they are in potential danger from flooding, although floods of sufficient magnitude may only take place at long intervals of time.

At all places constant care is required to see that sea walls and banks are kept in good condition. Rat holes, drying cracks, and untended weaknesses of any sort may lead to disaster. On an open beach where there are dunes every effort should be made to foster their growth. The making of paths through them may, in vulnerable places, lead directly to incursions of sea water. This kind of danger is especially troublesome in places where dune growth is small and where holiday makers are plentiful. It is difficult to convince people who only visit the coast for a summer holiday that the making of tracks through dunes may be a very real danger.[1] At other places if a thick growth of vegetation can be encouraged on marshes in front of walls it may have a beneficial effect. That this is so is evident in parts of Essex. It may not be without significance that some of the major breaks at the northern end of Foulness occurred where the width of marsh in front of the wall was small. Although a good deal is known about the ecology of dunes, shingle, and marshes, much more long-term experimental work is required. This can be done at places on the north Norfolk coast where experiments can be allowed to continue unmolested for years if necessary. Nor need these experiments in any way spoil the natural beauty of places like Scolt Head Island and Blakeney Point.

Conclusion

The amount of defence work already accomplished since the flooding is impressive and illustrates most effectively what can be done when necessity arises. It was imperative to protect as far as possible existing settlements since whatever future policies may be adopted it is obvious that, for example, the whole population of Canvey cannot be rehoused in a day! Moreover destruction of dunes and walls has led to the flooding of great areas of good agricultural land, and in the present state of national economy we must win back that land as soon as possible. The restoration of the *status quo* however, even if it can be accomplished, is not the only or even the most desirable end. Long-term planning must take much more into consideration:—the known subsidence of south-eastern England, the travel of beach material and the effects of groynes and harbour-works over long stretches of coast, the growth of dunes and how that growth can be aided or stimulated by human action, the more extensive experimental planting of areas of salt marsh and, still more, of bare sand, as well as the building of protective walls and other defences. It is partly with problems of this type in mind that the Government has appointed a Committee[2] under the Chairmanship of Viscount Waverley. It will be appropriate to end this paper with a statement of the terms of reference of that Committee:

"i. To examine the causes of the recent floods and the possibilities of a recurrence in Great Britain;

[1] "Trampled dunes may well have helped the attack of the sea on parts of the Lincolnshire coast." D. N. Robinson, *Survey*, Nottingham University, 3, 37, 1953 (March).

[2] Dr. G. M. B. Dobson, Sir Donald Fergusson, Mr. R. D. Gwyther, Sir Claude Inglis, Mr. R. G. Leach, Major Sir Basil Neven Spence, Professor J. Proudman, Mr. A. S. Quartermaine, Lord de Ramsey, Professor J. A. Steers, Sir Miles Thomas, Sir John Wrigley, Mr. T. Yates.

ii. To consider what margin of safety for sea defences would be reasonable and practicable having regard on the one hand to the estimated risks involved and on the other to the cost of the protective measures;
iii. To consider whether any further measures should be taken by a system of warning or otherwise to lessen the risk of loss of life and serious damage to property.
iv. To review the lessons to be learned from the disaster and the administrative and financial responsibilities of the various bodies concerned in providing and maintaining the sea defences and replacing them in the event of damage;
and to make recommendations."

These terms imply a comprehensive review of the disaster and the lessons to be learned from it. Whilst there is every reason to hope that a storm like that of last January will not recur for many years, there is always the possibility that meteorological and tidal conditions may combine to produce a similar crisis any winter. The careful sifting of evidence placed before the Committee, and the appraisal of the points of view of those best able to give advice take some time. The Government and the country should, however, have a report which will set out clearly the dangers that threaten and the best ways of confronting them.

[*Editor's Note:* A discussion of this paper has been omitted.]

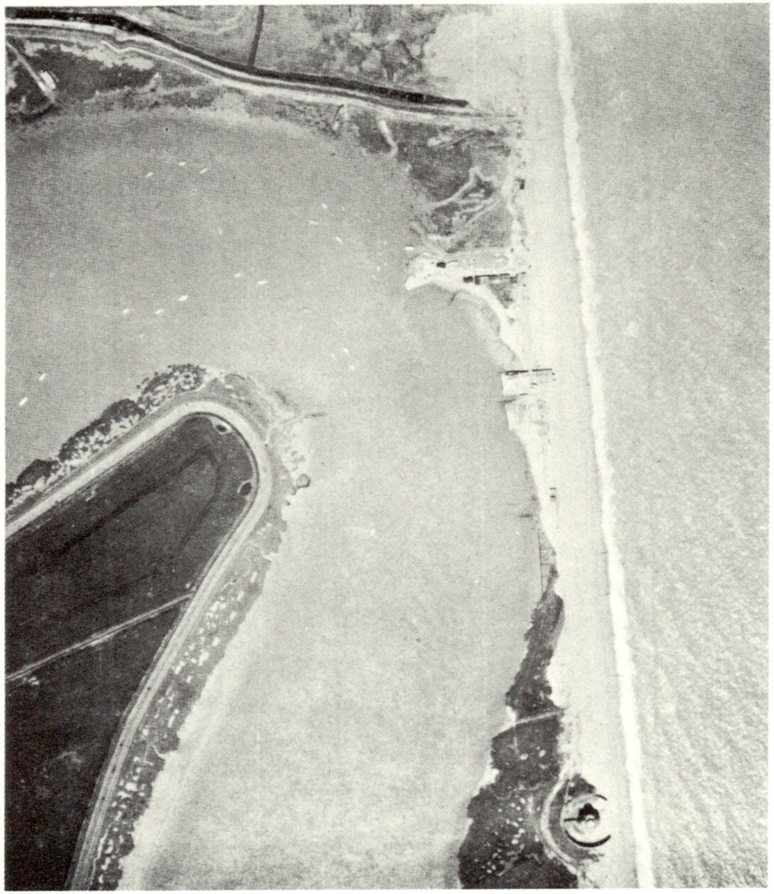

The bend of the Alde at Slaughden, before the storm, groynes just visible

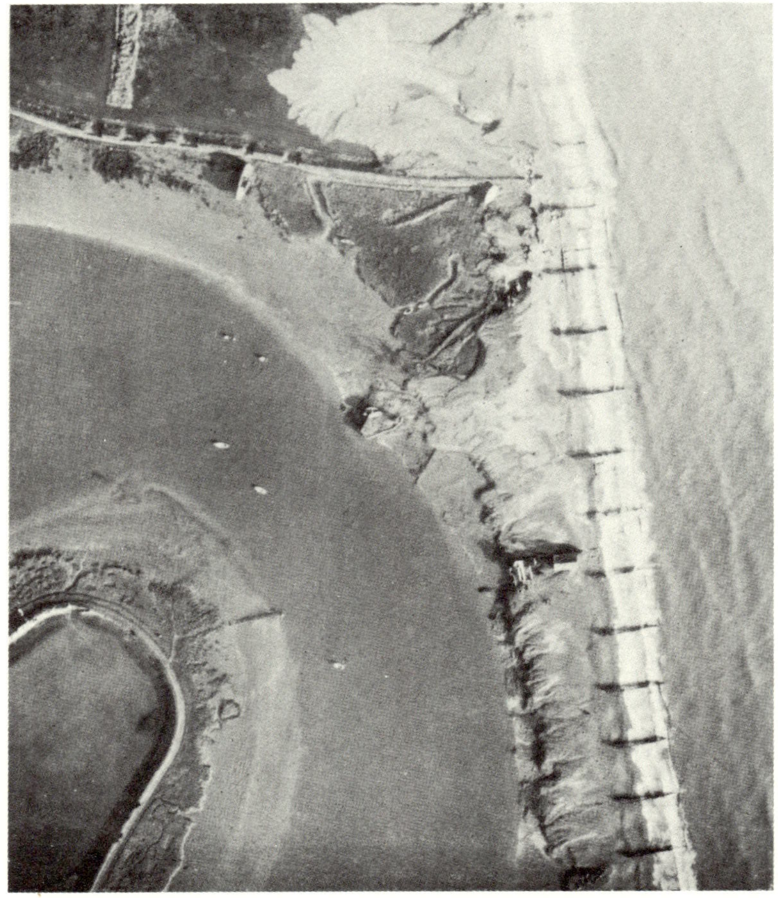

The bend of the Alde at Slaughden, after the storm. Note breaches in the wall, driving-in of shingle along beach, and exposure of groynes

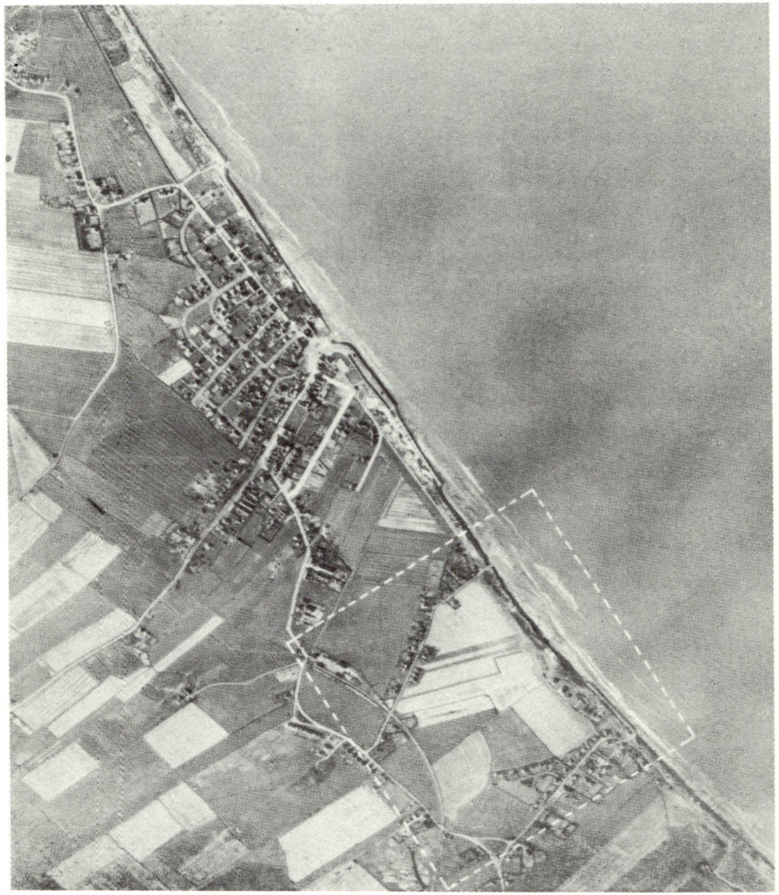

Sutton-on-Sea, before and after the storm. Possibly the largest single breach (about 300 yards wide) made by the storm

Scolt Head, Norfolk: the foreshore dunes were cut away the distance between the two figures and the dune foot

Gap at landward end of the Great Wall along western side of Wells Harbour

East from Waxham: dunes have been much reduced in width. Low ledge between third and more distant groynes was covered by dunes before the storm. Brushwood groynes are new

Beach at Sutton-on-Sea after the storm

11

Reprinted from *The Canadian Surveyor,* **19**(4):372–381 (1965)

Coastal Studies by Sequential Air Photography

H. L. Cameron

Department of Geology, Acadia University, and
Photogrammetry Division, Nova Scotia Research Foundation

With the advent of air photography, many problems of mapping coast lines, such as those caused by surf action and underwater obstructions, were solved, as the vertical air photo shows the exact configuration of the coast and reveals many of the shallow-water obstructions. This paper shows that aerial photographs, obtained at intervals ranging from 2 to 20 years, contain important information on long-term coastal changes, brought about by wave and current action over a period of time. Two examples of such changes, revealed by sequential aerial photography, are given. It is concluded that a program of systematic study of existing photographs and supplementary rephotography would more than justify its cost.

Avec l'arrivée de la photo-aérienne plusieurs problèmes de cartographie des côtes, comme ceux causés par l'action de vagues et les obstacles submergés, furent résolus puisque la photo montre la configuration exacte de la côte, et révèle plusieurs obstructions en eau peu profonde. L'article montre que des photos, prises à intervalles allant de 2 à 20 ans, renseignent sur les variations à long terme de la côte, apportées par l'action prolongée des vagues et des courants. On donne deux exemples de changements révélés par la photographie séquentielle. On conclut que les dépenses d'une étude systématique des photos existantes et de la reprise de photos seraient tout a fait justifiées.

Introduction

The mapping of coastlines has always presented unique difficulties to the chart maker. Inaccessibility, surf action, underwater obstructions and constant changes in the local topography, all add to his difficulties. With the advent of air photography, many of these problems were solved, as the vertical air photo shows the exact configuration of the coast, and many of the shallow-water obstructions are revealed. Wave and surf action are also shown, limited, however, to the particular day on which the photo cover was obtained.

The one field, still largely untouched, is that of coastal changes, brought about by wave and current action over a period of time. As usual, the experts do not agree on how these changes come about, but they do admit, from observations made at varying lengths of time, that major changes do occur. The present study indicates that data on

Published with the permission of Acadia University and the Nova Scotia Research Foundation.

such changes exist in the form of air photos, obtained at intervals ranging from 2 to 20 years, which can be used to indicate actual changes. From the changes that have taken place, deductions can be made on the causes and their operations. The National Air Photo Library, Department of Mines and Technical Surveys, is a mine of such information, which should be exploited without delay.

Photography

Until quite recently, coastal air-photo cover was the result of photography taken for topographic mapping for forestry studies. In the last five years, some medium to large-scale cover has been obtained where coastal works were contemplated and some extremely small scale cover for tidal-current surveys. This type of work should expand greatly in the near future.

Coastal photography has been taken mainly to obtain an exact knowledge of existing conditions, such as the position of rock outcrops and their underwater reef extensions, and sand beaches with their offshore bars. Because of the tides, it has been the practice to obtain two covers at low and high tides, as close together as tide times and daylight conditions would allow. Infrared film has been used to obtain exact high and low-water lines and color film to obtain bottom detail down to 25 feet or more.

Studies of coastline changes, based on wave refraction, current action and littoral drift over a long period of time, are few to non-existent. Those that have been made are often of changes brought about by a disastrous storm or earthquake and cannot be regarded as typical of long range coastal change. This paper is an attempt to demonstrate the possibilities of such studies, based on existing photo covers, and to urge the periodic rephotography of our coast lines as an economic means of study and control of shore-line processes.

The work, described in this paper, is based on photo covers obtained at intervals, ranging from 2 to 20 years, mainly in the Advocate Harbour and Rose Bay areas of Nova Scotia. With the great range of sequential photography available in the National Air Photo Library, it is suggested that studies should be initiated covering the coastal areas of Canada, including the Great Lakes. The practical aspects of such studies are well illustrated, for instance, in the Rose Bay area.

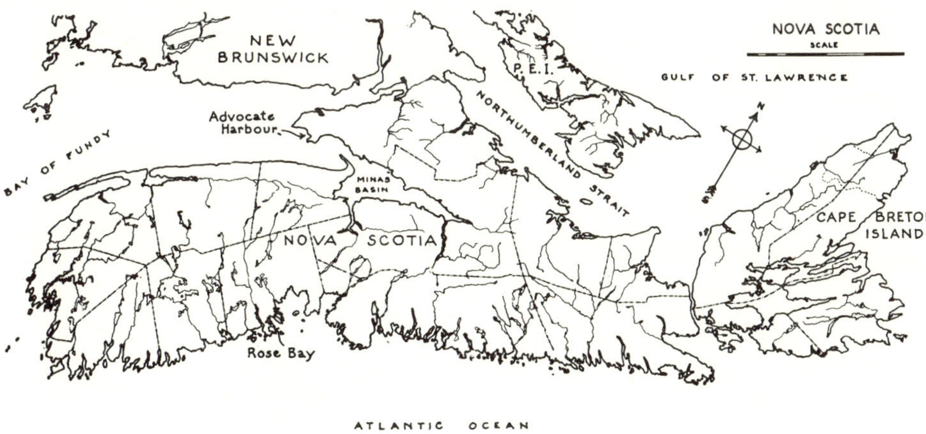

FIG. 1.

The photo covers used in these studies were taken by the RCAF for the federal Department of Mines and Technical Surveys, by the Province of Nova Scotia for topographic mapping and forestry surveys and by the RAF for water-current research.

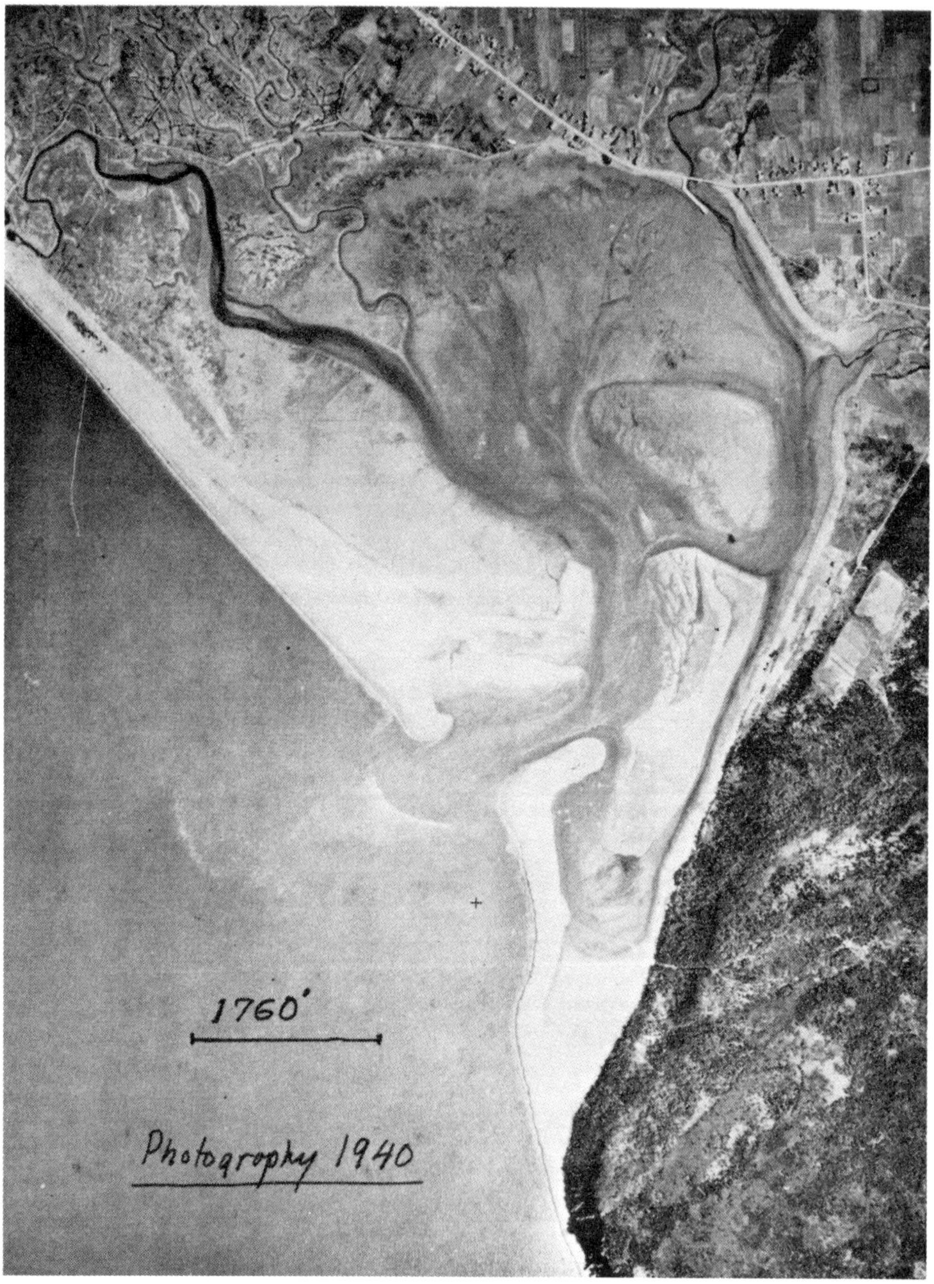

FIG. 2

Advocate Harbour

Advocate Harbour is just north of Cape d'Or, at the mouth of Minas Channel (Figure 1). It is a shallow barachois-type inlet, formed by the building of a sand bar between the Cape Chignecto coast line and the Cape d'Or headland by littoral drift from the west. This bar has never been continuous, probably because of the high tides and strong tidal currents; the southeast portion is a stubby section tangent to the Cape d'Or headland. The variations in the shape and position of the ends of the two harbour-mouth sand bars is shown in detail by the sequential photos in Figures 2, 3, 4a, 4b and 5.

Figure 2 is a photograph, taken in 1940, at a scale of one inch to 1,760 feet. The northwest bar terminus shows a hook on the inside of the harbour, with several overlapping bar segments. The outer hook is less prominent, probably because of the sweep of the westward currents on a rising tide. The southside bar also shows a hook, and the scour channel produced by the tidal currents is well defined.

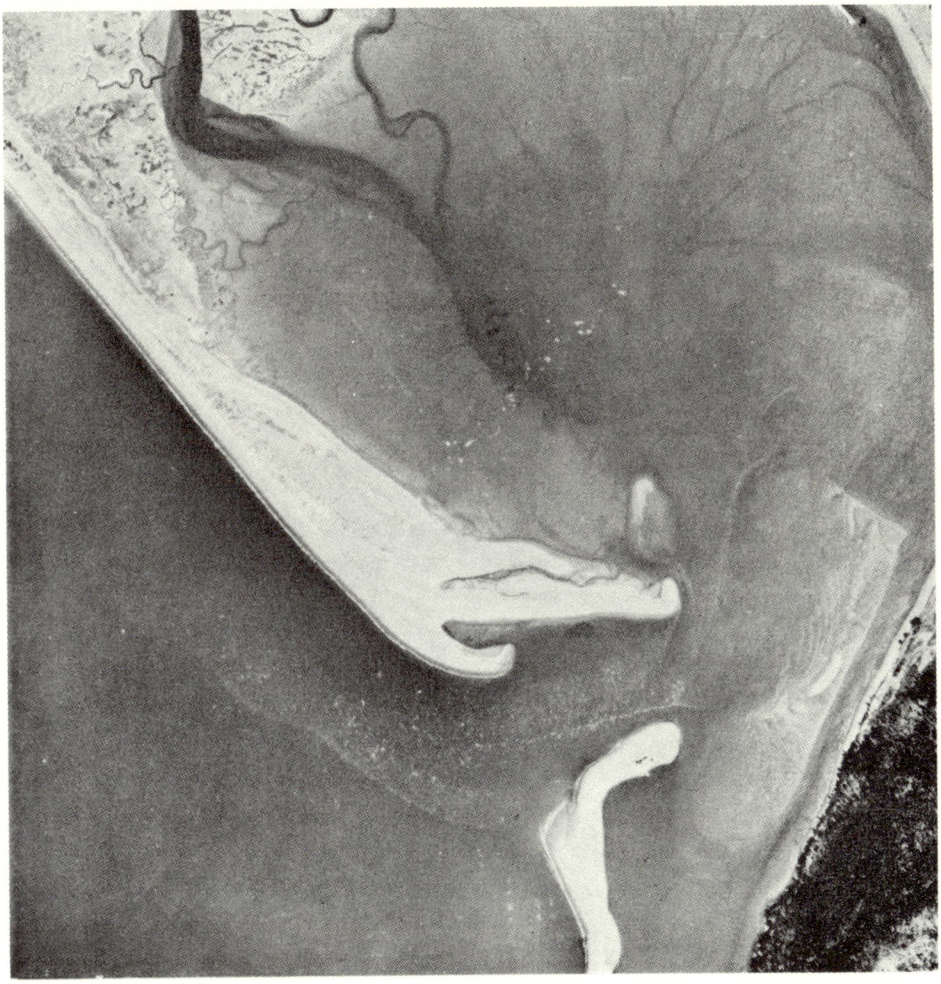

FIG. 3

Figure 3 is a photograph of the same area taken in 1948 at one inch to 1,320 feet. The changes in the north bar terminus in the eight-year period are obvious. The channel remains approximately the same, but the north bar terminal spits have changed shape. The outer spit, the immediate extension of the northwest beach, has lengthened and curved inward. The second spit has been extended and a short north hook developed. The third spit has lost part of its height, but is essentially the same shape as in 1940. The south beach has lost part of its westward extension, and the width of the spit has decreased. The inner sand bar, or delta, has changed shape. These descriptive remarks take cognizance of the fact that the tide is higher in the 1948 photos than in those of 1940.

Figures 4(a) and 4(b) were taken in 1954 as part of the joint federal-provincial forest inventory photography of Nova Scotia. They show approximately high tide and an interesting item in the extension of diking in the northwest end of Advocate Harbour, carried out by the Maritime Marshlands Rehabilitation Administration. The changes in the spits on the northwest bar are obvious. The harbour mouth channel appears to have widened and the southern spit to have straightened and narrowed. The wave refraction pattern in Figure 4(a) is particularly suggestive of the forces at work. Figure 4(b) shows the shape and position of the south bar, which is obscured by sun reflection in Figure 4(a). It is suggested that water-volume and current-flow studies be made to see what difference, if any, the diking off of the area at the northwest end of the Harbour made to the current at the harbour mouth.

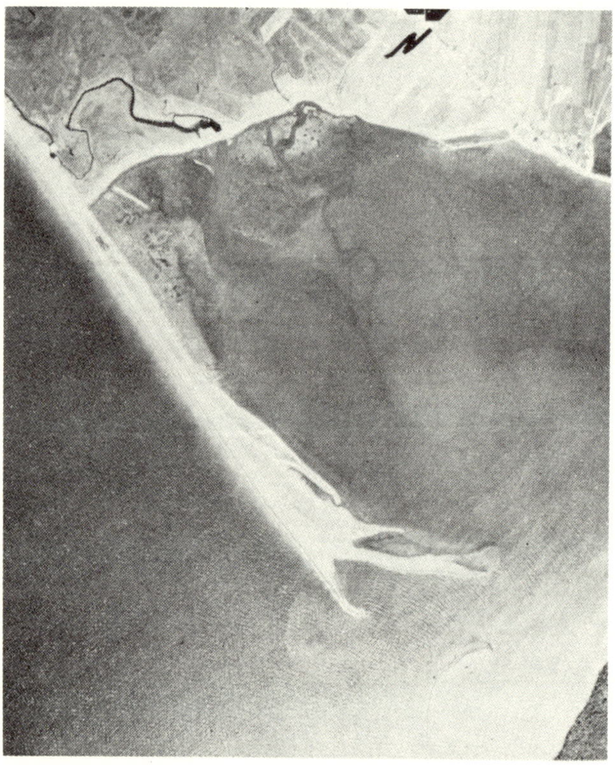

Fig. 4(a)

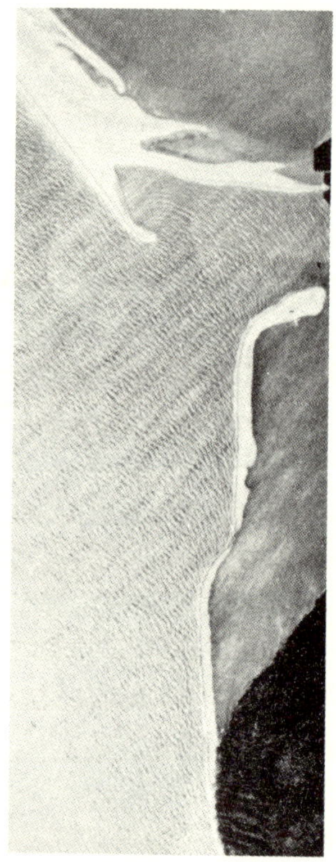

Fig. 4(b)

FIG. 5

In 1959 at low tide, the entire spit system and channel were revealed (Figure 5). The most remarkable features are the eastward extensions of the terminal hooks beyond the trend line of the two bars. The small basins on either side of the channel between the hooks are similar to the surge basins, built by the Department of Public Works at Dingwall Harbour. They were designed to keep the channel clear of sediment by providing an expansion area in the channel, which would provide a surge, or pulsation, to sweep the channel clear. Nature appears to have duplicated them at Advocate Harbour with a similar end result.

The extremely strong tidal current in this area can be seen setting eastward past Cape d'Or. Earlier studies [*Cameron* 1961] indicate that the main tidal wave sweeping up the Bay of Fundy splits west of Ile Haute and part of the Chignecto Bay segment rebounds past Cape Chignecto into Advocate Bay. This would explain the relative activity of the northwest beach bar as compared with the more static southeast beach bar.

FIG. 6

Rose Bay, Lunenburg County, Nova Scotia

The area of interest comprises the Conrad Island and adjacent main-shore section. Figure 6 is a photograph taken in 1945 at one inch to 1,320 feet. The island is a dual drumlin, often referred to as "Paramecium twins". This double drumlin is a common form in Nova Scotia, and comprises clay and boulders picked up by the glaciers and built into an elongated oval form under the ice. The island is in process of being cut off on the north by the prevailing currents in Rose Bay, which set to the west and north. The resultant debris has formed a bar with multiple lobes on the west and northwest of the island. The longshore drift southward from the mainland has built a mass of sand south from the mainland toward Conrad Island, which curves eastward. It would appear that eventually this bar will reach the island, or its west bar, and form a tie bar similar to many others along the same coast.

Fig. 7

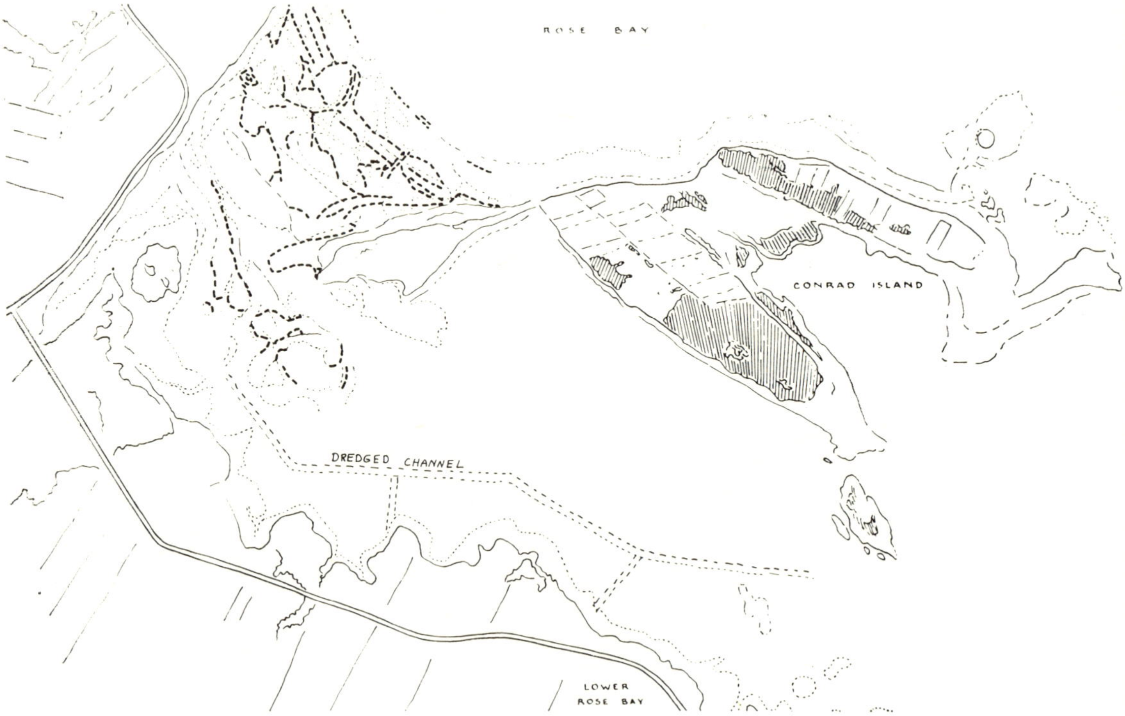

Rose Bay Area

Lunenberg County, Nova Scotia

—Shore line and Sand Bars — 1945

—Shore line and Sand Bars — 1955

FIG. 8

Another feature to be noted is the dredged channel along the southwest coastal section from Rose Bay south. This channel, with its local branches, is dredged to permit inshore fishing boats to use the numerous small fishing stations dotted along the coast.

Figure 7 is the same area at the same scale, but taken in 1955, 10 years after the cover described above. The changes are obvious and striking. The west bar of Conrad Island has been extended for 330 feet and small lobes and hooks formed at the end. The chaotic mass of sand with several channels, which extended southeast from the mainland, has formed an almost continuous bar with one well defined channel. The dredged channel still shows in the southwest. These changes are shown in Figure 8. This particular example of coastal change illustrates a possible practical application of such studies. If the sand bar, built up and consolidated northwest of Conrad Island, were to be aided by the placement of concrete tetrahedrons, or heavy rock fill, then dredging operations in the southwest sector would be eliminated, or much reduced.

Conclusion

Two examples have been given of coastal changes revealed by sequential air photography. There are probably hundreds of others which would be revealed by a search of the photography available in the National Air Photo Library in Ottawa. It is strongly urged that a study should be made of all existing photo covers of coastlines in Canada and a program set up to supplement them by rephotography at regular intervals. The assistance rendered by such a program to government departments, concerned with coastal works and protection, would more than justify the cost involved.

Bibliography

Cameron, H. L. Mapping tidal currents in the Bay of Fundy area by high altitude photography. Paper presented at March 1961 ASP Annual Meeting.

Interpretation of high altitude photography, *The Canadian Surveyor,* Vol. XV, November 1961.

Water current and movement measurement by time lapse air photography—an evaluation. *Photogrammetric Engineering,* Vol. XXVIII, No. 1, March 1962.

Ice cover surveys in the Gulf of St. Lawrence by radar. *Photogrammetric Engineering,* Vol. XXX, No. 5, September 1964.

Forrester, W. D. Plotting of water current patterns by photogrammetry, *Photogrammetric Engineering,* Vol. XXVI, No. 5, December 1960.

Keller, M. Tidal current surveys by photogrammetric methods, *Photogrammetric Engineering,* Vol. XXIX, No. 5, September 1963.

Sonu, C. J. Study of shore line processes with aid of aerial photogrammetry, *Photogrammetric Engineering,* Vol. XXX, No. 6, November 1964.

12

Reprinted from *Photogramm. Eng.* **33**(2):184–189 (1967)

SHORELINE FEATURES AND THEIR CHANGES

M. T. El-Ashry*
Prof. H. R. Wanless†
University of Illinois at Urbana

Sequence photos show long-term trends as well as alterations by storms.

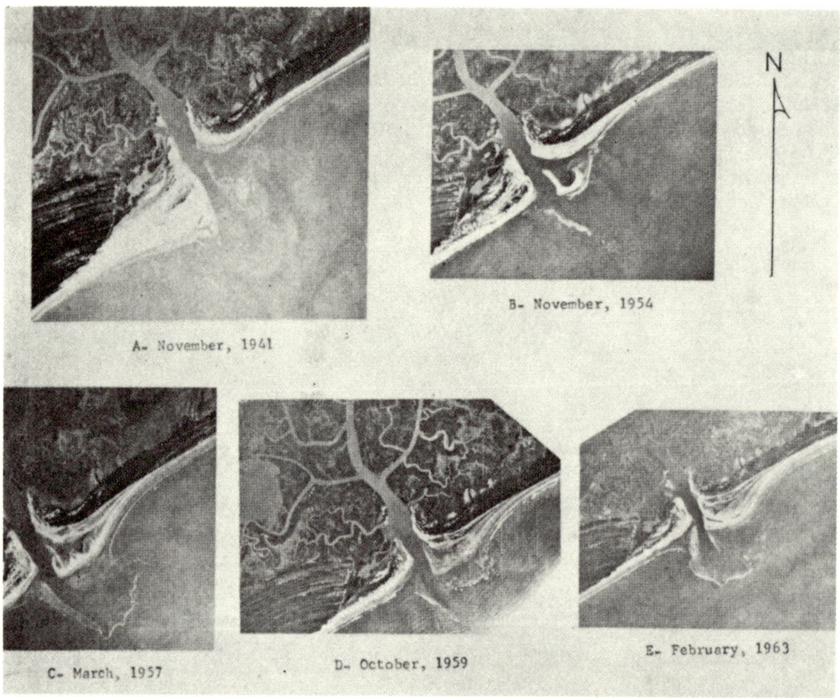

FRONTISPIECE. Shoreline development at Price Inlet, Charleston County, South Carolina. (See text page 187).

INTRODUCTION

SINCE WORLD WAR II aerial photographs have been used in the study of coastlines and in coastal engineering to an increasingly large extent. An aerial photograph provides an excellent record of the pattern of features on a coastline that existed at the time of photography. The size of waves, direction of wave fronts, distribution of shallow water sediment accumulations, and the direction of littoral drift may easily be determined. More detail of configuration of coastal and near-shore features is observable than is usually recorded in careful field surveys of a coastline.

From sequential aerial photography it is possible to observe and measure changes resulting from sediment shift during the inter-

* Submitted under the title "Sequential Aerial Photography and the Study of Shoreline Features and Shoreline Changes," based in part on research supported by a grant from the U. S. Coast & Geodetic Survey (Environmental Science Services Admin., Institute of Cartography) to the University of Illinois.

Now on staff of Univ. of Cairo, Cairo, U.A.R.

Abstract: *Sequential aerial photography is an excellent tool for recording long-term trends in shore line changes and for studying shoreline features. Such minor features as wave-cut scarps, sluiceways, washover fans, and underwater sediment patterns are better represented than on charts or maps. The photographs provide an exact basis for measuring net changes in dimension or shape between photographic dates. Changes in the width and position of inlets can be measured directly from the photographs and the amount of sediment eroded or added to the beach can be estimated. At the same time, an estimate of the amount of sediment carried by longshore currents or deposited in the form of shoals or tidal deltas can be determined.*

val between dates of photography. The greatest difficulty encountered in interpreting the changes results from the lack of a continuous day-by-day photographic coverage of the coastline. If the photographs indicate that a new inlet has been opened, an old one closed, a new beach ridge formed, or a spit lengthened or cut away, all that can be said is that these changes have occurred between the two dates of photography. The greatest advantage of sequential aerial photography, especially when supplemented by older surveys and charts, is in recording long-term trends in shoreline changes that help in planning shore protection programs.

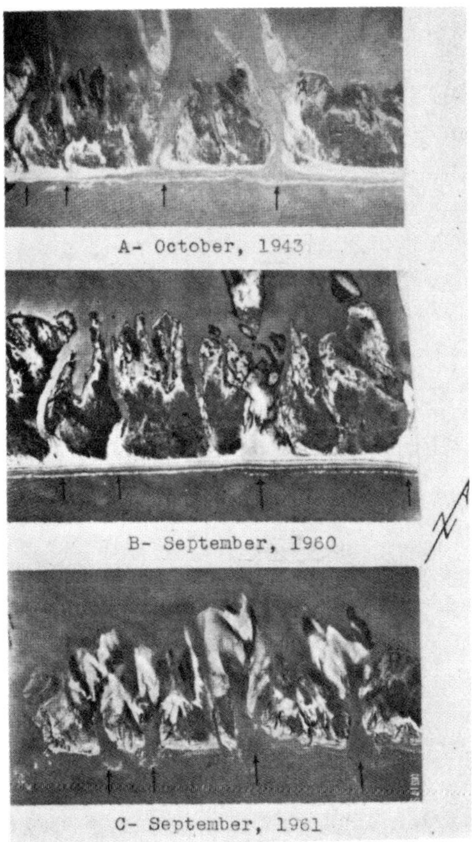

Fig. 1. Portion of Matagorda Peninsula, west of the Colorado River in Texas, showing some of the effects of hurricanes on narrow barrier islands.

Studies of hurricane effects and shoreline changes in selected areas along the Atlantic and Gulf Coasts of the United States by the authors in the past four years have demonstrated that sequential aerial photography is an indispensable tool, and those studies have suggested various other geologic and engineering applications of this tool. Examples of these applications include: river flood effects; changing outlines of deltas and their distributaries; shifting sand dunes; glacial fluctuations; landslides; earthquake effects; volcanic eruptions; and man-made changes.

The principal sources of aerial photography used by the authors are the U. S. Coast and Geodetic Survey and the U. S. Department of Agriculture. The number of dates of photography for the different coastal areas ranges from 3 to 10. Six examples are used in this paper to illustrate some of the phenomena and features discussed.

It should be mentioned that the two parts of the paper—the first dealing with shoreline changes, and the second with shoreline features—are interrelated and are only separated for the sake of the discussion.

Shoreline Changes

A preliminary study of sequential aerial photographs of the United States coast between Massachusetts and Texas shows that many parts have experienced notable changes in outline since the early 1930's when systematic aerial photography of the United States began. Detailed studies were carried out in North Carolina, South Carolina, the west coast of Florida, and Texas. Other

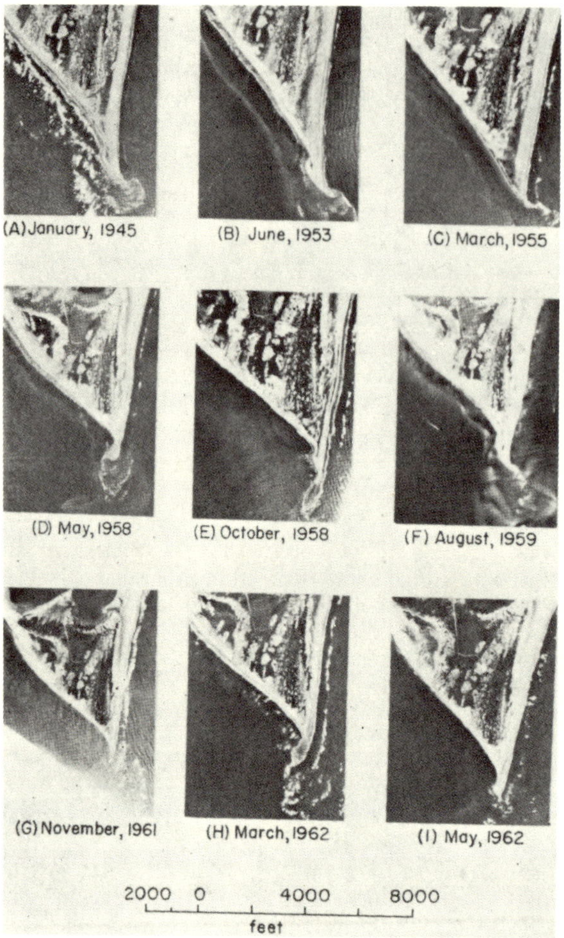

FIG. 2. Aerial photographs of Cape Lookout, North Carolina, 1945 to 1962.

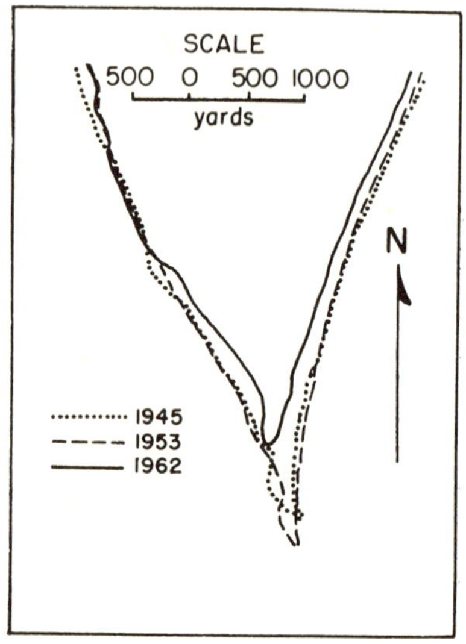

FIG. 3. Sketch diagram, traced from aerial photographs, showing the outline of Cape Lookout, North Carolina, in 1945, 1953, and 1962.

studies are underway by colleagues at the University of Illinois in Massachusetts, New Jersey, Maryland, and Virginia.

Careful examination of the photographs shows that the width of the beaches increases and decreases with the seasons and that a single storm may alter the configuration of a shoreline to such a great extent that a long period of time may be required for a shoreline to regain its normal outline and width.

The U. S. Coast and Geodetic Survey commonly photographs a coast just at the end of a violent storm (or in its later phases) to assess storm damage and to revise nautical charts. The photographs show that the greatest changes in coastal form are catastrophic, accomplished by high winds, waves and tides. Photographs taken during an interval of calm weather after a past major storm show that storm-induced changes are gradually smoothed and obliterated; thus, the coast may appear to have reassumed the form it had before the storm. Measurements on photographs show whether the beaches are at the same or different locations.

Figure 1 illustrates the effects of two hurricanes on a portion of Matagorda Peninsula, west of the Colorado River in Texas, and the changes that took place during a period of non-hurricane weather. Two inlets were opened by storm waves of the hurricane of August, 1942 (Figure 1A), washing across low and narrow parts of the barrier island. Spits growing in a southwesterly direction, across the mouths of the inlets, were formed during the year following the hurricane by littoral drift. Figure 1B shows the inlets completely closed, but they were reopened by hurricane Carla of September, 1961 (Figure 1C). Hur-

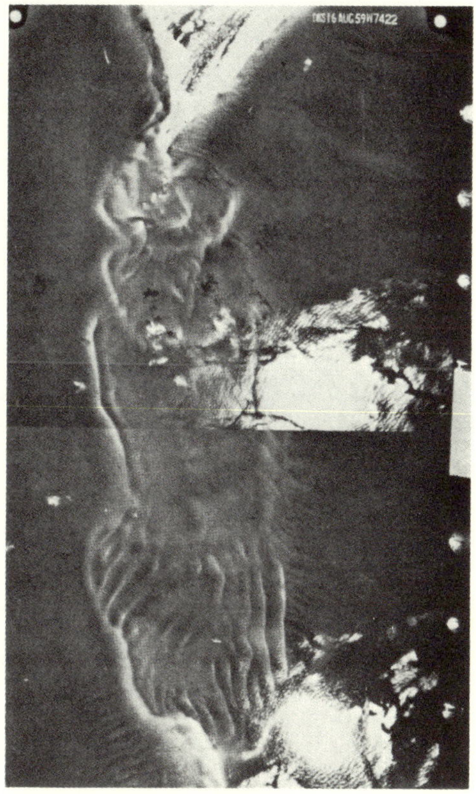

FIG. 4. Aerial photo mosaic of Cape Lookout Shoals, North Carolina.

ricane waves and surges associated with Carla eroded the beaches to a great extent and cut a number of sluiceways through the foreshore dunes. Material eroded from the outer beaches was deposited as washover deltas in the lagoon behind the barrier.

Price Inlet is the outlet of Price Creek in Charleston County, South Carolina (Frontispiece). Erosion of the southwest side and progradation on the northeast side of the inlet took place during the period of 1941 to 1963. A headland on the southwest side of the inlet (*A* in the Frontispiece) was eroded by concentrated wave action due to wave refraction by the headland. At the same time, the northeast side was prograding by the addition of a series of recurved spits built by the southwesterly littoral drift. The position of older spits on this side of the inlet is now marked by beach ridges that help in restoring the original outline of the shore *E* in the Frontispiece.

The Capes of North Carolina provide the largest number of photographic dates of any of the coastal areas studied. Nine dates of photography are available for Cape Lookout (Figure 2). The presence of old beach ridges help in restoring older outlines of the cape. From 1945 to 1962 approximately 100 yards of beach were eroded along the entire stretch of the east shore of the cape (Figure 3), and the spit extending southward from the cape point was cut back by about 650 yards, averaging about 40 yards per year (Figure 3). The orientation of the spit provides information as to the wind direction during the month preceding photography. From 1945 to 1955 the spit was oriented eastward from the cape point (Figure 2A, 2B, 2C); whereas, in most of the other photographs it curves slightly westward from the cape point.

Shoreline Features

Depositional features (such as bars, spits, sandkeys, washovers, barriers and beach ridges) and *erosional* features (such as wave-cut terraces, sea cliffs, and stacks) show more clearly and are easier to study from aerial photographs than from the ground. From sequential aerial photography it is possible sometimes to date individual beach ridges or to record the period of their formation within a decade or so. Changes in width and position of inlets can be measured directly from the photographs, and an estimate of the amount of sediment deposited in the form of shoals or tidal deltas, on either the lagoon-side or the ocean-side of the inlets, can easily be determined. Studies of the amount of sediment carried by littoral currents into shoaling areas, as, for example, off the southern tips of the North Carolina Capes, can be made by contouring the shoals at different dates, using a stereo plotter and estimating the amount added or removed during a given period of time. Figure 4 gives an excellent picture of the form of sand deposits composing Cape Lookout Shoals in North Carolina. The shoals extend about 9 miles south-southwestward from the cape point, reaching 2 miles in width and 2 to 18 feet in depth. These shoals were formed by sediment eroded from beaches north of the cape and carried southward into the ocean by littoral currents, particularly during times of severe storms.

The formation of new *barrier islands* and *sandkeys* from submerged sand deposits by storm waves and their subsequent changes can also be studied from sequential aerial photographs. Figure 5 shows LaCosta Island, which is located in Lee County, on the west coast of Florida. A strong outflow from Charlotte Harbor (on the north side of the island) drifts sediment in a southerly direction to

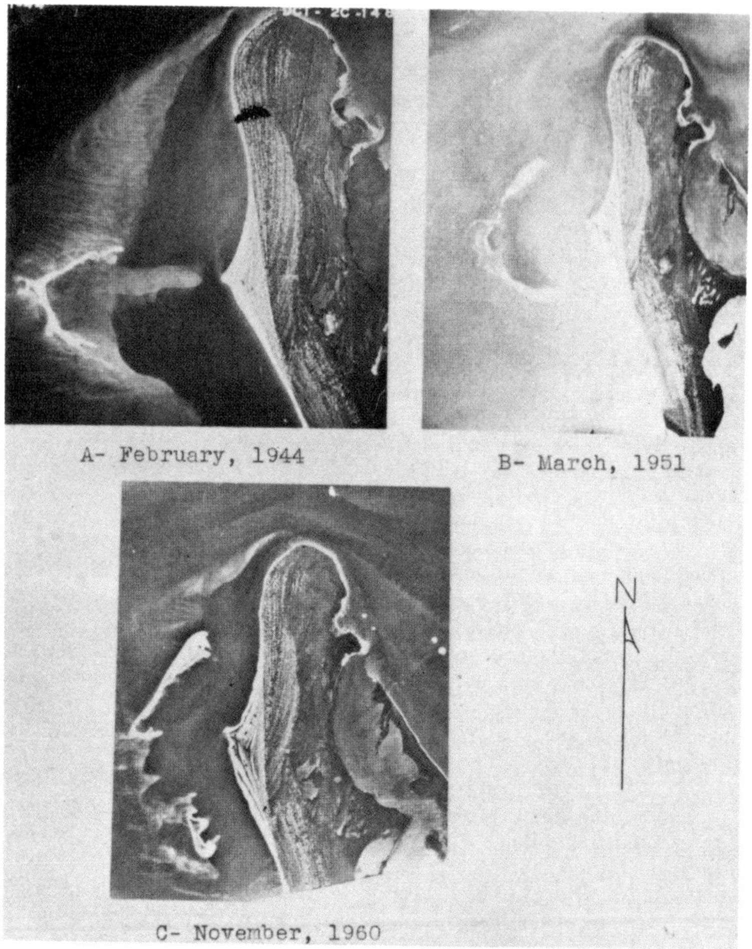

FIG. 5. Aerial photographs of LaCosta Island, Lee County, Florida, 1944 to 1960.

form a big shoal that resembles an arc surrounding the island (Figure 5A). A cuspate sandkey was formed by storm waves scouring the sediments of the shoaling area and a tombolo grew eastward towards the island. The presence of the shoal resulted in a wave refraction pattern that constantly brought material towards the island, and a cuspate foreland was formed at a point facing the sandkey. By 1951 (Figure 5B) the sandkey was split into two. By 1960 (Figure 5C) it migrated eastward, and its northern portion grew in length and developed a form close to that of 1944.

The building of a cone-shaped *delta* at the new outlet of the Brazos River in Texas and the alternate formation of barrier islands around the delta are illustrated in Figure 6. A diversion channel, cut by the U. S. Corps of Engineers in the late 1920's, allowed the river to enter directly into the Gulf after an older bay approximately 6 miles northeast of the present outlet had become filled. Sediments carried by the river were deposited to form an arcuate delta, and its seaward limit was marked by two barrier islands, one on each side of the river channel (Figure 6A).

From 1938 to 1952 the delta had extended a distance of about one mile into the Gulf, and the marshy areas had appeared above water due to accumulation of fine sediment from the river (Figure 6B). In 1952 the east side of the delta was marked by a barrier island, and the southwest side had two lines of emergent barrier islands. These islands were formed by hurricane surges from submerged longshore bars just below the intertidal zone off the river outlet. By 1960 a third line of barriers was built on the southwest side of the outlet (Figure 6C). After hurricane Carla of September 1961, the southwest side of the delta was cut back by a distance of about one-

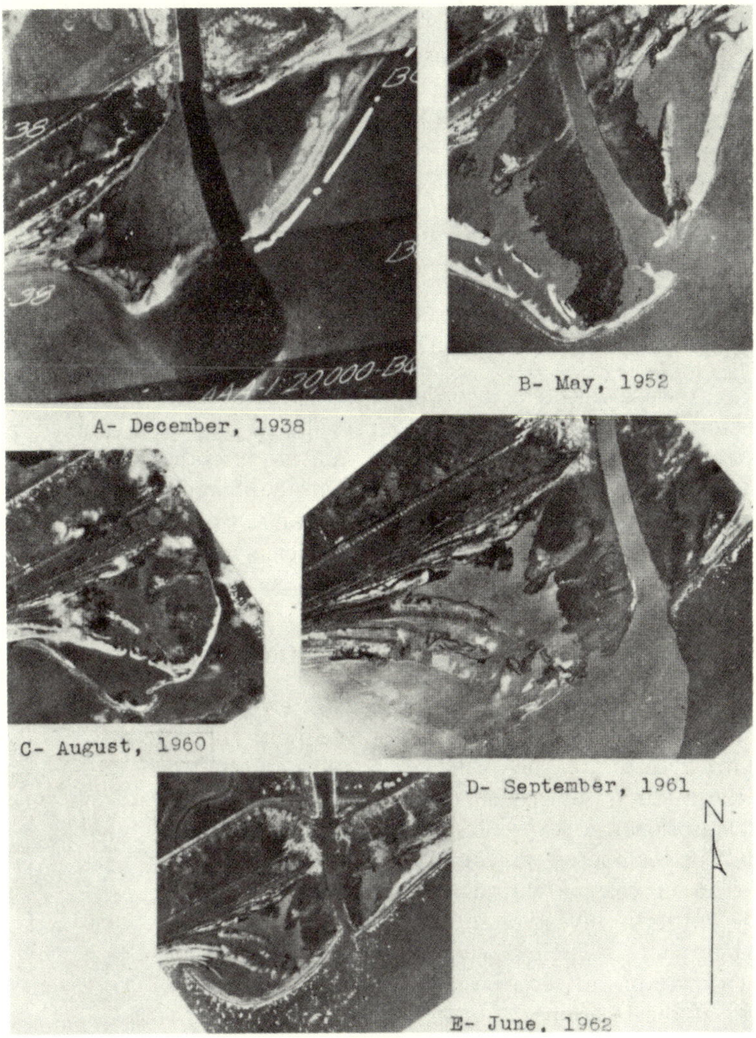

FIG. 6. Aerial photographs showing changes in the size and form of the Brazos River delta, Texas.

third of a mile; the barriers were almost entirely eroded and a large shoal was formed in their place (Figure 6D). About nine months after the hurricane, the sediments of the shoal were reworked and a large lunate bar was built enclosing remnants of the destroyed barriers (Figure 6E).

CONCLUSION

Sequential aerial photography is an excellent tool for the study of shoreline features and shoreline changes. Such minor features as wave-cut scarps; sluiceways; washover fans; and underwater sediment patterns in lagoons, inlets, or the seas are better represented than on any charts or maps. The photographs provide an exact basis for measuring net changes in dimension or shape between photographic dates. However, they do not provide information as to the exact time of origin of a new coastal feature, nor its impelling cause.

13

Reprinted from *Coastal Eng. Japan* **10**:67–83 (1967)

A STUDY ON EROSION OF COASTAL CLIFFS BY USING AERIAL PHOTOGRAPHS

*Kiyoshi Horikawa**
*Tsuguo Sunamura***

ABSTRACT

Two areas were selected, in order to investigate erosion of coastal cliffs of the short term by using aerial photographs. At one area annual amount of sand lost from coastal cliff was estimated, and at the other the relationship between eroded distance of cliff and the strength of rocks or the frequency of wave occurrence, and the future eroded distance were examined. In studying erosion of coastal cliff, especially of a short term, aerial photographs seem to be a very useful means as compared with previous ones.

I. INTRODUCTION

Japan is suffering from beach and coastal erosion, and this problem has become acute from the viewpoint of land protection. In recent years, erosion of coastal cliffs, which are the principal sources of sand drift in some areas, has attracted the attention of coastal engineers. Figure 1 shows areas where beach and coastal erosion are severe in Japan. Locations numbered in Fig. 1, where erosion of coastal cliffs is severe, were hitherto investigated by coastal geographers. Table 1 shows mean annual rate of cliff erosion and the geology of the locations.

This paper describes general conception of erosion of coastal cliffs, and instructions of use of aerial photographs in studying erosion of coastal cliffs, and also shows two examples of study by using aerial photographs.

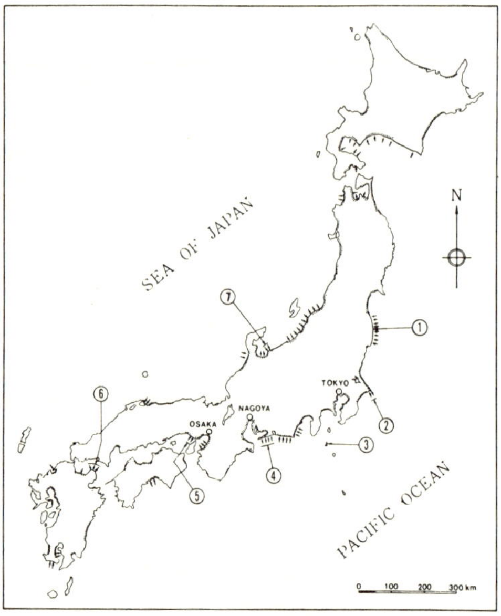

Fig. 1 Areas where beach and coastal erosion are severe in Japan.

II. GENERAL CONCEPTION OF EROSION OF COASTAL CLIFFS

A coastal cliff is "a nick or scarp resulting from wave erosion[8]" and erosion of coastal cliffs implies a phenemenon of recession of coastal cliffs due mainly to wave action.

* Professor of Civil Engineering, University of Tokyo.
** Post-graduate Student, Department of Civil Engineering, University of Tokyo.

Table 1 Mean annual rate of cliff erosion and the geology of locations numbered in Fig. 1.

Location	Mean annual rate	Geology
1. Joban Coast	0.3-0.7 m/year[1]	Sandstone, Siltstone, Mudstone
2. Byobugaura Coast	0.4-1.1 m/year[2]	Diluvial deposits, Siltstone
3. Habuseura Coast, Niijima Is.	5.5 m/year (mean)[3]	Dacitic pumiceous sand
4. Southern Coast of Atsumi Pen.	0.6 m/year (max.)[4]	Diluvial deposits
5. Akashi Coast	0.1 m/year (mean)[5]	Diluvial deposits
6. Northern Coast of Kunisaki Pen.	2.2 m/year (mean)[6]	Unconsolidated volcanic breccia
7. Toyama Coast	0.5-1.0 m/year[7]	Alluvial fan deposits

Important factors affecting erosion of cliffs are the erosive force of waves at cliff base and the resistibility of rocks exposed there. The erosive force of waves has the connection with wave energy in deep water, and also it is affected by many following factors while waves reach at cliff base: (1) characteristics of submarine topography, (2) width and slope of beach at cliff base, (3) grain size of beach deposits, (4) tidal range, etc. The erosive force of waves consists mainly of actions such as compression, cavitation and abrasion[9]. The resistibility of rocks against erosion is due to not only mechanical properties such as compressive strength, tensile strength, abrasive hardness, etc. but also rock structures such as stratifications, joints, faults, etc. In some areas, a factor of weathering that modifies properties of rocks must be taken into account.

Another important factor is landslide which is caused by erosion of cliff base due to wave action. Landslide includes various forms such as slump, debris fall, rockfall, etc.[10], which are controled by the structures and mechanical properties of rocks, especially shear strength of rocks. Factors that accelerate landslide mentioned above are rainfall, its intensity, its duration, etc.

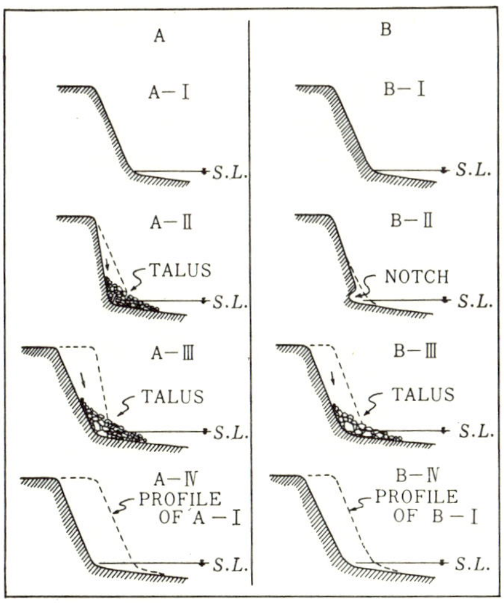

Fig. 2 Schematic diagram illustrating processes of coastal cliff erosion.

Rill erosion, that is erosion of cliff face due to rainfall, is less important as compared with many factors described above. But it would be necessary to take rill erosion into account in some places.

Figure 2 is a schematic diagram illustrating processes of coastal cliff erosion. A in Fig. 2 indicates a process described below. Landslide occures when cliff base (A-I) is eroded and talus developes (A-II). The cliff base is escaped from wave erosion while the talus is not removed by wave action. Removal of the talus (A-II) enables waves to erode the cliff base. Consequently landslide occures again and then new talus develops (A-III). After the talus is removed (A-IV), waves attack the cliff base. B shows another process. According to erosion of cliff base (B-I), notch develops there (B-II). Landslide follows development of the notch and builds talus (B-III). Waves remove the talus (B-IV), and then erode the cliff base.

In short, the process A has the characteristics of immediate occurrence of landslide after erosion of cliff base, and the process B has the characteristics of formation of notch, and of existance of time lag between erosion of cliff base and occurrence of landslide. An interposed process between A and B would be seen in many places. Differences which are in these two processes are due mainly to mechanical properties and structures of rocks of coastal cliffs.

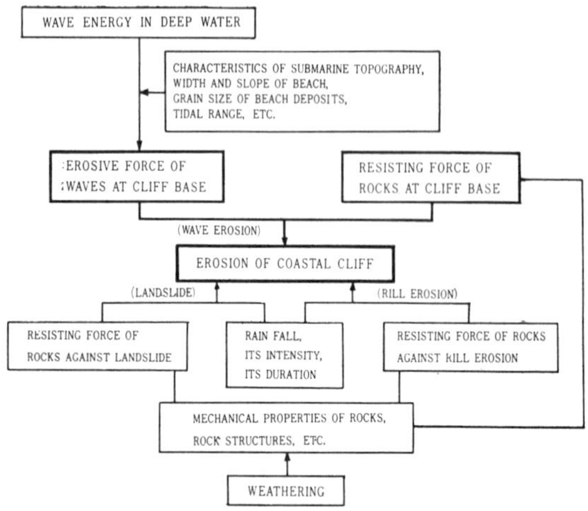

Fig. 3 Factors affecting coastal cliff erosion.

Processes described above possess in common the following cyclic process: wave erosion of cliff base → landslide → talus → removal of talus due to wave action → wave erosion of cliff base.

Figure 3 shows factors affecting erosion of coastal cliffs.

III. HOW TO USE AERIAL PHOTOGRAPHS

Eroded distance perpendicular to coastal cliff is called "the eroded distance" in this paper. The eroded distances described in previous papers were examined by using the following means: the Ordnance Survey maps (ab. 1/10,000)[11)], cadastral maps (scale unknown)[12)], distribution maps of national forest (1/50,000)[13)], surveying[14),15)], horizontal photographs[16),17)], etc. And the eroded distances examined by using these means were those of relatively long term of 10–10^2 years. It would be difficult to examine the eroded distance at cliff base with accuracy and to draw profiles of coastal cliffs, using the above-mentioned means except surveying. Even surveying, however, is not effective in places where slope of coastal cliffs is steep and also waves reach at cliff base in normal weather conditions.

In order to study erosion of coastal cliffs of a short term in connection with the erosive force of waves and the resisting force of rocks, it is necessary to examine the eroded distance at cliff base as accurately as possible. In order to obtain amounts of sand lost from coastal cliffs, on the other hand, it is important to draw profiles of cliffs. A method using aerial photographs would be powerful for these purposes. Instructions of use of aerial photographs will be described below.

Generally speaking, aerial photographs have tilt displacement and relief displacement (Fig. 4). The tilt displacement may be corrected by rectification that is a process of projecting a tilted photograph onto a horizontal datum plane. The rectified photographs enable the eroded distance to be examined by using stereo measurements. Accurate measurement, however, cannot be expected in

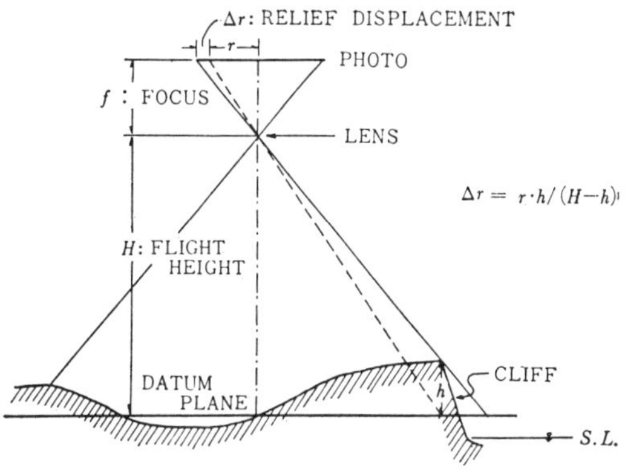

Fig. 4 Diagram showing relief displacement.

proportion to cost and time which are expended in using this method. Consequently, photogrammetric mapping seems to be the most effective method in measuring the eroded distance and in drawing profiles of cliffs, as accurately as possible. Large-scale mapping (more than 1/2,000) is desirable for these purposes. In order to diminish errors in protting from photographs, it is desirable to adopt the scribing method which is to draw scratching the surface of an emulsion-coated film with needle instead of pencil attached to a plotting machine. The smallest contour interval accurately plottable is due to the C-factor that varies according to the elements and conditions of photogrammetric instruments. The accuracy of map compilation is determined by the scale of a photograph, the scale of a drawn map, and the characteristics of the photogrammetric system. For further information about rectification, mapping, photogrammetric instruments, accuracy of map compilation, the C-factor, etc., please refer to textbooks[18] of photogrammetry.

IV. TWO EXAMPLES OF STUDY BY USING AERIAL PHOTOGRAPHS

1. Annual Amounts of Sand Lost from Coastal Cliff at Taito-Misaki District, Chiba Prefecture

a) Physiographical Conditions

The studied area is shown in Fig. 5. Coastal cliff in this area is of steep slope, and its height is 10–70 m above M.S.L. Rocks exposed to the cliff are silt-

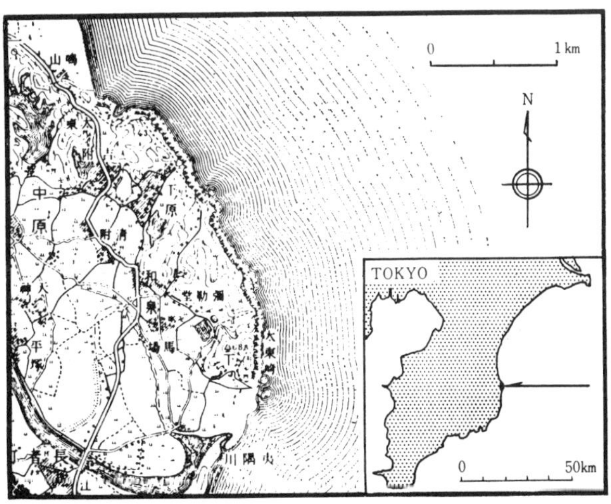

Fig. 5 Studied area.

stone and mudstone of the Pliocene, Tertiary, which are of fractures such as joints and faults. The extent of a coast-line of this studied area is about 2.5 km. Submarine contours run nearly parallel to the shoreline and the submarine slope is about 1/100 up to a water depth of 5 m and about 1/200 for further depth. The mean tidal range is about 1.2 m in this area.

b) A Method Estimating Annual Amounts of Sand and its Result

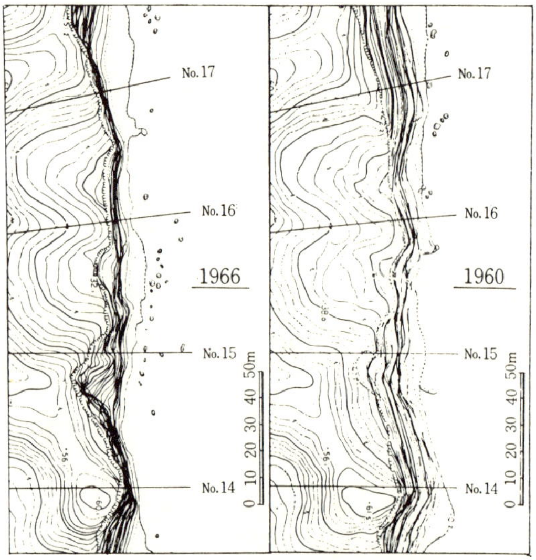

Fig. 6 A part of topographic maps compiled from aerial photographs. (Contour interval is 2 m.)

Aerial photographs were taken on March 19, 1960 and October 30, 1966 and their scale is about 1/12,000 and about 1/8,000 respectively. Autograph A7 and Stereoplotter A8 were used for map compilation. The scale of topographic maps plotted from these photographs is 1/1,000 each (Fig. 6). Based on these drawn maps, the authors estimated annual amounts of sand lost from the coastal cliff.

In the first place, the authors drew 37 cliff profiles on straight lines common to each map (e.g. No. 14, 15, 16, 17 in Fig. 6), which were fixed as perpendicularly as possible to the contour lines of each cliff, generally at intervals of 50 m (Fig. 7). In the next place, area of section S (Fig. 8) was calculated from each

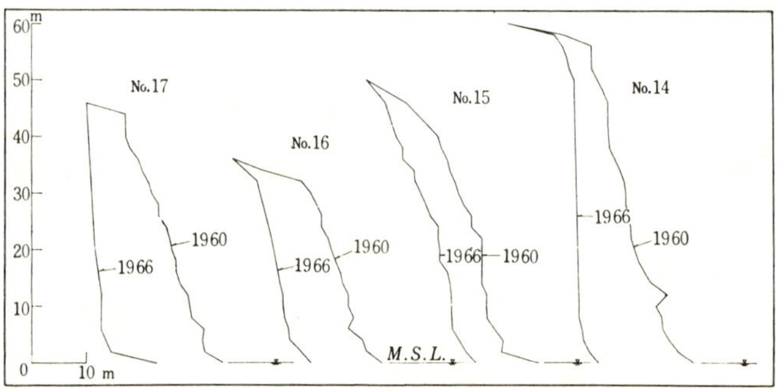

Fig. 7 Cliff profills. (Location numbers refered to Fig. 6.)

of these 37 profiles. In this paper, D_h is defined as the eroded distance at h m above M.S.L. and also H as height of nickpoint which is shown by the symbol **mm** on the map of 1960, that is, as cliff height (Fig. 8). And then the authors examined correlation between S and $D_h \cdot H$ which is area of parallelogram drawn by a dotted line in Fig. 8, for various h. As a result of its examination, $h=10$ m gave the best correlation. Figure 9 indicates the correlation between S and $D_{10} \cdot H$. Based on the results in Table 2, the authors show the relationship between S and H, for various H/D_{10}, in Fig. 10. Curves in Fig. 10 indicate the results

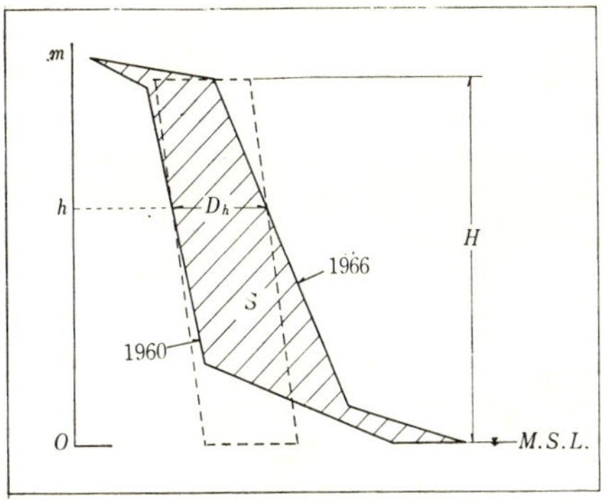

Fig. 8 Explanation of definitions.

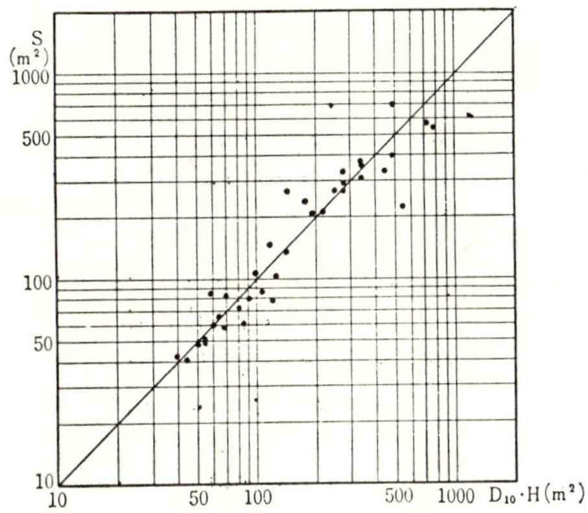

Fig. 9 Correlation between S and $D_{10}\cdot H$.

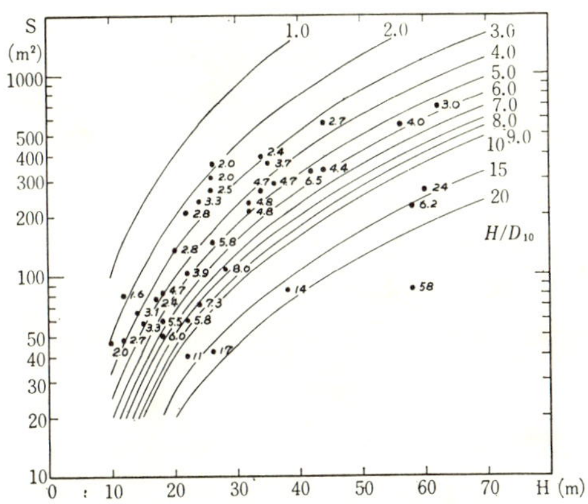

Fig. 10 Relationship between S and H, for various H/D_{10}.

Table 2 Results of investigation.

Location (No.)	S (m²)	H (m)	D_{10} (m)	H/D_{10}
1	230.7	32	6.7	4.8
2	692.2	62	7.7	8.0
3	83.9	58	1.0	58.0
4	266.7	60	2.4	24.0
5	221.6	58	9.4	6.2
6	390.7	34	14.1	2.4
7	101.5	22	5.7	3.9
8	208.0	22	8.6	2.8
9	237.6	24	7.2	3.3
10	48.6	12	4.5	2.7
11	71.9	24	3.3	7.3
12	42.1	26	1.5	17.0
13	364.9	35	9.6	3.7
14	544.2	56	14.0	4.0
15	328.0	42	6.5	6.5
16	358.2	26	12.9	2.0
17	562.2	44	16.3	2.7
18	332.0	44	10.0	4.4
19	135.4	20	7.1	2.8
20	286.9	36	7.6	4.7
21	84.2	38	2.8	13.6
22	209.4	32	6.7	4.8
23	107.1	28	3.5	8.0
24	40.1	22	2.0	11.0
25	78.2	17	7.1	2.4
26	82.1	18	3.8	4.7
27	267.0	34	7.3	4.7
28	148.4	26	4.5	5.8
29	265.3	26	10.6	2.5
30	79.8	12	7.5	1.6
31	304.8	26	13.0	2.0
32	60.3	22	3.8	5.8
33	47.9	10	5.0	2.0
34	50.3	18	3.0	6.0
35	59.6	18	3.3	5.5
36	58.3	15	4.5	3.3
37	65.0	14	4.5	3.1

of calculation which was conducted on the basis of assumption that recession of cliff is completely parallel. The agreement between the calculated curves and the plotted data seems to be appropriate for the present purposes. Therefore the amounts of sand lost from the cliff are shown as follows:

$$V \fallingdotseq D_{10} \cdot H \cdot L\,, \qquad (1)$$

where V is amount of sand, and L is alongshore length of the cliff. Using Eq. (1), we may be able to estimate the amount of sand as described below*.

D_{10} is determined from Fig. 11. The straight lines in this figure are fixed at intervals of 10 m, and as perpendicularly as possible to each 10 m-contour line which is drawn from two maps of 1960 and of 1966. And also H at corresponding location to D_{10} is determined from the map of 1960. As a result of calculation by using Eq. (1), the amount of sand lost from the coastal cliff from March 19, 1960 to October 30, 1966 were estimated at about 409,000 m^3. In calculating this, however, were excluded locations where H was not measurable, that is, pocket beaches, gully mouths and locations of cliff where nickpoint was not shown. Therefore mean annual amount of lost sand is about 62,000 m^3/year. As the alongshore length of the cliff is about 2 km, mean annual amount of sand per kilometer is about 31,000 m^3/year/km.

The relationship between the eroded distance of the cliff and the strength of rocks or the frequency of wave occurrence is now being investigated.

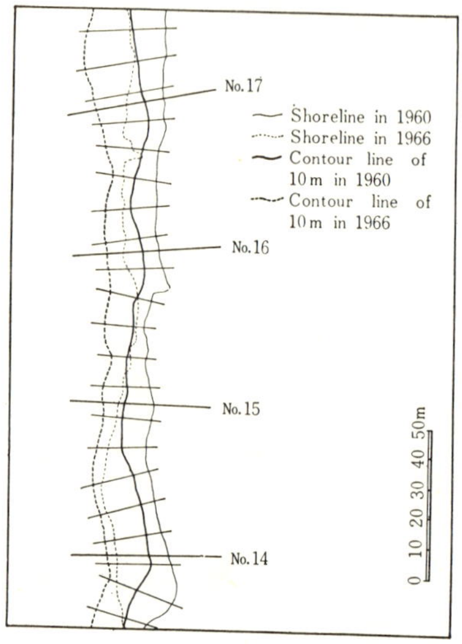

Fig. 11 Map showing the change in shoreline and 10 m-contour line. (The same location as shown in Fig. 6.)

2. Erosion of Coastal Cliff at Okuma District, Fukushima Prefecture

a) Physiographical Conditions

The studied area (Figs. 12 & 13) is marked by an uplifted marine terrace. The extent of a coast-line in this area is about 900 m, and the height of coastal cliff is 30-35 m above M.S.L. Rocks exposed to the cliff are nearly horizontal sedimentary rocks of the Pliocene, Tertiary, which consist of unconsolidated sandstone and relatively massive mudstone (Figs. 14 & 15). The submarine slope is about 1/40 to a seaward distance of 450 m from the shoreline and about 1/130 further out; and submarine contours deeper than a water depth of 5 m run nearly parallel to the shoreline. In this area the mean tidal range is about 1.3 m and a narrow beach at the cliff base is covered by water at high tide, even in normal weather conditions.

b) Relationship between the Eroded Distance and the Strength of Rocks

Nineteen locations were established along the coast-line, generally at intervals of 50 m (Fig. 13). The authors examined the eroded distance at 13 locations except gully mouths and a promontory. It was examined, based on aerial photographs which were taken on April 3, 1947, September 18, 1961 and October 9, 1963, and

* Estimation by using Eq. (1) increases errors in places where cliff height is above 40 m (Fig. 10).

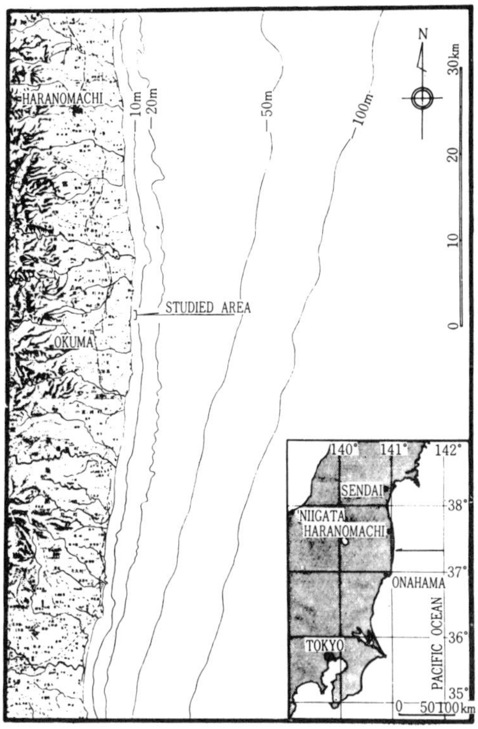

Fig. 12 Studied area.

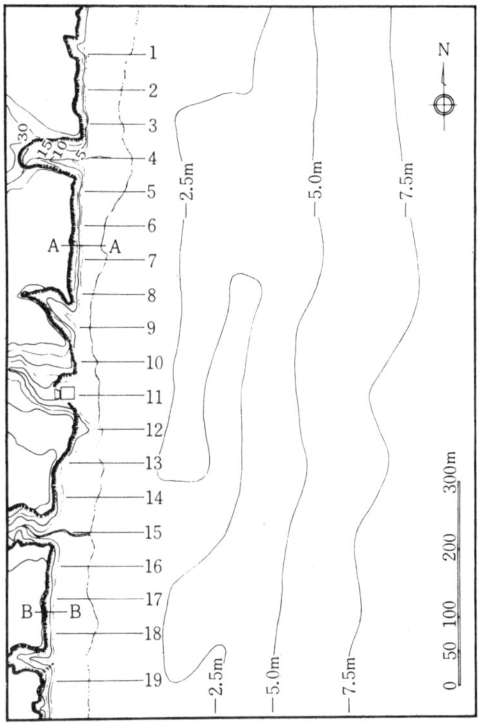

Fig. 13 Topographic map of studied area and location numbers.

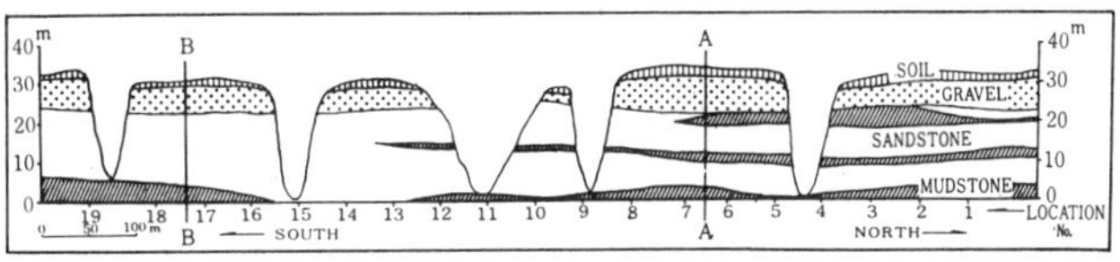

Fig. 14 Alongshore section of coastal cliff.

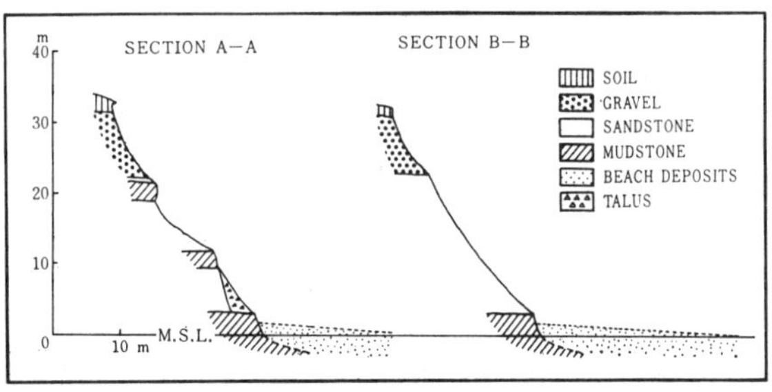

Fig. 15 Geological cross sections of coastal cliff.

on a topographic map of August 31, 1965; the scale of these aerial photographs is about 1/10,000, 1/12,000 and 1/12,000 respectively, and that of the topographic map is 1/200. The scale of maps compiled from these photographs is 1/2,000, 1/1,000 and 1/1,000 respectively. Autograph A7 and Stereoplotter A8 were used

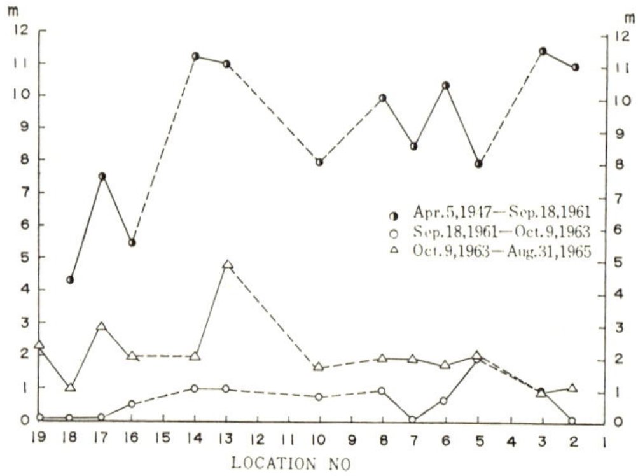

Fig. 16 Eroded distance along coast-line for each term.

for map compilation.

Figure 16 shows the eroded distance along the coast-line, and indicates the existance of a considerable regional variation even along the coast-line of only 900 m in length.

In this paper, the strength of rocks*, that is one index showing the resistibility of rocks against wave erosion, is expressed in terms of the unconfined compression strength** of rocks in the vicinity of the cliff base (Table 3). Figure 17(A) shows the relationship between the eroded distance and the strength of rocks for three different terms, while Fig. 17(B) indicates the relationship between the eroded distance during the period of about 18 years from 1947 to 1965 and the strength of rock. This figure shows also the general tendency that the eroded distance decreases with the increase of the strength of rocks, while it indicates a considerable scattering of the data. This scattering is due probably to the following reasons:

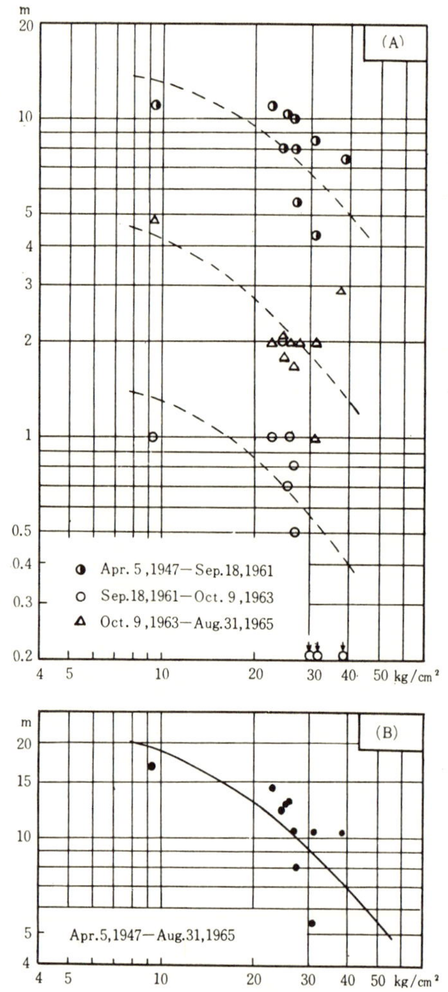

Fig. 17 Relationship between the eroded distance and the strength of rocks expressed in terms of the unconfined compression strength.

(1) The erosive force of waves at the cliff base has not been constant in time or place.

(2) The resistibility of rocks against wave erosion was represented by the unconfined compression strength only.

(3) The unconfined compression strength of rocks in this paper was not always equivalent to that of rocks in the past.

Figure 18 is drawn by the authors, based on the data[19] of Dr. H. Yamanouchi who studied erosion of coastal cliff of 3 km in length at coast near Harano-

* The strength of rocks in this paper is equivalent to mechanical strength of rocks.
** The rate of loading was 0.2 kg/cm²/sec, and wet test pieces were used for testing.

Table 3 Eroded distance, mean annual rate of erosion and characteristics of rocks in the vicinity of cliff base.

Location (No.)	Apr. 5, 1947 ~ Sep. 18, 1961		Sep. 18, 1961 ~ Oct. 9, 1963		Oct. 9, 1963 ~ Aug. 31, 1965		Apr. 5, 1947 ~ Aug. 31, 1965		Rocks	Unconfined compression strength* (kg/cm²)
	14.5 years		2.1 years		1.8 years		18.4 years			
	m	m/year	m	m/year	m	m/year	m	m/year		
1	Gully mouth		—	—	—	—	—	—	—	—
2	11.0	0.76	0.1	0.05	1.1	0.58	12.2	0.66	Mudstone	—
3	11.5	0.79	1.0	0.48	1.0	0.53	13.5	0.73	Mudstone	—
4	Gully mouth		—	—		—	—	—	—	—
5	8.0	0.55	2.0	0.95	2.1	1.11	12.1	0.66	Mudstone	24.9
6	10.4	0.72	0.7	0.33	1.8	0.95	12.9	0.70	Mudstone	25.3
7	8.5	0.59	0.1	0.05	2.0	1.05	10.6	0.58	Mudstone	31.6
8	10.0	0.69	1.0	0.48	2.0	1.05	13.0	0.71	Mudstone	26.7
9	Gully mouth		—	—	—	—	—	—	—	—
10	8.0	0.55	0.8	0.38	1.7	0.90	10.5	0.57	Mudstone	27.0
11	Gully mouth		—	—	—	—	—	—	—	—
12	Promontory		—	—	—	—	—	—	—	—
13	11.0	0.76	1.0	0.48	4.8	2.52	16.8	0.91	Sandstone	9.4
14	11.2	0.77	1.0	0.48	2.0	1.05	14.2	0.77	Sandstone	22.9
15	Gully mouth		—	—	—	—	—	—	—	—
16	5.5	0.38	0.5	0.24	2.0	1.05	8.0	0.44	Mudstone	27.2
17	7.5	0.52	0.1	0.05	2.9	1.53	10.5	0.57	Mudstone	38.3
18	4.3	0.30	0.1	0.05	1.0	0.53	5.4	0.29	Mudstone	30.6
19	—	—	0.1	0.05	2.3	1.21	—	—	Mudstone	—
	8.90*	0.62*	0.65*	0.31*	2.05*	1.08*	11.7*	0.63*	* .... Mean value	

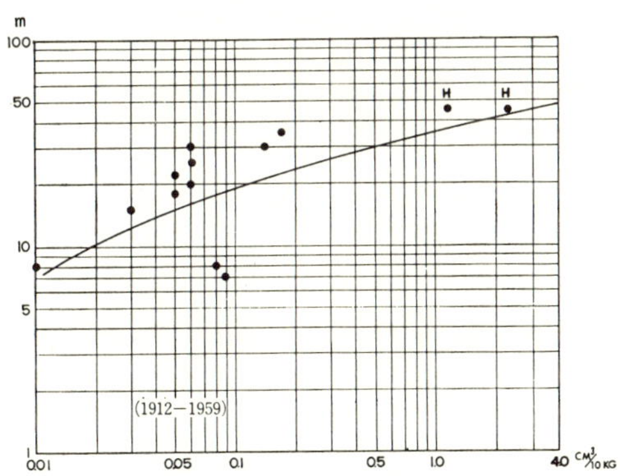

Fig. 18 Relationship between the eroded distance and the strength of rocks expressed in terms of the penetration value, at coast near Haranomachi City (*H* means humussoil, and the others mudstone).

machi City, about 20 km north of the present studied area (Fig. 12). In Fig. 18, the ordinate shows the eroded distance during the period of 47 years from 1912 to 1959, and the abscissa shows the penetration readings on mudstone and humus soil exposed at the cliff base, which were examined by using a penetrometer. The penetration strength, as well as the unconfined compression strength, seems to be one index showing the resistibility of rocks against wave erosion. Unlike the compression strength in Fig. 17, however, the penetration values in Fig. 18 show that the strength of rocks decreases with the increase of values along the abscissa. In order to compare with Fig. 17, Fig. 19 is drawn, the abscissa of which shows the inverse of the penetration values in Fig. 18. Figure 19 shows the same tendency as shown in Fig. 17, and also a considerable scattering of the data.

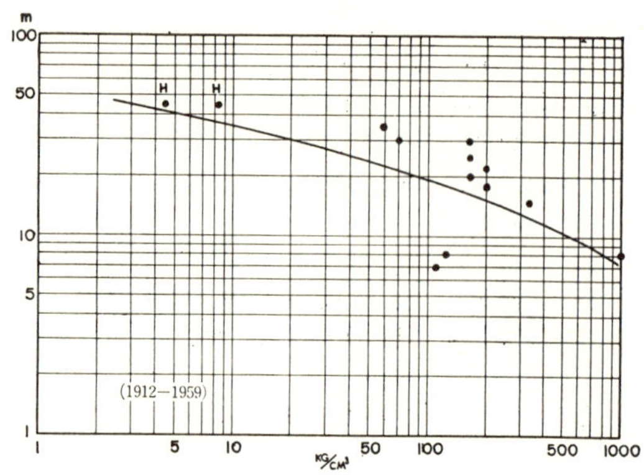

Fig. 19 Redrawing of Fig. 18. (The abscissa shows the inverse of the penetration value.)

c) Relationship between the Rate of Erosion and the Frequency of Wave Occurrence

Figure 20 shows the relationship between the strength of rocks and the mean annual rate of erosion, which is expressed by the eroded distance per year. As shown in Table 3, the rate of erosion from 1947 to 1961, from 1961 to 1963, and from 1963 to 1965 is about 0.6 m/year, 0.3 m/year and 1.1 m/year on the average, respectively. These values fluctuate considerably according to the terms for consideration[20]. It is unquestionable that the rate of erosion from 1963 to 1965 is high as compared with the rate before 1963.

The frequency of occurrence of wave height in Fig. 21 was examined in relation to visual observation data*[21] at Onahama port at about 55 km southward from this area, because it was impossible to obtain wave data of such long term

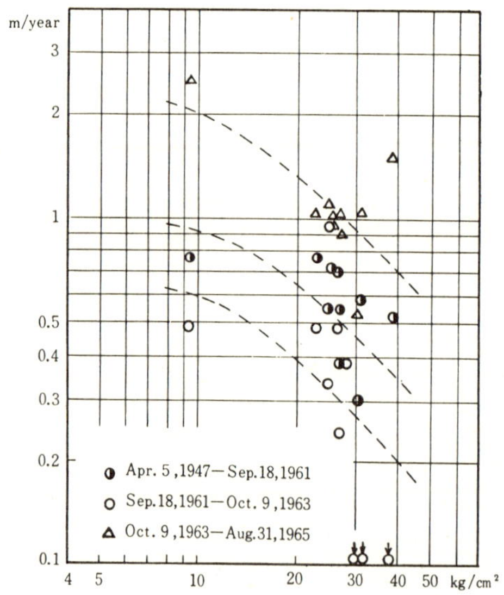

Fig. 20 Relationship between the mean annual rate of cliff erosion and the strength of rocks.

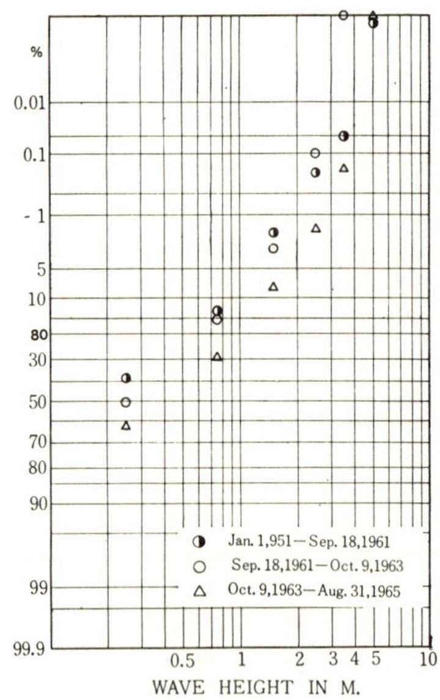

Fig. 21 Frequency of occurrence of wave height. (After visualobservati on data at Onahama port.)

* Records of wave period were not included in these data.

as mentioned above at this coast. Because of the lack of data before 1951 the authors examined the frequency of occurrence by using data for about 15 years from 1951 to 1965. As shown in Fig. 21, waves observed from 1963 to 1965 have high frequency of occurrence of wave height as compared with waves before 1963, and this fact explains well the difference in the rate of erosion mentioned above.

d) Estimation of the Future Eroded Distance

It seems to be possible to estimate the eroded distance in the future, if using such data that indicate wave characteristics of storm which causes the cliff to be cut back, and also the eroded distance for this storm duration. However, the authors could not obtain such convenient data, so they estimated the future eroded distance by a simplified method described in the following.

The authors consider only two of many factors affecting erosion of the cliff; one is the erosive force of waves and the other is the resisting force of rocks. Assuming that there exists the linear relation between the eroded distance of the cliff X and the erosive force f, and that the eroded distance becomes dX in dt by f, the following equation may be written:

$$dX = C_R \cdot f \cdot dt \,, \tag{2}$$

where C_R is a coefficient expressing the erosibility of rocks, and also has relation to the inverse of the resisting force of rocks.

The erosive force of waves consists mainly of actions such as compression, cavitation and abrasion, as described previously. In this paper, however, considering compression only, we will assume the following expression for f:

$$f = k \cdot H \,, \tag{3}$$

where H is wave height (at the cliff base) after breaking and k a coefficient. And we will further assume the following expression[22),23)] for H:

$$H = H_b \cdot e^{-\alpha x} \,, \tag{4}$$

where H_b is wave height of breaker, x the distance from the breaking point to the cliff base (Fig. 22), and α a coefficient*. Substituting Eq. (4) into Eq. (3), we have the following equation:

$$f = k \cdot H_b \cdot e^{-\alpha x} \,. \tag{5}$$

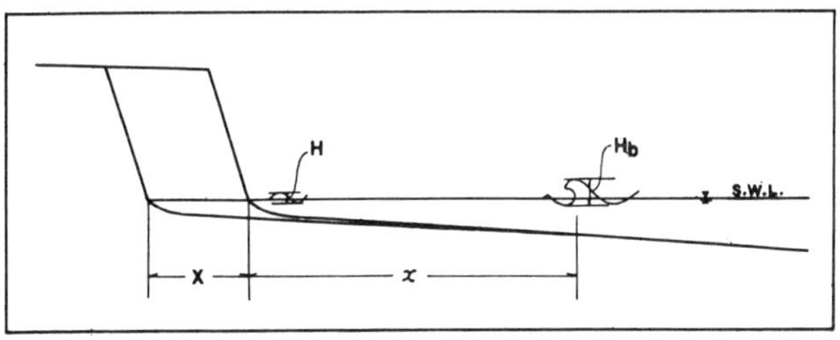

Fig. 22 Explanation of definitions.

* A coefficient α must generally be expressed as a function of wave period and water depth, but α is assumed to be a constant in this paper.

Differentiation of Eq. (5) with respect to time t reduces to

$$df/dt = -\alpha \cdot f \cdot dx/dt \,. \tag{6}$$

Only two conditions will be considered in this paper. One is dx/dt=const., that is, the distance from the breaking point to the cliff base increases at a constant rate, and the other is $dx/dt=0$, that is, the distance from the breaking point to the cliff base is unvaried.

(i) If dx/dt=const., Eq. (6) becomes

$$df/dt = -\beta \cdot f \,, \qquad \beta = \alpha \cdot dx/dt\,(=\text{const.}) \,. \tag{7}$$

Integrating Eq. (7), we have the following equation:

$$f = C \cdot e^{-\beta t} \,, \tag{8}$$

where C is an integration constant. Substituting Eq. (8) into Eq. (2), we have the following equation:

$$dX/dt = C_R \cdot C \cdot e^{-\beta t} \,. \tag{9}$$

Integration of Eq. (9) reduces to

$$X = \int_0^t C_R \cdot C \cdot e^{-\beta t} dt = C_R \cdot C \cdot (1 - e^{-\beta t})/\beta \,. \tag{10}$$

Equation (10) seems to account for the experimental result[24] on the transformation of reservoir shores and of faces of earth dams investigated by using two-dimentional wave flume, that is, this equation may present the relationship between the eroded distance and time at coasts where rock debris resulted from cliff erosion are transported not along the coast-line but perpendicular to it, and also the profile of equilibrium develops in a short time. Accordingly, it is questionable to apply Eq. (10) to natural coasts where, in general, there exists not only onshore-offshore movement of material by wave action but also alongshore movement, and the profile of equilibrium does not develop in a short time.

(ii) If $dx/dt=0$, Eq. (6) reduces to

$$f = C \,, \tag{11}$$

where C is an integration constant. However, C is a function of H_b, because f becomes a function of H_b, if x is constant, as indicated by Eq. (5). Substituting Eq. (11) into Eq. (2), we have

$$dX/dt = C_R \cdot C \,. \tag{12}$$

Integration of Eq. (12) reduces to

$$X = \int_0^t C_R \cdot C dt = \gamma t \,, \qquad \gamma = C_R \cdot C \,. \tag{13}$$

As a consequence, we can find a linear relation between the eroded distance and time.

On the assumption that erosion of cliff is accompanied with submarine erosion of bedrocks and the distance from the breaking point to the cliff is kept constant at coasts where cliff erosion is severe, the authors apply Eq. (13) to their studied area. Evaluation of γ for this area will be described below.

Because C in Eq. (13) is a function of H_b as mentioned above, and also because C_R is considered to be constant, γ is also a function of H_b. Assuming that there exists a linear relation between H_b and the observed wave height, γ may become a constant for a certain wave height. We should notice that t and X in Eq. (13) are duration of action of the erosive force which causes cliff erosion and the eroded distance, respectively, during the period of T years. Here we will introduce a certain wave height A which is to produce cliff erosion effectively. If the occurrence frequency of waves higher than A m in height is $\delta\%$, t may be expressed as follows:

$$t=\delta T/100\,. \tag{14}$$

Substituting Eq. (14) into Eq. (13), we have

$$\gamma=100X/\delta T\,. \tag{15}$$

The frequency of occurrence of wave height δ was examined based on wave data at Onahama port. Due to the lack of data before 1951, as previously mentioned, data of two terms, that is, September 18, 1961 to October 9, 1963 and October 9, 1963 to August 31, 1965 were used. The value of δ of two terms is shown in Fig. 23. The eroded distance X of each term is expressed by the mean value shown in Table 3, because the eroded distance varies considerably from place to place as indicated in Fig. 16. In order to determine two unknown factors such as A and γ

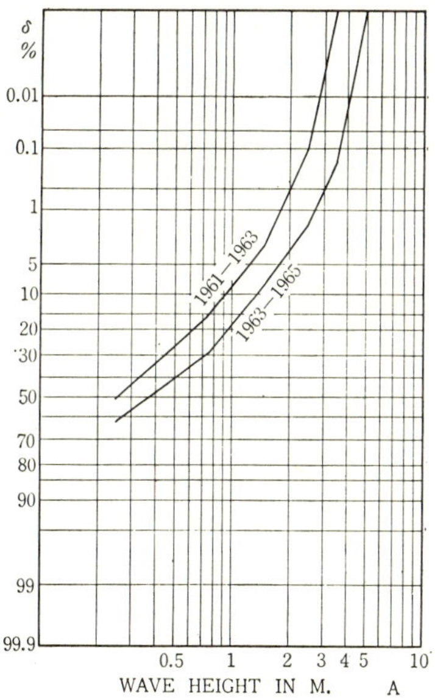

Fig. 23 Frequency of occurrence of wave height of two terms. (Essentially the same as shown in Fig. 21.)

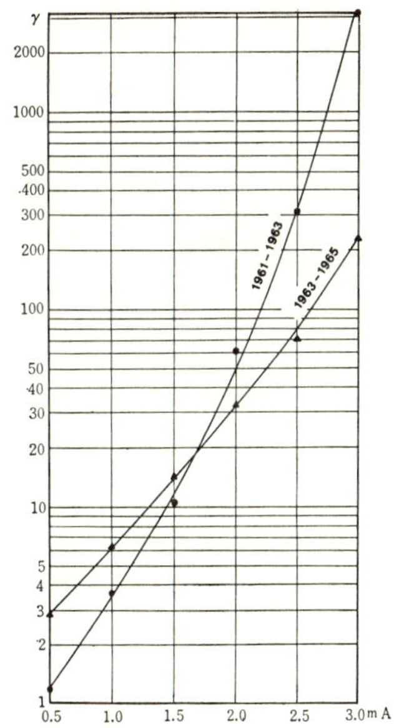

Fig. 24 Relationship between γ and wave height A.

Table 4 Results of calculation.

Wave height A (m)	Term	T (year)	X (m)	δ (%)	γ
0.5	Sept. 18, 1961～Oct. 9, 1963	2.1	0.65	26.0	1.19
	Oct. 9, 1963～Aug. 31, 1965	1.8	2.05	40.0	2.85
1.0	1961 ～ 1963	2.1	0.65	8.5	3.63
	1963 ～ 1965	1.8	2.05	18.0	6.33
1.5	1961 ～ 1963	2.1	0.65	2.9	1.07×10
	1963 ～ 1965	1.8	2.05	8.0	1.42×10
2.0	1961 ～ 1963	2.1	0.65	0.5	6.19×10
	1963 ～ 1965	1.8	2.05	3.5	3.26×10
2.5	1961 ～ 1963	2.1	0.65	0.1	3.10×10^2
	1963 ～ 1965	1.8	2.05	1.6	7.12×10
3.0	1961 ～ 1963	2.1	0.65	0.01	3.10×10^3
	1963 ～ 1965	1.8	2.05	0.05	2.28×10^2

in this area, the authors determined δ of each term for various wave height A from Fig. 23 and then calculated γ by using Eq. (15). Using the results of calculation (Table 4), the authors drew a graph showing the relationship between γ and wave height A for each term (Fig. 24). An intersecting point of two curves indicates that $\gamma \doteqdot 20$, and the effective wave height A equals to 1.7 m. Accordingly, substituting $\gamma=20$ into Eq. (15), we have the following tentative equation:

$$X=0.20\delta T. \qquad (16)$$

Equation (16) shows the future eroded distance of the coastal cliff in this area for T years, for which waves of more than 1.7 m in height occur with probability of δ% (Fig. 25).

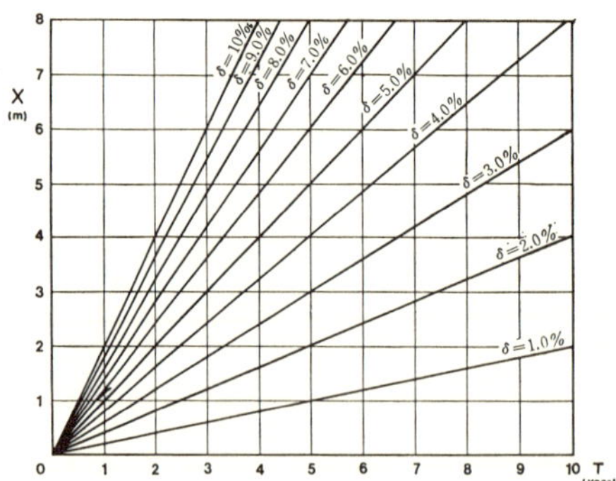

Fig. 25 Relationship between the future eroded distance and time, for various occurrence frequency of wave height of more than 1.7 m.

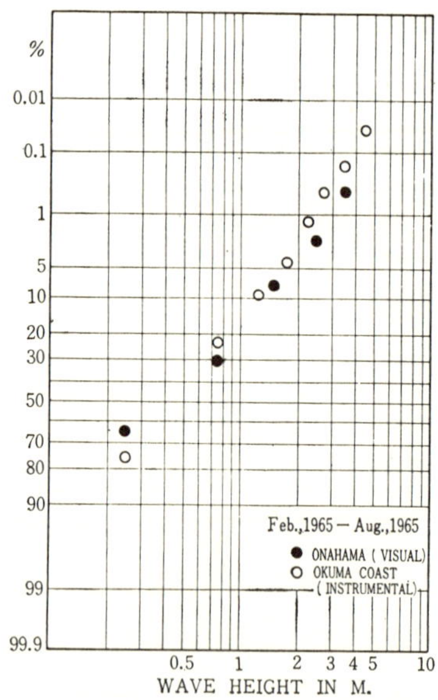

Fig. 26 Comparison of occurrence frequency of wave height at Onahama port and that at Okuma coast.

For reference, Fig. 26 shows a comparison of the frequency of occurrence of wave height at Onahama port and that at this coast, during the period of 6 months from February to August in 1965. The former was examined in relation to visual observation data and the latter to records by wave gauge of pressure type.

V. CONCLUDING REMARKS

A method using aerial photographs enabled to examine the eroded distance and the transformation of coastal cliffs with higher accuracy as compared with previous means. It is difficult at present, however, to collect data of waves which are main factor affecting cliff erosion. In studying erosion of coastal cliffs not only from the engineering viewpoint but also from the scientific one, it may be necessary to comprehend the erosive force of waves and the mechanical properties of rocks.

ACKNOWLEDGEMENTS

The authors express their appreciation to Dr. M. Hom-ma, Professor Emeritus of Civil Engineering, University of Tokyo, for his guidance, and also to M.S. in Civil Eng. R. Kajima, Central Research Institute of Electric Power Industry, who assisted in the field investigations at Okuma district. Finally it should be noticed here that this study has been supported by a part of the Research Grant of the Ministry of Education.

REFERENCES

1) Yamanouchi, H.: On the Retreat of the Sea Cliff along the Pacific Coast at Haranomachi, Japan, Geogr. Rev. Vol. 37, pp. 138-146, 1964. (in Japanese)
2) Kawasaki, I.: Geomorphological Study of the Byobugaura Sea-Cliff in the Vicinity of Iioka-machi, Chiba Prefecture, Geor. Rev. Vol. 27, pp. 213-217, 1954. (in Japanese)
3) Yajima, A.: A Study on Recession of Coastal Cliff at Habuseura, Niijima Island, Graduation Thesis, Tokyo Univ. of Education, 1965. (in Japanese)
4) Yamanouchi, H.: Geomorphological Study on the Retreat of the Sea Cliff along the Southwestern Coast of the Atsumi Peninsula, Doctorate Thesis, Tokyo Univ. of Education, 1965. (in Japanese)
5) Yoshikawa, T.: On the Formation of Wave-Cut Bench, Geogr. Inst., Univ. of Tokyo, Bull. Vol. 1, pp. 99-113, 1951. (in Japanese)
6) Komukai, R.: Beach Erosion of Omori-hama, Hakodate City, Hokkaido, Japan, Hydrogr. Office, Bull. Vol. 13, Introduction, 1959. (in Japanese)
7) Yoshikawa, T.: Shore Erosion and its Bearing on Topography of Kurobe Alluvial Fan, Toyama Prefecture, Geogr. Inst., Univ. of Tokyo, Bull. Vol. 2, pp. 92-109, 1952. (in Japanese)
8) Thornbury, W. D.: Principles of Geomorphology, John Wiley & Sons, New York, p. 433, 1960.
9) Yatsu, E.: Rock Control in Geomorphology, Sozosha, Tokyo, pp. 34-35, 1966.
10) Sharpe, C. F. S.: Landslides and Related Phenomena, Pegeant Books, New Jersey, p. 67, 1960.
11) Steers, J. A.: Notes on Erosion along the Coast of Suffolk, Geol. Mag. Vol. 88, pp. 425-439, 1951.

12) Kawasaki, I.: op. cit., 2)
13) Yamanouchi, H.: op. cit., 1)
14) Valentin, H.: Der Landverlust in Holderness, Ostengland, von 1852 bis 1952, Die Erde, Bd. 6, SS. 296-315, 1954.
15) Yajima, A.: op. cit., 3)
16) Shepard, F. P., and U. S. Grant: Wave Erosion along the Southern California Coast, Geol. Soc. Amer., Bull. Vol. 58, pp. 919-926, 1947.
17) Byrne, J. V.: Coastal Erosion, Northern Oregon, Essays in Marine Geology in Honor of K. O. Emery, Univ. of Southern Calif. Press, pp. 11-13, 1963.
18) Hallert, B.: Photogrammetry, McGraw-Hill, New York, 1960.
Thompson, M. M.: Manual of Photogrammetry, Amer. Soc. of Photogrammetry, 1966.
19) Yamanouchi, H.: op. cit., 1)
20) Steers, J. A.: Coastal Cliffs: Report of a Symposium, Geogr. Jour. Vol. 128, pp. 303-320, 1962.
21) Japan Meteorological Agency: The Results of Marine Meteorological and Oceanographical Observations, 1951-1965.
22) Ijima, T., and Others: Wave Characteristics in the Surf Zone Observed by Stereophotography, Report of Transportation Tech. Research Inst., Report No. 31, 1958.
23) Horikawa, K., and C. Kuo: A Study on Wave Transformation inside Surf Zone, Proc. of 10th Conf. on Coastal Engineering, 1967.
24) Popov, I. J.: Experimental Research in Formation by Waves of Stable Profile of Upstream Faces of Earth Dams and Reservoir Shores, Proc. of 7th Conf. on Coastal Engineering, 1961.

ERRATA

A STUDY ON EROSION OF COASTAL CLIFFS BY USING AERIAL PHOTOGRAPHS

By Kiyoshi Horikawa and Tsuguo Sunamura

Pages		
67	2nd line from the top	At one area... → In
67	3rd line from the top	at the other... → in
67	2nd line from the base	phenemenon → phenomenon
68	7th and 13th lines from the base	occures → occurs
71	Explanation of Fig. 18	profills → profiles refered → referred
73	14th line from the top	were → was
75	18th line from the top	rock → rocks
76	Explanation of Fig. 18	humussoil → humus soil
77	Ordinate of Fig. 21	take out the figure " 80 " between " 10 " and " 30 ".
77	Explanation of Fig. 21	visualobservati on → visual observation
79	18th line from the top	two-dimentional → two-dimensional
81	Bottom of the colum δ (%) in Table 4	0.05 → 0.5

14

Reprinted from *U.S. Geol. Survey Prof. Paper 542-B*:1–21 (1965)

EFFECTS OF THE EARTHQUAKE OF MARCH 27, 1964, AT WHITTIER, ALASKA

Reuben Kachadoorian

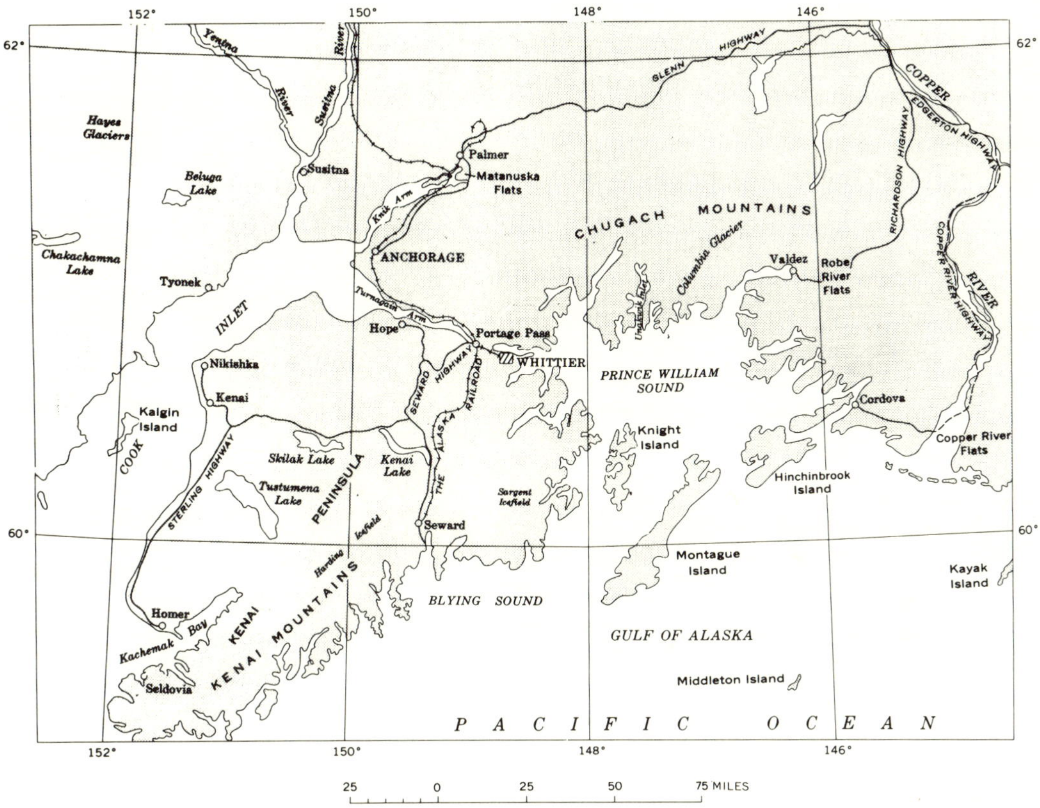

1.—Index map showing location of Whittier area.

ABSTRACT

Whittier, Alaska, lying at the western end of Passage Canal, is an ocean terminal of The Alaska Railroad. The earthquake that shook south-central Alaska at 5:36 p.m. (Alaska Standard Time) on March 27, 1964, took the lives of 13 persons and caused more than $5 million worth of damage to Government and private property at Whittier.

Seismic motion lasted only 2½–3 minutes, but when it stopped the Whittier waterfront was in shambles and the port facilities were inoperable. Damage was caused by (1) a 5.3-foot subsidence of the landmass, sufficient to put some of the developed land under water during high tides, (2) seismic shock, (3) fracturing of fill and unconsolidated sediments, (4) compaction of fill and unconsolidated deposits, (5) submarine landslides which generated waves that destroyed part of The Alaska Railroad roadbed and other property, (6) at least two, but probably three, waves generated by landslides, which completely wrecked the buildings of two lumber companies, the stub pier, the small-boat harbor, the car-barge slip dock, and several homes, and (7) fire that destroyed the fuel-storage tanks at the Whittier waterfront.

Many buildings and other facilities were totally wrecked, others were damaged to lesser degrees. For example, the 14-story reinforced-concrete Hodge Building, which rests upon at least 44 feet of sandy gravel, was moderately damaged by seismic shock, but the six-story reinforced-concrete Buckner Building, which rests upon bedrock, was only slightly damaged.

INTRODUCTION

The port of Whittier (fig. 1) was constructed under the supervision of the Corps of Engineers in 1942–43 to provide an all-weather terminal for The Alaska Railroad. During World War II, Whittier and Seward served as the two all-weather railroad ports that safeguarded the flow of military supplies, equipment, and personnel from tidewater to Anchorage and Fairbanks. The town of Whittier is owned and operated by the U.S. Government—specifically by The Alaska Railroad of the Department of the Interior and by the U.S. Army of the Department of Defense. Some of the land has been leased to private enterprise.

Whittier was hard hit by the earthquake of March 27, 1964. Tragically, 13 persons were lost during the earthquake. At the time, only 70 people were living at Whittier. The official 1960 census lists a population of 800; however, this figure included military personnel who were subsequently transferred when the Army closed its Whittier operation. Only one body was recovered; the remaining 12 persons were presumed dead. In addition, the earthquake destroyed or made inoperable a major part of the port facilities. Total damage to the Federal and privately owned facilities at Whittier was in excess of $5 million.

The loss of the Whittier port facilities, coupled with destruction of those at both Seward and Valdez, left Alaska without any all-weather port for unloading supplies for movement either by rail or highway to the metropolitan areas of Anchorage and Fairbanks.

FIELDWORK

This report is based on data collected by U.S. Geological Survey personnel during visits to Whittier, and supplemented by data supplied by The Alaska Railroad, the U.S. Army-Alaska, the U.S. Coast and Geodetic Survey, and many citizens of Alaska, particularly those of Whittier. A 10-day study of south-central Alaska was made by the Geological Survey immediately after the earthquake, and the results were published on April 27, 1964 (Grantz, Plafker, and Kachadoorian, 1964). During the summer of 1964, Whittier was again visited by the author and by George Plafker, D. S. McCulloch, and L. R. Mayo—all of the U.S. Geological Survey. Plafker collected additional data on the onshore effects of the earthquake; McCulloch and Mayo made offshore submarine geologic investigations.

ACKNOWLEDGMENTS

Most of the logistic support for the initial study by the Geological Survey between March 29 and April 9, 1964, was provided by the U.S. Army-Alaska (USARAL); this support was provided under the difficult conditions that existed in Alaska immediately after the earthquake. During the followup studies in the summer of 1964, the U.S. Army again provided support. The author appreciates the efficient and enthusiastic support of Col. M. L. Fallwell and Lt. Col. H. E. Nolde, their staffs, and the officers and men of the USARAL Aviation Company of the 19th Aviation Battalion at Fort Richardson, Alaska.

Officials of The Alaska Railroad also cooperated with the Survey by providing preearthquake maps and photographs of the Whittier area as well as postearthquake information. Among the officials who generously supported the investigation were I. P. Cook, C. L. Griffith, T. C. Fugelstad, B. E. Cannon, and J. A. Morrison.

The author also thanks the many residents of the State of Alaska, especially of Whittier, who supplied eyewitness descriptions and other firsthand information about the earthquake.

CLIMATE

Whittier lies in the coastal rain belt, but is surrounded by high mountains. Thus, the climate is intermediate between that of coastal mountains and open coast. Residents and The Alaska Railroad personnel state that Whittier is frequently buffeted by high westerly winds blowing off Portage Pass. Mr. Jack Farnsworth (oral commun., 1965) stated that winds locally have been so strong that house trailers were blown over by them. The high winds are not restricted to the winter months, as one might expect, but occur the year-round. Climatological data for Whittier are shown on table 1.

TABLE 1.—*Climatological data for Whittier, Alaska*

[U.S. Weather Bureau, 1964]

Month	*Average temperature* [1] *(°F)*	*Average total precipitation* [2] *(inches)*
January	25.7	9.45
February	28.3	21.40
March	24.5	8.18
April	31.3	11.05
May	43.7	14.72
June	46.3	16.17
July	53.8	8.84
August	53.8	6.45
September	48.1	16.40
October	38.4	[3] 24.40
November	22.8	[3] 6.13
December	22.9	[3] 15.00
Annual average	37.2	[3] 157.89

[1] Based on 14-year record.
[2] Based on 9-year record.
[3] Amount is wholly or partly estimated.

GEOGRAPHIC SETTING

Whittier is at the head of Passage Canal on the northeast side of the Kenai Peninsula in south-central Alaska (fig. 1). It is at lat 60°47′ N., long 148°40′ W., about 50 miles southeast of Anchorage and 40 miles southwest of the epicenter of the March 27 earthquake. Passage Canal, a west-southwest-trending fiord, is a western arm of Prince William Sound and is about 11 miles long and 1½ miles in average width.

The Kenai Mountains form the backbone of the Kenai Peninsula and rise sharply from tidewater to an altitude of more than 3,500 feet within 2 miles of Whittier. The large Harding icefield and much smaller Sargent icefield cover much of the higher part of the mountains. The Chugach Mountains to the east do not have a large icefield, but do have several large glaciers. North of Passage Canal the mountains rise abruptly to 4,600 feet, 1.5 miles from the shore.

The boundary between the Chugach and Kenai Mountains is Portage Pass west of Whittier. The pass is 800 feet above sea level; it was used extensively by prospectors during the gold rush as a route from Passage Canal west to Turnagain Arm. The pass is in part occupied by Portage Glacier and Portage Lake.

The main port facilities of Whittier are on the south shore of Passage Canal about 1 mile east of its western end. The town is built on a fan-shaped delta formed by Whittier Creek—a creek fed by Whittier Glacier. The delta has a maximum width of approximately 2 miles, is about 1½ miles long, and rises from sea level to an altitude of 90 feet with a fairly uniform slope of 60 feet per mile.

At the head of Passage Canal is a low-lying deltaic area about 2.3 miles wide and 2.3 miles long having an average seaward slope of 140 feet per mile. The southern part of this lowland consists of coalescing deltas of creeks flowing from Portage Pass, Shakespeare Glacier, and Learnard Glacier. The northern part of the lowland was overridden by a minor advance of Learnard Glacier. When the glacier retreated, it left behind a semicircular moraine with irregular hummocky topography. The margin of the moraine is 10–15 feet high, and the surface of the interior part contains depressions as much as 5 feet deep and small hillocks as much as 10 feet high.

After the ice that sculptured Passage Canal had retreated, the canal remained as a steep-walled U-shaped fiord locally more than 1,000 feet deep. However, immediately after the retreat of the ice, melt water began to construct deltas at the margins of the fiords—Whittier delta and the delta at the west end of Passage Canal being the largest. Another large delta was formed by Billings Creek on the north side of the canal about 3 miles northeast of Whittier.

Prior to the earthquake, Passage Canal was approximately 300 feet deep 0.1 mile from shore at its western end. Along the north shore, the depth within 1.5 miles of the head of the canal and within 0.7 mile of shore was 600 feet. Submarine slopes in front of the unconsolidated delta deposits were steeper than the average slope of the Whittier delta which was about 30° at the time of the earthquake.

GEOLOGY

The rocks in the Whittier area are slate and graywacke of probable Cretaceous age, locally overlain by unconsolidated Quaternary deposits consisting of glacial moraine, reworked outwash and stream gravel, and artificial fill (pl. 1). The slate and graywacke were intruded by quartz diorite and diorite during Cretaceous or Tertiary time. Barnes (1943) has given a more detailed description of the geology of the Whittier area.

BEDROCK

The Cretaceous(?) rocks that underlie the Whittier area are predominantly slate containing minor amounts of graywacke locally interbedded. Commonly the slate grades into the graywacke. Barnes (unpub. data, 1939) stated that locally, along the shore of Passage Canal, graywacke that appears massive above high-tide level actually has a platy structure below. This structure is apparently emphasized by the weathering of the graywacke in the intertidal zone.

The slate and, to a lesser extent, the graywacke have a pronounced cleavage generally parallel to the bedding. The attitude of the cleavage planes ranges from N. 20° E. to N. 30° E., the dips from N. 65° W. to N. 80° W. The strike of the bedding is from N. 20° E. to N. 30° E., and the dips are from N. 60° W. to N. 65° W.

Along the south shore of Passage Canal the slate and graywacke have been intruded by four quartz diorite or diorite dikes or sills, 3, 4, 7, and 100 feet thick, that are approximately parallel to the bedding of the host rock.

UNCONSOLIDATED DEPOSITS

Unconsolidated deposits consisting of outwash and stream gravels form the delta upon which Whittier rests and also the southern part of the delta at the head of Passage Canal.

The Whittier delta is composed predominantly of coarse, subangular to subrounded gravel in a matrix of coarse sand. The maximum thickness of the gravel is unknown but is at least 44 feet beneath the Hodge Building (figs. 2 and 3).

The sediments composing the delta at the head of Passage Canal are generally similar to those of the Whittier delta. However, the outwash and stream deposits in the northern part of the delta are overlain by a moraine of the Learnard Glacier. The thickness of the gravel deposits is unknown.

The moraine that overlies the delta deposits at the head of Passage Canal consists primarily of jumbled heaps and ridges of coarse angular blocks of slate and graywacke. Locally, patches of sand and gravel are interspersed with coarse blocks.

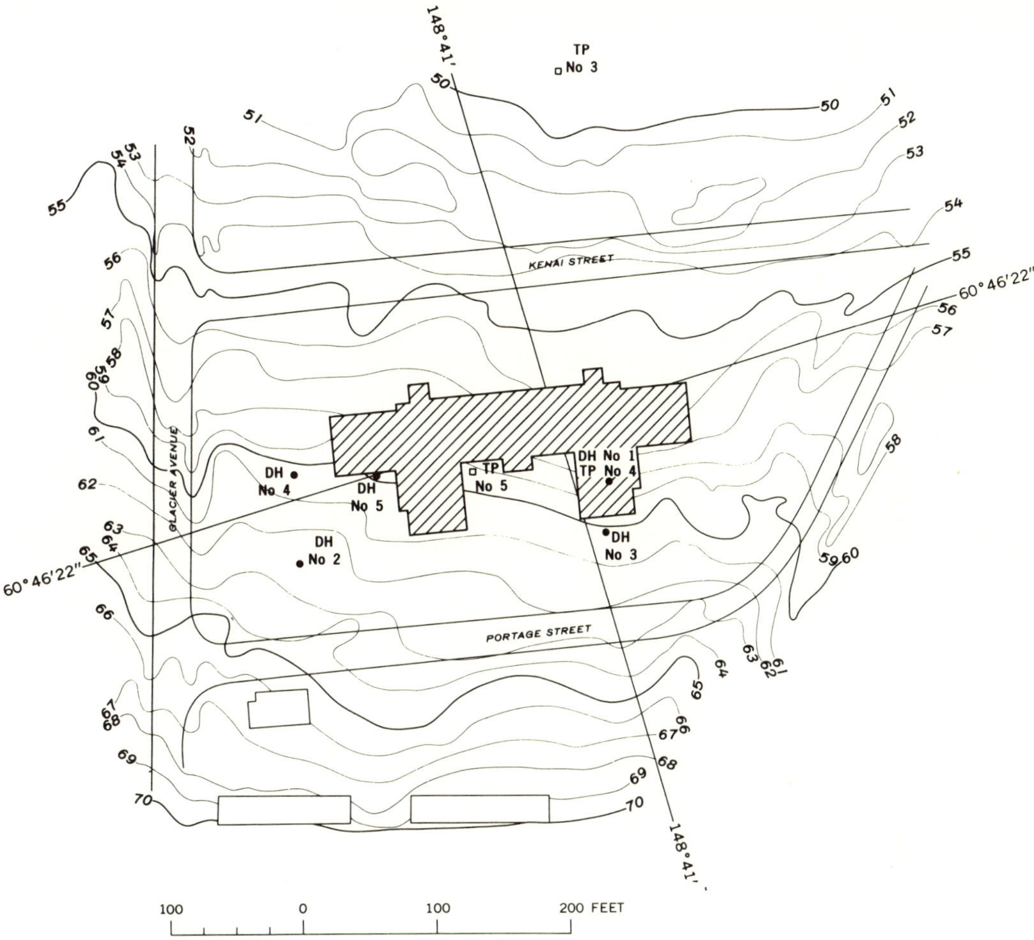

2.—Map showing Hodge Building, surrounding area, and location of drill holes. Graphic logs of these holes are shown in figure 3.

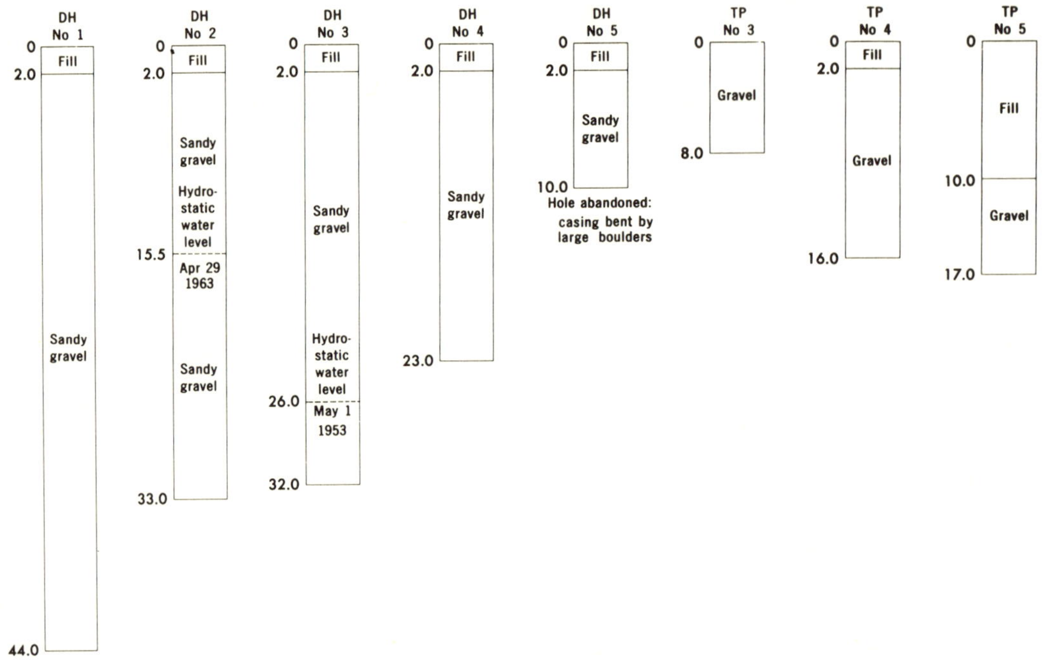

3.—Graphic logs of drill holes (DH) and test pits (TP) in the Hodges Building area. Location of holes and pits are shown in figure 2.

ARTIFICIAL FILL

Areas of artificial fill are shown on plate 1. The easternmost 500 feet of the airstrip, an area near the Union Oil Co. fuel tanks, and an extensive area by the marginal wharf are the largest of these areas. The fill was obtained locally and thus consists predominantly of delta deposits, or glacial outwash and stream deposits.

DESCRIPTIONS OF THE EARTHQUAKE

On Good Friday, March 27, 1964, at 5:36 p.m. (Alaska Standard Time) the earthquake struck Whittier. A light snow was falling, the surface of Passage Canal was "glassy," and the tide was at 1 foot above mean lower low water. Seismic motion accelerated to its maximum intensity in about one-half minute, maintained this intensity for 1–1½ minutes, and then gradually subsided. Thus, seismic motion lasted for a total of approximately 2½–3 minutes. The eyewitness estimates of the time agree closely with charts from automatic recording devices in the powerhouse and salt-water pumphouse at Whittier. The most reliable time record—a chart from the outside air-temperature recording gage in the salt-water pumphouse—shows that seismic shaking lasted for about 4 minutes (fig. 4).

Residents of Whittier said that the motion was generally east-west. However, Mr. W. P. LaRue, powerhouse superintendent, stated that the motion was generally north-northwest–south-southeast. People who were on bedrock at the time of the earthquake said that the motion was of a jarring nature; those on fill or unconsolidated deposits reported more of a rolling or "round-and-round" motion. Most people felt no vertical motion. Mrs. D. Keat-

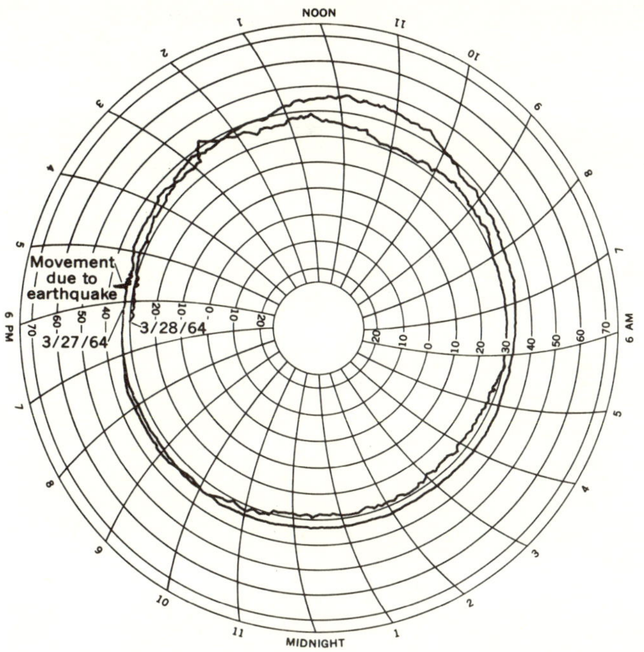

4.—Chart of outside-air temperature from automatic recording device in salt-water pumphouse in Whittier.

ing, who was in the Hodge Building during the earthquake, said, however, that midway through the violent phase of the shaking, the Hodge Building seemed to have been lifted up and set down.

At least three waves were observed during and immediately after the earthquake. Mr. M. J. Dixon and Mr. Farrian Kolb reported that about 1 minute after the earthquake started, the water in Passage Canal rose rapidly to a maximum altitude of 25–26 feet [1] in the vicinity of the town. The water was still glassy and did not contain any debris.

The water immediately receded, and 1–1½ minutes later a large breaking wave came in. This wave was muddy, contained much debris, and was about 40 feet high when it reached The Alaska Railroad depot (pl. 1). Mrs. Farrian Kolb, who observed the wave from the second story of the depot, described it as a wall of water 30–50 feet high that rose near the center of Passage Canal. She stated that something seemed to be exploding underneath the water. Mr. L. McDonough said that the water in Passage Canal about 50 yards from shore appeared to be boiling. This wave struck the railroad depot 8–10 feet above the ground surface, or at an altitude of 34–36 feet.

From ½ to 1 minute after the first breaking wave (second rise of sea level), another wave struck Whittier. This wave was similar in nature to the first breaking wave. It reached an altitude of about 30 feet near The Alaska Railroad depot.

There were no eyewitnesses to the waves that struck the shore in other parts of the area, but high-water marks on snow and trees and deposition of debris indicate that waves in Passage Canal reached their maximum altitude not at Whittier but along the shore to the northwest (as measured by Plafker and Mayo). There the waves reached as high as 104 feet. At the extreme north end of the delta at the head of Passage Canal, a wave reached an altitude of 82 feet.

During the earthquake a fire started in the Union Oil Co. and U.S. Army tank-farm area (fig. 5) and burned until March 31; it completely destroyed the fuel-storage tanks.

Thirteen lives were lost at Whittier as the result of wave action. Eyewitnesses state that 10 of the persons lost and presumed dead had gone into the home of one of the Columbia Lumber Co. workers a few minutes before the earthquake struck. The waves totally destroyed both the company's industrial buildings and the homes of the workers (figs. 6 and 7). Two other persons associated with the company are missing and presumed dead. None of the 12

[1] Unless otherwise indicated, altitudes given in this report are based on a pre-earthquake datum and do not reflect the 5.3-foot regional subsidence of the Whittier area during the earthquake because it is not known when the subsidence occurred.

Datum for subaerial contours on all plates is mean sea level; datum for submarine soundings and contours is mean lower low water (MLLW) or 6.3 feet below mean sea level. Height of waves is the actual distance between trough and crest. For example, a 40-foot wave would have an altitude of 34.7 feet because the earthquake occurred at a 1-foot tide or 5.3 feet below mean sea level.

The tide range at Whittier is from a low of −3.8 feet to a high of 15.4 feet (datum is MLLW).

5.—Aerial view, looking northeast, of Union Oil Co. and U.S. Army fuel-storage tanks burning. Note high-water mark where waves washed away snow cover. Photograph by U.S. Army, March 28, 1964.

6.—Preearthquake aerial view, looking south, of Union Oil Co. fuel- storage tanks and Columbia Lumber Co. Photograph by U.S. Army, June 7, 1963.

7.—Postearthquake aerial view, looking north, of destruction of Columbia Lumber Co. Photograph by U.S. Army, March 28, 1964.

bodies was recovered from the wreckage; the only traces found of the missing persons were pieces of children's clothing. It is presumed that all were washed out into Passage Canal. The 13th victim, and the only one whose body was recovered, was a child who lost her life near The Alaska Railroad depot.

Several residents of Whittier commented that on the morning after the earthquake, red snapper fish, which normally inhabit waters deeper than 400 feet, were floating on the surface of Passage Canal and were also strewn along parts of the shoreline. Everyone who saw the fish remarked that they were larger than those normally caught by local fishermen. Dr. N. J. Wilimovsky, a marine biologist from the University of British Columbia, stated that 26 genera of red snappers are caught commercially in Prince William Sound. Two of these—*Seabastolobus* and *Sebastodes*—increase in abundance with depth. Most of the red snappers found dead along the shore and on the surface of Passage Canal were probably fish of these two genera. The fish apparently were killed by a rapid change of pressure in the water and were then brought to the surface by upwelling of the water or by turbidity currents triggered by the seismic shaking.

EFFECTS OF THE EARTHQUAKE

Seven different factors related to the earthquake were responsible for the extensive damage in the Whittier area. These include (1) a change in land level, (2) seismic shock, (3) fracturing of fill and unconsolidated deposits, (4) differential subsidence due to compaction, (5) landslides, (6) waves generated by submarine landslides, and (7) fire. A large avalanche fell on the moraine at the head of Passage Canal; it consisted chiefly of snow debris and did no harm. Plate 1 indicates the nature and extent of damage by seismic shock, compaction, landslides, waves, and fire. Figures 8 and 9 are prequake and postquake aerial photographs of much of the Whittier area.

LAND-LEVEL CHANGES

Studies by the U.S. Geological Survey and the U.S. Coast and Geodetic Survey determined that tectonic subsidence of the landmass at Whittier was 5.3 feet during the earthquake. Residents of Whittier estimate that only 20 percent of the prequake tidal zone is now exposed during low tides (compare pls. 2 and 3). High tides now inundate some of the delta area at the head of Passage Canal, much land leased by the Two Brothers Lumber Co., and about 350 feet of the eastern end of the airstrip (fig. 10). The 5.3-foot subsidence is not wholly responsible for the change in exposure of the tidal zone; landslides

8.—Vertical aerial photographs of Whittier. Top, Prequake; composite of photographs by Bureau of Land Management, September 23, 1963. Bottom, Postquake; photograph by U.S. Army, March 28, 1964.

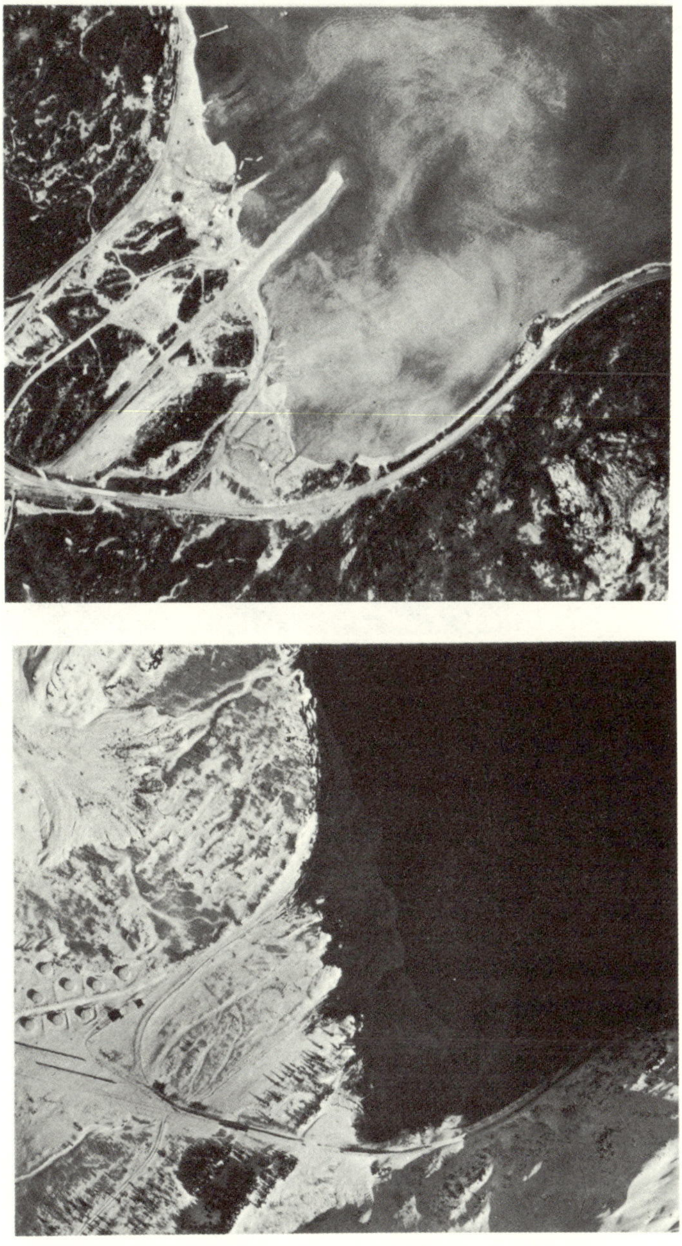

9.—Vertical aerial photographs of delta at west end of Passage Canal. Top, Prequake; photograph by Bureau of Land Management, September 23, 1963. Bottom, Postquake; photograph by Air Photo Tech, Anchorage, April 1, 1964.

and compaction (p. B15) also were, in part, responsible.

The most pressing problem at Whittier was to put back into service the port and railroad facilities that were destroyed or damaged by land subsidence. Much of the reconstruction in Alaska depended on immediate use of the car-barge slip dock, which was the only connecting link for transferring supplies and equipment from ships to The Alaska Railroad after the earthquake. To permit maximum use of the dock, which is stationary at the land end and free to rise and fall with the tide at the sea end, it was necessary to know the actual amount of subsidence at Whittier. Once the amount of subsidence was determined, the dock was reconstructed and railroad traffic to Anchorage was reestablished in a remarkably short time.

10.—Aerial photograph, looking south, of delta at head of Passage Canal. Remains of Two Brothers Lumber Co. in right foreground. Projecting into the water is the airstrip. Note high-water mark indicated by snow. Photograph by U.S. Army, March 28, 1964. ⟶

SEISMIC DAMAGE

In the downtown area of Whittier away from the waterfront, most of the damage was caused by seismic shock. Within the area inundated by waves (pl. 1), seismic shock severely damaged one of the slip towers of the car-barge slip dock (fig. 11), part of the marginal wharf, one of the storage tanks at the U.S. Army tank farm, and the highway and rail-

11.—View, looking west, of remains of car-barge slip dock. Slip tower only part of structure left. Photograph by The Alaska Railroad, 1964. ⟶

12.—View, looking southeast, of Hodge Building. Photograph taken April 4, 1964.

13.—Aerial view, looking south-southeast, of part of postquake Whittier. 1 Hodge Building, 2 gymnasium, 3 composite shop, 4 fire station, 5 station storage warehouse, 6 telephone building, 7 ACS building, 8 intransit storage shed, 9 marginal wharf, 10 The Alaska Railroad depot, 11 FHA housing, and 12 Buckner Building.

road bridges crossing Whittier Creek. Beyond the inundated zone there was moderate damage to the Hodge Building (fig. 12), the composite shop, and part of the station storage warehouse (fig. 13). The Buckner Building (fig. 14), gymnasium, fire station, telephone building, ACS (Alaska Communication System) building, powerhouse, and the railroad bridge near the airstrip were slightly damaged.

The Hodge Building, a reinforced-concrete 14-story structure 265 feet long and 50 and 110 feet wide (figs. 2 and 3), rests on at least 44 feet of unconsolidated sand and gravel. The building was constructed as three structural monolithic units connected by 8-inch expansion joints. In the corridors, metal jumper plates spanned the 8-inch joints; on the exterior, the joints were secured by a weather seal. The long dimension of the building was parallel to the predominant east-west direction of the seismic motion; the expansion joints were perpendicular to the motion. During the earthquake the individual units banged against each other, the weather seals popped out, and the jumper plates were curled. Damage to the structural connections (weather seals, jumper plates, and spandels) was progressively more intense upward in the structure owing to the larger horizontal component of movement. Jumpers on the top floor were shortened 4 inches by curling, whereas on the third floor they curled only a quarter of an inch.

The interior partition walls of the Hodge Building were constructed of concrete block. Damage to the walls was also greater in the upper floors. The walls on the 13th and 14th floors either collapsed or were cracked. On the fourth floor the blocks were loosened, except near the expansion joints where some of the blocks were dislodged. The exterior of the building was not substantially damaged, even where the units banged against each other.

The interior and roof of the composite shop were moderately damaged by seismic shaking. Most of the interior blocks and some of the concrete posts and beams were cracked; in numerous places the roof purlins were twisted or separated from the beams. Much of the internal bracing in the building was either broken or strained.

The largest building in Whittier, probably the largest in Alaska, is the Buckner Building (figs. 14, 15, and 16). It is a six-story reinforced-concrete structure about 500 feet long and 50–150 feet wide. The long dimension of the structure runs northeast-southwest, oblique to the seismic motion of the earthquake. The building rests on bedrock and did not receive any significant structural damage. The steam powerhouse near The Alaska Railroad depot also rests on bedrock and was not damaged by seismic motion.

Several small structures on un-

14.—Aerial view, looking southeast, of Buckner Building; building in foreground is intransit storage shed; long structure at left of photograph is FHA housing; structure north of FHA housing is depot. Note high-water mark of wave on snow. Photograph taken April 4, 1964.

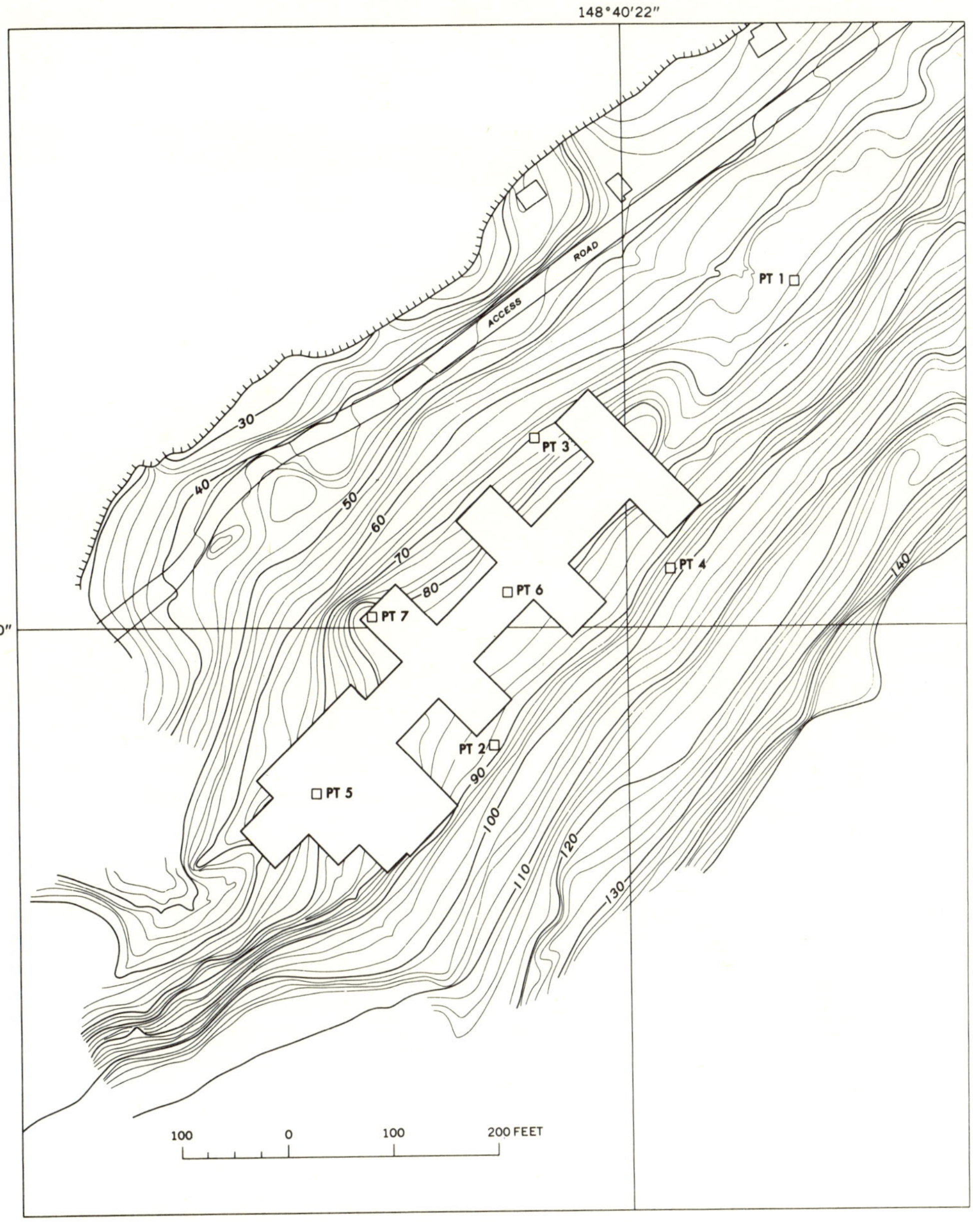

15.—**Map showing Buckner Building, surrounding area, and location of test pits** (Pt). **Graphic logs of these pits are shown in fig. 16.**

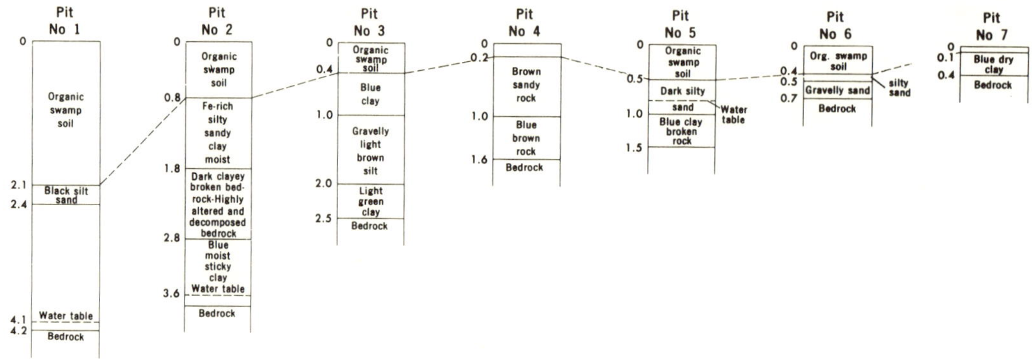

16.—Graphic logs of test pits (Pt) in the Buckner Building area. Locations of the pits are shown in figure 15.

consolidated sandy gravel which were only slightly damaged by seismic activity include (1) the fire station, (2) the telephone building, which cracked at its corners, (3) the ACS building, (4) the gymnasium, which had some cracks at the top of the structure, and (5) the tunnel between the Hodge Building and the school.

The intransit storage shed, which rests upon several feet of fill on top of delta sediments, was slightly to moderately damaged by the combined effects of seismic shaking, subsidence, and wave action. In places the walls of the structure moved as much as 7 inches; subsequently, the building appeared to have been stretched by the earthquake effects.

In addition to surficial damage, seismic shock broke two 10-inch pipelines from the water-storage reservoir. In order to operate the steam powerplant, water was pumped from the salt-water pumphouse through a firehose.

FRACTURING

Only a few fractures occurred in the fill and the unconsolidated sediments and these caused no damage. A ground fracture formed during the earthquake near The Alaska Railroad depot. This fracture extends beneath the structure and appears to follow the contact between the cut and fill made during the construction of the depot. Minor fracturing in fill occurred near the intransit storage shed. Near the FAA station, a series of cracks parallel to the shore extends about 100 feet inland.

COMPACTION

Compaction of unconsolidated sediments and differential subsidence of the land damaged four structures: the pumphouse near the FAA station, the DeLong Dock, the intransit storage shed, and the marginal wharf. These structures were also damaged by waves. In addition, the pumphouse is now inundated by high tides.

The marginal wharf and the intransit storage shed were constructed on fill overlying delta sediments. The station storage warehouse and the pumphouse near the FAA station are underlain by delta sediments. How much fill, if any, underlies the DeLong Dock is not known.

The maximum recorded subsidence (about 1½ feet) was measured in the area near the marginal wharf and intransit storage shed. Subsidence may have been greater (an estimated 2–3 feet) in the delta area at the head of Passage Canal. The pumphouse here is a concrete structure enclosing two 40-foot wells. The concrete floor subsided 6 inches relative to the well casings during the earthquake.

LANDSLIDES

The landslides triggered by the earthquake (pl. 3) formed in delta sediments at the head of Passage Canal, in delta sediments and fill at Whittier waterfront, and possibly in submarine talus or in a submarine lateral moraine along the north shore of Passage Canal. It is not known if any of the slip surfaces of the landslides are in bedrock.

The profiles of the slides and submarine slopes are shown on plate 3, and the changes in degree of slope are shown in table 2 (next page). The landslides occurred on slopes ranging from 20° offshore of the airstrip to 31° at the marginal wharf. The largest change of sub-

TABLE 2.—*Effect of landslides on average submarine slopes at Whittier, Alaska*

Landslide	Submarine slopes (degrees)	
	Pre-landslide	Post-landslide
Marginal wharf	21	23
Airstrip	25	15
Tank-farm area	25	24
FAA station	20	20
Passage Canal, north shore	31	27

marine slope, 10°, was produced by the airstrip landslide. The marginal-wharf landslide effected the only increase in slope. The only landslide scarp exposed is at the FAA station landslide.

The only structures damaged by landslides (pl. 1) were the FAA station (fig. 17), a section of the railroad, a small pier near the FAA station, and the extreme eastern end of the airstrip (fig. 10). The slip tower at the car-barge slip dock may have been destroyed by the landslide at the marginal wharf.

The landslides in the Whittier area did not all occur simultaneously. There were two, and possibly three, different periods of sliding during the earthquake.

LANDSLIDE-GENERATED WAVES

Submarine landslides in Passage Canal generated at least two, and probably three, waves. Plate 1 shows the area inundated and destroyed by waves. The direction and altitude of the highest waves are indicated on plate 3. Only the direction of the highest wave in a particular area can be indicated, because a higher wave would destroy any evidence of an earlier lower wave. A lower wave that followed a higher wave would merely rearrange debris. The maximum altitude of the waves was 104 feet along the north shore of Passage Canal (pl. 3). About a quarter of a mile north of the FAA station, the wave reached an altitude of 82 feet. Near the airstrip the wave reached 20 feet, and at the extreme south shore of Passage Canal the wave topped the railroad, reaching an altitude of 52 feet. In Whittier the maximum altitude reached by a wave was 43 feet at the small-boat harbor northeast of The Alaska Railroad depot. On the waterfront area the altitude was 26 feet.

At the Whittier waterfront, submarine landslides apparently generated two different types of waves. Figure 18 illustrates schematically the sequential formation of the waves.

The first wave formed in the scarp area of the submarine landslide. As the landmass slid, a wave was generated by the water rushing in to fill the void formed behind the sliding mass, and a drawdown of water occurred. This wave was directional and its form was chiefly a function of (1) the size and shape of the landslide, (2) the distance of the landslide scarp offshore, (3) the velocity of the sliding mass, and (4) the configuration of the coastline.

17.—View, looking north, of FAA station. Track destroyed by landslide. Bow of boat resting on scarp. Photograph by G. D. Hanna, 1964.

The second type of submarine landslide wave was generated when the landslide came to rest at the bottom of Passage Canal. The volume of the landmass displaced the water and created an upwelling near the margins of the depositional area (fig. 18). When the upwelling water surfaced at Passage Canal, it generated a wave system. This wave system was not as directional as the wave generated at the scarp of the landslide. However, water moved primarily in two directions: (1) toward the scarp area, and (2) toward the landslide movement.

The size and force of the second wave were controlled chiefly by (1) the size and shape of the sliding landmass, (2) the velocity of the sliding mass, (3) the submarine configuration of Passage Canal, (4) the distance offshore at which upwelling occurred, (5) the depth of water in which the slide came to rest, and (6) the configuration of the coastline. The character of the waves may also have been influenced by other landslide-generated waves in nearby areas.

Eyewitness accounts indicate that probably both types of waves inundated parts of Whittier. There were no witnesses to the waves that struck the delta at the west end of Passage Canal or its north shore. The apparent submarine explosion described by Mr. Kolb and the "boiling out" of water in Passage Canal described by Mr. McDonough no doubt, preceded waves of the second type. Eyewitnesses reported that one of the slip towers of the car-barge slip dock collapsed before any wave struck the dock. The author believes that the tower was at the head of the marginal-wharf slide (pl. 3) and that the tower collapsed as the land slid during the early violent phase of seismic shaking.

There are several possible explanations of the glassy wave reported by witnesses. If the sliding mass moved with sufficient velocity, the volume of drawdown could be large enough to form a wave that would reach an altitude of 25 to 26 feet when water rushed into the vacated area yet which probably would not be highly turbulent.

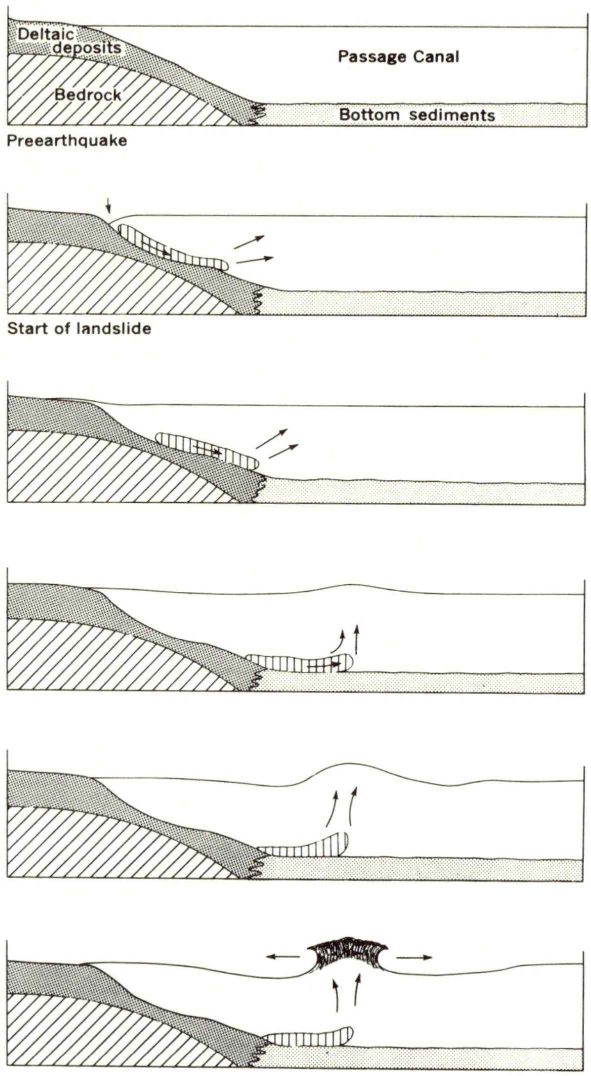

18.—Schematic sections showing generation of waves associated with submarine landslides.

Another possible explanation for the glassy wave is that it was caused by seiching in Passage Canal. Inasmuch as Passage Canal is in open communication with a large body of water, the period (T) of seiching would occur according to the formula

$$T=\frac{4l}{\sqrt{(gh)}}$$

where T is period in seconds, l is length of Passage Canal in meters, h is the average depth of Passage Canal in meters, and g is the acceleration of gravity. Assuming l is 15,000 meters and h is 300 meters, the period of seiche waves in Passage Canal would be 1,110 seconds or about 18.5 minutes. The earthquake occurred when the tide was at 1 foot and the land subsided 5.3 feet; thus the seiche waves would have to be at least 18.7 feet high to reach the 25-foot altitude (preearthquake datum) attained by the glassy wave. If such seiching occurred, evidence of a high-water level would be found near this altitude in the cove into which Cove Creek drains. Such evidence was not found in the cove. If seiche waves in Passage Canal had been generated by the earthquake, they would have continued at an 18.5-minute interval throughout the night of March 27, 1964. However, residents of Whittier who worked all night after the earthquake did not report any unusual waves in Passage Canal other than the three waves that struck Whittier during or immediately after the earthquake. Therefore, the glassy wave and the following two breaking waves must have been generated by submarine landslides. The absence of waves throughout the night as reported by the eyewitnesses also rules out the possibility that small seiche waves occurred but did no damage at Whittier.

19.—View, looking south, of The Alaska Railroad depot. Wave destroyed waiting room of the depot. Photograph taken April 7, 1964.

It is remotely possible that submarine landsliding near Billings Creek was responsible for the glassy wave. However, any landmass failure there would have had to be immediate—within 45 seconds of the earthquake—to allow sufficient time for the wave to reach Whittier by the time the glassy wave was observed. The area was not included in the study for the present report.

Wave damage in the Whittier area is shown on plate 1. The wave completely destroyed (1) the small-boat harbor, (2) the stub pier, (3) the car-barge slip dock, (4) the waiting room of The Alaska Railroad depot (fig. 19), and (5) the Columbia Lumber Co. structures (fig. 5 and 7). Many facilities were damaged by a combination of destructive forces: (1) the Army tank farm by fire, seismic shaking, and waves, (2) the Union Oil Co. fuel tanks by fire and waves, (3) the DeLong Dock and part of the marginal wharf by wave action and subsidence, and (4) the intransit storage shed by waves, subsidence, and seismic shaking.

The waves that struck Whittier had sufficient force to carry heavy objects many feet inland. A 4-cubic-yard barnacle-covered boulder was carried about 125 feet and deposited on The Alaska Railroad track near the depot at an altitude of about 26 feet. Boulders as much as 6 feet across were strewn on the road between the small-boat harbor and the Union Oil Co. fuel tanks. One of the Union Oil Co. fuel tanks containing ballast was moved at least 40 feet south by waves, and another tank containing 1 million gallons of fuel was moved 15 feet south (fig. 5).

The waves that struck the shoreline at the head of Passage Canal damaged practically all the structures near the shore. The waves completely destroyed three unoc-

20.—Photograph of 2- by 6-inch plank driven through a 10-ply tire by wave.

cupied homes on the shore, 200–400 feet south of the FAA station, and the buildings of the Two Brothers Lumber Co. The homes were washed away, and the nearby FAA station was partially inundated by the waves but not significantly damaged. The Two Brothers Lumber Co. buildings were carried inland in a southwest direction; the company's trimmer and conveyor chain were moved 200 feet, the bucking machine 40–50 feet, and the 2,300-pound mill about 100 feet to the southwest. The waves struck the Two Brothers Lumber Co. installation with such force that a 2- by 6-inch plank was driven through a 10-ply forklift tire (fig. 20).

The waves (or single wave) that damaged the west coastline of Passage Canal originated along the north coastline of the canal and traveled southwest. The wave (or waves) was apparently diverted by the eastern end of the airstrip—the structure about 300 feet south of the airstrip was only moderately damaged (pl. 1).

The southwest- and south-traveling wave struck the point along the south shore, about 4,000 feet west of Whittier Creek, with tremendous force. The wave reached an altitude of more than 50 feet. It carried a 1-ton winch and a boulder weighing 2–3 tons 120 feet south and deposited them on The Alaska Railroad tracks, which are about 30 feet in altitude.

FIRE

Fire was chiefly responsible for destruction of the Union Oil Co. and the U.S. Army fuel-storage tanks at the Whittier waterfront. Fire, together with damage by wave and seismic action, completely destroyed the tanks. The cause of the fire is not known, but there are at least two possible explanations: (1) seismic motion may have snapped a live powerline in the area, or (2) fuel from a damaged tank may have flowed into a boiler that was operating in the area.

CONCLUSIONS

The chief conclusion that can be drawn from the Geological Survey's investigation of the effects of the Alaska earthquake of March 27, 1964, on the Whittier area is that damage can be correlated with the local geology.

At Whittier all facilities built on the slate and graywacke bedrock were damaged slightly, or not at all, by seismic shaking. These structures include (1) the six-story Buckner Building, (2) the three-story FHA building, (3) the 1½-million-gallon water tank, (4) the powerhouse, and (5) the salt-water pumphouse. All are concrete except the FHA building, which is a wood frame structure.

In contrast to the facilities on bedrock, most of the buildings on unconsolidated sediments or fill were damaged significantly by seismic activity. These buildings include (1) the 14-story Hodge Building, (2) the gymnasium, (3) the composite shop, (4) the school, (5) the station storage warehouse, (6) the intransit storage shed, (7) the telephone building, (8) the ACS building, (9) the pipe shed, and (10) the fire station. All of the 10 buildings are of concrete construction except the composite shop, intransit storage shed, station storage warehouse, and the pipe shed, which are of various types of construction.

In any earthquake the structural response of a building to a seismic wave is a function of (1) the amplitude, acceleration, and frequency of the seismic wave, (2) the duration of shaking of the building, (3) the shape, height, and width of the building, (4) the type of construction used in the building, and (5) the orientation of the building to the seismic-wave direction.

Richter (1958) stated that there is an elementary relation between amplitude, acceleration, and frequency. He further stated that "Duration of shaking is the single most important factor in producing excessive damage. It takes time to break buildings up once damage starts. Long duration, reasonably high acceleration, and considerable amplitudes are the combination which does most damage in buildings."

Gutenberg (1957) reported that strong seismic shaking may last several times longer in unconsolidated deposits than in bedrock. He also stated that seismic amplitudes in unconsolidated deposits may be as much as 10 times the amplitudes in bedrock. Therefore, because there is a relation between amplitude, acceleration, and frequency and because of the longer duration of shaking, buildings on unconsolidated sediments are subjected to greater damaging forces than buildings on bedrock. The structural damage at Whittier clearly reflected this relationship.

Not only is there a correlation between damage and the underlying bedrock or unconsolidated deposits, but damage appears to be controlled also by the type, size, and thickness of the unconsolidated sediments. The Hodge Building, a 14-story reinforced-concrete structure which rests on coarse sandy gravel, was just moderately damaged by seismic activity. The Whittier delta at this point is only 1,200 feet wide, and the east end of the building is within 300 feet of bedrock. Therefore, the thickness of the sandy gravel that underlies the structure is probably not more than 150 feet.

The Hodge Building received substantially less damage than buildings of comparable height at Anchorage—a city about twice as far from the epicenter as Whittier. At least three geologic factors are responsible for this marked difference of damage: (1) the Hodge Building rests on a small area of unconsolidated deposits as compared to the structures at Anchorage, where a plain several miles wide is underlain by at least several hundred feet of clay, silt, sand, and gravel; (2) the deep fine-grained sediments that underlie Anchorage magnified the seismic amplitudes substantially more than the shallow coarse-grained sediments that underlie Whittier; (3) because the higher frequency components of ground motion attenuate with distance more rapidly than the lower frequency components, the tall structures at Anchorage were more nearly attuned to long-period ground motion. Thus, these structures suffered a proportionately greater degree of damage than the 14-story Hodge Building at Whittier.

The duration of ground motion differs not only between bedrock and unconsolidated deposits, but also with the type of unconsolidated deposits. For example, the longest duration of motion reported was at Portage, where ground shaking lasted about 15 minutes. Portage (fig. 1) is on the eastern shore of Turnagain Arm about 10 miles west of Whittier and approximately 40 miles southeast of Anchorage. The longer duration of ground motion at Portage than at either Anchorage or Whittier can be attributed to the finer grained sediments underlying the area, the shallow water table, and the greater size, shape, and depth of the sedimentary basin.

Ground fractures in the coarse-grained unconsolidated delta and moraine deposits at Whittier are uncommon, especially in the Whittier delta. Although the sediments were laterally extended toward free faces during the earthquake, the movement of material was not expressed as surface fractures. Extensive ground fracturing occurred at (1) Valdez, (2) Robe River flats, (3) Copper River flats, (4) Matanuska flats, (5) Anchorage, (6) Seward, and (7) the Portage area. In these places, the unconsolidated deposits are finer grained and thicker, and they cover a more extensive area than the unconsolidated sediments at Whittier. Apparently, the coarse-grained texture and, in part, the thinness of the sediments were responsible for the lack of extensive ground fractures at Whittier.

Even the coarse unconsolidated deposits at Whittier were compacted by the seismic shaking. Compaction and subsequent subsidence should be expected in most unconsolidated sediments during an earthquake of large magnitude.

Landslides formed on submarine slopes in unconsolidated material inclined 20° to 25°. The three submarine landslide-generated waves that struck Whittier were of two types. The first type of wave was generated in the scarp area of the submarine landslide, was unidirectional, and damaged only the area behind the landslide scarp. The glassy rise of water may have been associated with this wave.

The second type of wave was formed by the volume displacement of water offshore. The upwelling water moved primarily in two opposing directions: (1) toward the scarp area, and (2) in the direction of the landslide movement. These waves were not as directional as the waves generated at the scarp of the landslide, but they were far more destructive.

The landslide-generated waves were responsible for the loss of life and were the chief instrument of damage in the Whittier area. Because the submarine slopes in Passage Canal were not significantly decreased by the landsliding during the earthquake, another earthquake of comparable magnitude would probably trigger more submarine landslides. Destructive waves would inevitably follow.

The hazard of fire is always associated with earthquakes. The towns of Seward, Valdez, and Whittier had fires that originated at the fuel-storage tanks in the waterfront area. An immediate power failure spared Anchorage from electrically ignited fires.

REFERENCES CITED

Barnes, F. F., 1943, Geology of the Portage Pass area, Alaska: U.S. Geol. Survey Bull. 926–D, p. 211–235.

Grantz, Arthur, Plafker, George, and Kachadoorian, Reuben, 1964, Alaska's Good Friday earthquake, March 27, 1964, a preliminary evaluation: U.S. Geol. Survey Circ. 491, 35 p.

Gutenberg, Beno, 1957, Effects of ground on earthquake motion: Seismol. Soc. America Bull., v. 47, no. 3, p. 221–250.

Richter, C. F., 1958, Elementary seismology: San Francisco, Calif., W. H. Freeman and Co., 768 p.

U.S. Weather Bureau, 1964, Climatological data, Alaska, *in* Annual Summary, 1963: v. 49, no. 13, p. 219–230.

Part III

COASTAL ENGINEERING PROBLEMS

Air photograph (1960) of jetties and groins at Cold Spring Inlet, New Jersey. Compare with photograph of 1940, page **233**. (U.S.C. & G.S. photo)

Editor's Comments on Papers 15 Through 19

The problems of beach erosion are of concern to governments, planners, and people throughout the world. As indicated in Part II of this volume, the problems of beach erosion, shoreline recession, coastal flooding, and shoaling in inlets and harbors are not unique to any particular section of a coastline and do not stop at national boundaries. Concern over these problems and research and engineering planning of mitigating measures have increased in the past few years and will continue in the future as a result of economic development, population growth, improvements in transportation systems, and increased demand for natural resources and recreational areas in coastal regions.

Short-term and long-term cycles in beach erosion and shoreline changes are recognized. Shepard (1950b), Bajorunas and Duane (1967), and Sonu and Van Beek (1971) reported on short-term cycles (daily and seasonal) resulting from tides, swell, and storm waves. It is usually assumed that sediment removed from a beach during the fall and winter seasons (storm wave condition) is moved offshore to be temporarily stored in the form of longshore bars and offshore shoals till the spring and summer seasons (fair weather swell conditions) when the sediment is moved back to the beach. If the volume of sand returned to the beaches equals the volume removed from the beaches, net changes over the long term, theoretically, would be zero. However, onshore and offshore movements of sediment are not the only factors in considering beach erosion and shoreline changes. Longshore currents also play an important role in sediment movement. Assuming uniformity in occurrence and magnitude of natural processes, without human interventions or modifications, we would expect then a condition of stability of beaches and coasts over long periods of time. However, weather conditions over land and water and attendant coastal processes vary from time to time, and severe storms and cyclones frequently interrupt fair weather conditions in many parts of the world. During such severe storms the volume of sediment transported by longshore currents may vastly exceed the volume transported offshore (Galvin, 1967; and Longuet-Higgins, 1970). Also, any activities that alter the volume of sediment contributed to the beaches by inland rivers or by longshore drift drastically affect the equilibrium and result in erosion in some sections and accretion in others. All of these factors have cumulative effects and the net result in the long term would be erosion of several hundred feet of the beaches.

El-Ashry (1971) discussed causes for recent increased erosion along shorelines of the United States. The conclusions are considered applicable to shorelines of tectonically stable regions in many parts of the world. Three major factors are considered responsible for the recent increased erosion along the shores. These are: (1) hurricanes (cyclones) and severe storms, (2) recent eustatic rise in sea level, and (3) man's interference with natural shore processes. Hurricane surges cut channels and inlets through foreshore dunes, beach ridges, and barrier islands and deposit the eroded sediments in back of the beaches as washover fans and deltas. This eliminates sediments that are normally available for beach and longshore drift. Inlets that remain open after a storm normally interrupt the supply of sediment carried by longshore currents to beaches on the downcurrent side of the inlets and result in increased erosion of these beaches. These interruptions usually occur when sediments are trapped in strong flood (tidal) currents that carry them through the inlets to be deposited in lagoons or sounds as tidal deltas. Also, during storms, finer portions of the sediments eroded from the beaches may be carried by strong return (rip) currents onto the continental shelf, representing a permanent loss of beach sediments.

Eustatic rise in sea level can explain some of the long-term and slow recessions of the shorelines. Fairbridge (1961) mentioned that some of the long-term eustatic changes observed involve 100 m (vertical) in 200,000 years, which averages 0.5 mm per year; current rise in sea level, measured by tide gauges, is 1.2 mm per year. A rise in sea level of up to 10 mm annually, continuing over many centuries is recorded on the coasts of northern Canada and parts of Scandinavia. In discussing the effect of sea-level rise on erosion along a coast, Bruun (1962) assumed that for any rise in sea level the bottom of the nearshore littoral drift zone will be raised simultaneously until it is covered with the same depth of water at the same distance from the (new) shoreline as it was before the rise. He concluded that the material needed to raise the bottom comes from the corresponding shore area through the movement of material by rip and diffusion currents. This was experimentally proven by Schwartz (1965; 1967) in scale-model and field studies on Cape Cod.

During the last decade, above average precipitation (33 cm above normal) has occurred in the Great Lakes region resulting in higher lake levels and causing extensive erosion and flooding. Of the 5,920 km of United States shoreline on the Great Lakes, erosion is occurring on about 2,080 km and another 536 km are being flooded. The increased erosion results from waves being able to break farther shoreward of former beaches and bars and directly against bluffs in addition to the fact that much eroded material is washed into deeper water rather than being carried along the shore (Hartford and Tanner, 1976).

Man's activities that interfere with normal shore processes and have adverse effects in the coastal zone include: (1) upstream dams on coastal rivers resulting in a reduction of sediment contributed to the beaches; (2) construction of jetties for inlet and harbor maintenance and improvements; (3) bulkheads and seawalls to protect property on the upper part of the beach from wave attack; (4) breakwaters for protecting harbors and providing shelter for boats; (5) groins for the interruption of longshore sand movement and accumulation of sand on the shore, or for retardation of sand loss; and (6) construction activities, in general, that result in the destruction of natural dune systems which are a major supply of sediment for the beaches.

Under natural conditions, shorelines develop their own natural defenses to cope with extreme sea conditions. Gradually sloping beaches dissipate most of the wave energy, dunes absorb the energy of storm waves and offer protection to the land behind them, and barrier islands act as a front line defense affording complete protection of lands along the main coastline. Continued encroachment in the coastal zone with man-made development has in many cases taken place without regard to existing natural defenses. Dunes are leveled and graded to make room for buildings, and barrier islands have been modified in many ways for resort and recreation facilities. Where such developments were left unprotected, great damage has resulted from storms and artificial protection methods were sought.

Coastal engineering planning considers knowledge of long-term trends in shoreline changes and the processes that caused them. Air photography has been instrumental in supplying such information (Magoon, Jarman, and Berg, 1971; and Stafford, 1972). It also provides a tool for monitoring construction activities and performance as well as for recording effects of structures on the shorelines. Rates of erosion or deposition may be determined from air photos and older maps and charts. Areal estimates of sediment removal or accumulation can be transformed to volumes from shoreline change maps, or from repetitive offshore profiles, over a number of years.

A major consideration in the planning, design, and maintenance of shore facilities is sand movement in the littoral zone (Saville and Watts, 1969). Wave action is the most important factor affecting sediment movement along United State shorelines. Tidal currents are generally negligible in influencing sediment transport in the United States but may be a major factor in other parts of the world. Due to changes in the direction of wave approach, sand transport along the shore may be reversed in terms of hours, but most reversals in the direction of longshore drift are seasonal. Observations and study of seasonal changes in the direction of littoral drift and resulting beach width over a complete year's cycle are necessary before stating erosion rates of an area to be protected by structures. Air photography can provide valuable information in this connection (Slaughter, 1973).

Calculations of rates of littoral transport are complicated by the added effect of sediment supply contributed by rivers to the shore zone and by onshore-offshore movement of sediment.

Shore Protection Methods and Their Effects

Coastal engineering problems may be classified* into four general categories: shoreline stabilization, backshore protection

* Based on information derived from the "Shore Protection Manual" by the U.S. Army Coastal Engineering Research Center.

(from storm waves and surges), inlet stabilization, and harbor protection.

Measures to stabilize the shore fall into two classes: structures to protect against wave attack such as seawalls, bulkheads, and revetments; and an artificial supply of beach sand to make up for a deficiency in sand supply because of natural processes or due to man-made causes.

Seawalls, bulkheads, and *revetments* are structures placed parallel to the shoreline to separate a land area from a water area. The primary purpose of a bulkhead is to prevent sliding of the land while offering protection from wave attack. A seawall or revetment protects the land from damage by waves. There are no exact distinctions between the three structures, and in many cases the same type of structure is given a different name in different localities. While seawalls may protect the land, they can create local problems. Downward forces of water created by waves hitting the wall, can rapidly remove sand from the seaward side of the wall, eventually resulting in excessive scouring and undermining. A stone apron is often necessary to protect against that. Also, the presence of walls sharply changes the nature of longshore drift. Zenkovich (1973) reported that as a result of oblique surf waves striking a wall along the Caucasian Black Sea coast, an unusually fast current develops with velocities of up to 8 m/sec, and carries with it sediments which include large stones up to 20 cm in diameter. Thus, the transporting capacity greatly exceeds that on beach sections without walls.

A *breakwater* is a structure protecting a shore area (beaches, bluffs, or dunes), harbor, anchorage, or basin from waves. All breakwaters reduce or eliminate wave action. Whether offshore or shore-connected, the breakwater's elimination of wave action reduces longshore transport, obstructs the free flow of sand along the shore, and deprives beaches of the downdrift side (Hartley, 1965). At a harbor breakwater, the sand supply can be restored by pumping sand from the area of accumulation to eroded areas on the downdrift side. This type of operation has been in use for many years at Santa Barbara, California (see Papers 15 and 16 in this volume). At an offshore breakwater, sand is deposited in the quiet water between it and the shore. The buildup serves as a barrier and completely dams the sand stream, depriving the downdrift beaches of sand and causing erosion. There is one case, however, in which that type of breakwater can be used to aid shore processes. When placed on the updrift side of inlets, the structure impounds sand, prevents it from entering the inlet channel, and provides shelter for a floating dredge to pump the impounded sand across the inlet into the longshore drift. This method is used at a harbor near Port Hueneme, California (Herron, 1960; and Cassidy, 1961).

A *groin* is a shore protection structure designed to build a protective beach or to retard erosion of an existing or restored beach by trapping littoral drift. Groins are usually perpendicular to the shore and vary in length from less than 30 m to over 100 m, but normally extend only to the breaker zone. Groins may be classified as permeable or impermeable, high or low, long or short, and fixed or adjustable. They are normally constructed of timber, steel, stone, or concrete. Impermeable groins have a solid or nearly solid structure which prevents littoral drift from passing through them. Permeable groins have openings through the structure and permit some of the wave energy and moving sand to pass through them. Low groins allow appreciable amounts of sand to pass over the top of the structure. Nearly all groins are permanent and fixed structures. However, in England and Florida, groins of adjustable heights have been used which consist of removable panels between piles. Groins could result in less material being transported along shore than before their construction, and a permanent adverse effect to the downdrift beaches may result. Adverse effects to adjacent beaches may also result from the diversion of littoral drift, by groins, into deep water.

A *jetty* is a structure extending into the water to direct and confine river or tidal flow into a channel, and to prevent or reduce the shoaling of the channel by littoral material. Jetties constructed at the entrance to a bay or river also serve to protect the entrance channel from wave action and cross currents. When located at inlets through barrier islands, they also stabilize the inlet location and prevent migration. A jetty represents a perfect and complete barrier to littoral drift. Accretion takes place at the updrift jetty thus preventing nourishment of downdrift beaches and causes erosion (see Papers 17, 18, and 19 in this volume; Gofseyeff, 1952; Hartley, 1965; Wong, 1970; Noble, 1971; and Komar, Lizarraga-Arciniega, and Terich, 1976). Before the construction of the jetty, nature supplied sand by transporting it across the inlet, intermittently along the outer bar, and hence to the downdrift beaches. One solution to the problem of downdrift erosion is to dredge the sand accumulated at the updrift jetty and to pump it through a pipeline (bypassing the inlet) to the eroding beach (Gofseyeff, 1952; and Cassidy, 1961). A more recent development for sand bypassing provides a low section or weir in the updrift jetty over which sand moves into a sheltered pre-dredged deposition basin. The basin is dredged periodically and the sand is normally pumped across the inlet to the eroding beaches on the down drift side. A weir jetty of this type at Masonboro Inlet, North Carolina is discussed in Paper 16.

Beach structures, when properly used, have a place in shore protection (Rankin, 1952). But research has shown that the best protection is provided by using methods similar to natural ones. To simulate natural protection, dunes and beaches are rebuilt artificially. To restore an eroded beach and to stabilize it at the new position, sand from sources behind the beach or offshore is placed directly along the eroded section, and then the beach is artificially nourished by the stockpiling of suitable beach material at the updrift sector of the problem area.

Engineering design of shore protection and improved navigation measures, in addition to considering the materials to be used and the natural forces that will affect them, must also consider the alterations that such measures will cause to the natural processes operating in the coast and nearshore zone. In addition, the effects of such measures on adjacent lands must be considered to the extent of providing the required protection with the least amount of disturbance to current and future land use, environmental factors, and aesthetics of the area.

15

Reprinted with the permission of the National Academy of Sciences from pp. 272–290 of *Applied Sedimentation*, P. D. Trask, ed., John Wiley & Sons, New York, 1950, 707 pp.

GEOLOGY IN SHORE-CONTROL PROBLEMS

Martin A. Mason

Chief, Engineering and Research Branch
Beach Erosion Board
Corps of Engineers, Department of the Army
Washington, D. C.

The development of knowledge of shore processes has been due traditionally to the interest of geologists and others working in related fields. However, in the application of that knowledge to the increasingly important practical problems of shore-control, contributions by geologists have been notable for their absence. This chapter purposes to call the attention of geologists to some of the challenging practical problems encountered by engineers concerned with shore control and the manner in which geologists may contribute to the solution of the problems.

SHORE-CONTROL PROBLEMS

The problems of stabilization and rehabilitation of beach areas are the most widespread of shore-control problems. Stabilization is of high economic significance in resort areas, such as the New Jersey coast or the Miami-Palm Beach coast of Florida, but the significance varies only in degree in the case of individual property owners. In the typical instance, highways, homes, hotels, business establishments, and other structures pertinent to beach utilization in American society are located, during the development period, immediately adjacent to the beach. Little, if any, allowance is made for the possibility of encroachment by the sea (Fig. 1). When erosion of the beach occurs, as it normally does, the structures are endangered or damaged, and the engineer is called in to prevent further erosion and rehabilitate the beach. Both planners, where they have participated in beach developments, and property owners are at fault in such situations for not having provided an adequate marginal area to allow for the inevitable encroachment of the sea. The shore-control engineer is consulted after the patient is ill, in many cases in extremis, and his problem is to plan and construct defenses or works that not only will resist the attack of the sea but also will regain at least a part of the shore already lost. Often the problem is presented as treatment of a limited area, say a few hundred feet of frontage, within a large region of erosion, the re-

FIG. 1. Atlantic City, New Jersey. The highly vulnerable position of structures built without allowance for an adequate margin for erosion is well illustrated. September 1944. (Photo by U. S. Navy.)

mainder of the region being endangered but the property owners being in economic circumstances precluding any corrective action.

Another typical problem frequently encountered is that arising from the provision of deep navigation channels from the sea or lakes into rivers or inlets. To illustrate this problem we may assume an inlet into a bay. Initially, under natural conditions, the inlet channel might have had a controlling depth of, say, 10 feet, and would have migrated within certain limits. As the channel approached one limit of its migration, a new channel would break out near the opposite limit, with the old channel closing as the new channel improved, the new channel then migrating to repeat the cycle. In the usual case there would be a submerged bar associated with the inlet channel. Now demands are made for a channel of greater depth, say 20 feet, and the stabilization in position of the inlet channel to serve the interests of navigation. The engineer recognizes that the natural conditions represent an equilibrium situation involving, among other factors, the rate of transport of material alongshore and the tidal or flow characteristics of the inlet. Equilibrium will be destroyed by dredging of a chan-

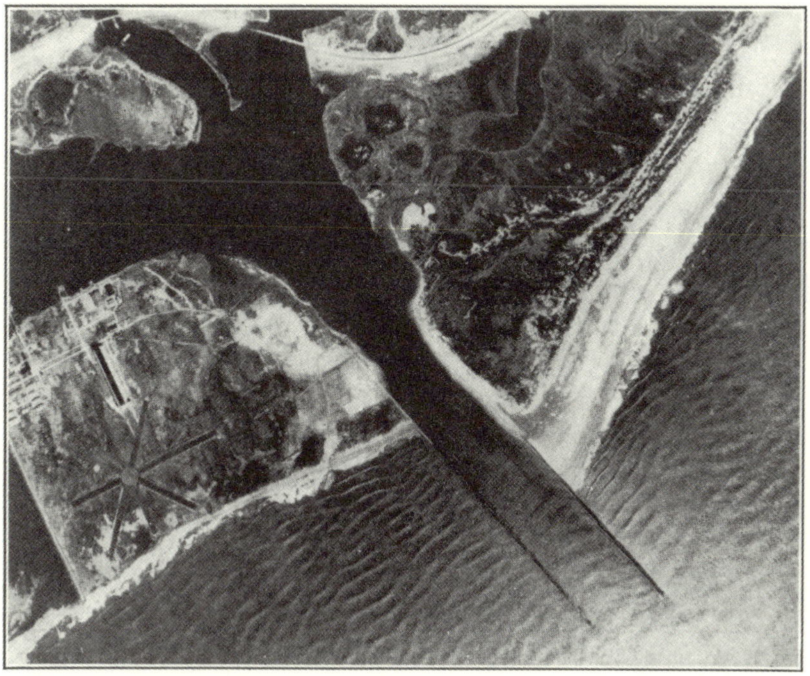

FIG. 2. Cold Spring Harbor, New Jersey. A wide beach has been formed by material trapped by jetties (right). The downdrift beach (left) is starved and subject to serious erosion. April 1940. (Photo by Aero Service Corp.)

nel, and natural forces will tend to re-establish equilibrium, in part at least, by filling the dredged channel. Jetties, or structures parallel to the channel axis and extending from shore seaward to the end of the dredged channel, are usually provided to eliminate refilling of the dredged channel. These structures, however, also inhibit or prevent the natural movement of material alongshore, thus establishing a new regimen which probably will involve starvation of the downdrift beaches due to interruption of the natural littoral supply (Fig. 2). The engineer's problem then is so to design the navigation works as to interfere as little as possible with the natural littoral transport of material, yet protect the navigation channel against filling, and to provide palliative measures for such deleterious effects of the works as may not be avoided.

A third type problem of shore control is associated with harbor-protective works. These works take the form of breakwaters and jetties extending seaward from the shore. Their function is to protect a given water area against wave action, thus providing a shelter for craft.

FIG. 3. Santa Barbara, California. Littoral material impounded by harbor breakwater and passing into harbor. June 1938. (Photo by Beach Erosion Board.)

Unfortunately the structures projecting from the shore act as barriers to the natural movement of material alongshore. The offshore structures, or breakwaters, reduce or destroy wave action, thus eliminating the principal force that causes material transportation. The problem here is to prevent starvation of the downdrift shore, and in some cases to control excessive accretion of material on the updrift side of the harbor. Santa Barbara Harbor, California (Fig. 3) is an excellent illustration of the problem. A dog-leg, or curved, breakwater was built from shore into deep water to provide a sheltered anchorage for small craft. Within a few years the entire updrift area of impoundment of the jetty was filled to capacity, and littoral drift passed into the harbor, shoaling it to such an extent that it became partially unusable. Concurrently, erosion of the downdrift shore occurred, finally extending its effects almost 10 miles downcoast from Santa Barbara.

The evidence at hand indicates that the net results of the action of shore processes over the period of record is an average annual loss of

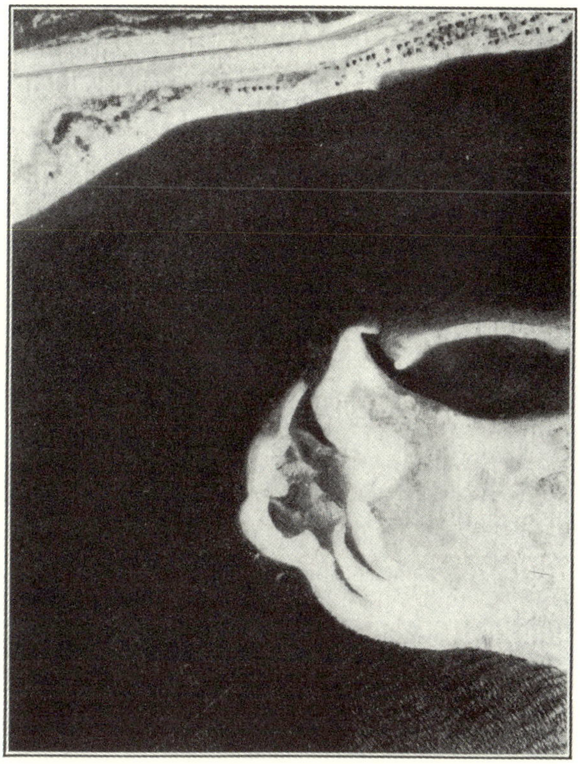

FIG. 4. Fire Island Inlet, New York, before construction of jetty, September 1939. (Photo by Beach Erosion Board.)

land in the United States equivalent to about a 1-foot strip over our entire 52,000 miles of shore line, a loss in land area approximating 6,400 acres per year. In some areas loss is general and excessive, for example, the Chesapeake Bay area, where entire islands have disappeared by erosion within the memory of man, and where the unprotected shores of both developed and agricultural lands are now being lost at rates as high as 15 feet or more per year. In other limited sections land is being gained, as at Fire Island Inlet on the south shore of Long Island (Figs. 4 and 5), where littoral drift is being trapped by a recently constructed jetty.

Shore-line modifications vary widely in scale. The larger changes are those involving considerable physiographic modification, for example, the truncation of headlands and the associated formation and growth of spits and bars; or the very important formation of barrier beaches and lagoons responsible for the present coastal features of New

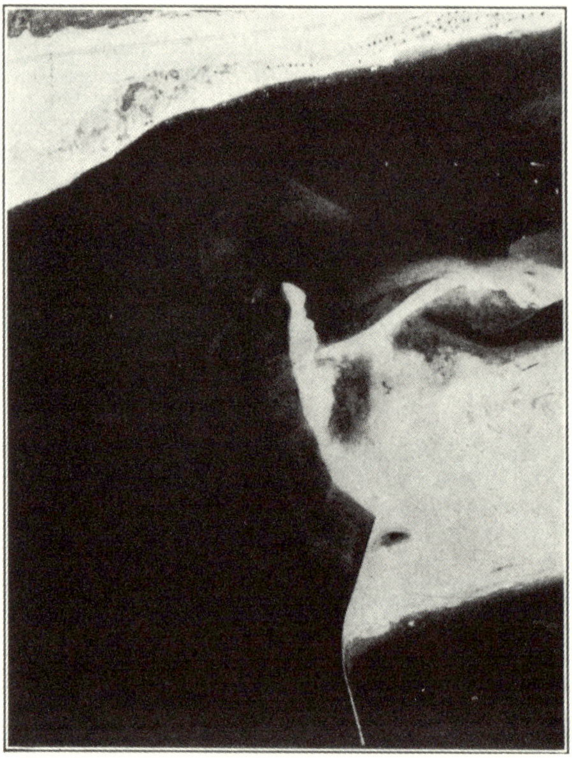

Fig. 5. Fire Island Inlet, New York, after jetty construction, 1941. (Photo by Beach Erosion Board.)

Jersey and the south shore of Long Island. It is in the study of these large-scale changes that geologists have made their greatest contributions to knowledge of shore processes in the past. The information developed is of great value as background for the solution of practical problems of control, but because of the long-term nature of the phenomena it is of lesser value in study of the relatively short-term problems of control. Shore-control engineers seek and need immediate control methods, effective for perhaps 20 to 50 years, and, justly, have little direct interest in or use for information about processes involving perhaps thousands of years for their consummation. It is in the study of small-scale, short-term shore processes that the need for assistance from geologists is most important.

The economic importance of shore control is largely a function of the development and use of the specific shore area considered. At Atlantic City, New Jersey, the shore-front property has a value of about $200,000,000, and this value is dependent in major part on the maintenance in satisfactory condition of some 10,000 feet of shore. Judged by this standard, each front foot of beach at that locality has a worth of $20,000. Other resort areas have similarly fantastic worth attaching to their limited sea boundary.

The value of beaches is strikingly illustrated by examination of the expenditures made to maintain or retain them. The State of New Jersey has spent $1,000,000 annually for 20 years in protecting its beaches. Harrison County, Mississippi, is spending federal and state funds in excess of $2,000,000 to repair a sea wall and provide beaches for the cities of Biloxi, Gulfport, and Pass Christian.

It is in navigation-shore-control problems that the economic significance of shore control is most striking. The annual reports of the Chief of Engineers, U. S. Army, covering the major portion of such problems, show that each year the federal government alone spends $48,000,000 to remove about 185,000,000 cubic yards of sediment deposited in navigation channels. In addition, millions of dollars damage to downcoast shores is caused by the trapping of these unwanted deposits and their resulting loss from the littoral supply.

It is difficult to assess the total economic significance of shore control to the people of the United States. Perhaps a conservative guess might be of the order of $100,000,000 annually, considering only actual expenditures and damage costs, and disregarding the intangible and difficult to evaluate costs associated with loss of recreational benefits and loss of shore area without immediate resulting damage.

FACTORS INFLUENCING THE PROBLEM

Two concepts are fundamental to the study of shore phenomena. The first is that of the physiographic unit. This unit may be defined as a shore area so limited that the shore phenomena within the area are not affected by the physical conditions in adjacent areas. Analysis of shore problems indicates that, for the purposes of studies directed toward protection or modification of the shore, all important phenomena depend on the material and energy available to the shore. What we seek in a physiographic unit, therefore, is a shore area so defined that the energy and material available within the area are not dependent on adjacent areas. Thus, for example, we may consider an island as a physiographic unit; it is probable that the entire Atlantic coast of New Jersey from Sandy Hook to Cape May is a unit; and on a smaller scale, Monterey and Emerald bays (Fig. 6) in California are units. The physiographic unit defines the boundaries of the area to be studied when considering shore problems.

FIG. 6. Emerald Bay, California. A physiographic unit.

The second basic concept is that of material-energy balance within the physiographic unit. Material balance concedes equality between the quantity of material supplied to and that lost from a shore area. Obviously a time factor must be introduced which requires that the volume of material considered be thought of as a rate of supply or loss. We are concerned chiefly with material balances pertaining to days, seasons, years, and storm durations. Since material is moved into and away from any given beach or shore section only by the application of forces, we must think also of rates of doing work, or energy.

Let us examine briefly the factors involved in material balance. If we assume some hypothetical shore stretch and consider the material situation therein, we find that formulation of a complete balance requires knowledge of (1) the rates of littoral drift to and from the area; (2) the rates of supply and loss from and to the adjacent sea bottom; (3) the rates of supply and loss from and to the adjacent land, including specifically dune areas and other marginal-land features; (4) the rate of supply of material contributed by drainage features. Unfortunately it is not possible, in the present state of the science, to evaluate these factors in other than a qualitative fashion.

Comparative study can indicate the probable relative importance of each of the factors, but only rarely are quantitative data available. In many areas only cursory examination is required to establish the littoral drift as the most important item, for example, along the New Jersey coast; whereas other areas are very complicated, for example, the Monterey Bay, California, coast where a submarine canyon and large supplies of material brought to the coast by streams greatly influence the coastal regimen and probably completely dominate the littoral phenomena.

Littoral drift is now evaluated by the amount of material trapped by shore structures, either natural or man-made, and from knowledge of the rate of depletion of sources of supply, for example, eroding cliffs or headlands.

Estimation of the material volume derived from or lost to the adjacent sea bottom is unsatisfactory at present. Experiments are in progress at reduced scale in the laboratory which give promise of confirming some conclusions reached from theoretical considerations regarding material transportation on the sea bottom. However, there is little, if any, evidence of a conclusive nature with respect to this phenomenon in nature.

The quantity of material derived from or lost to the land is now evaluated with fair accuracy by comparative study of the topography of the area at various times, coupled with knowledge of the character of the land and shore material.

The determination of the volume of material supplied to a shore area by streams, gully washes, or other drainage features is facilitated by knowledge of the suspended and bed load carried by the flow. Sometimes this information can be gained from study of the stream mouth delta, more easily from records of the rate of silting, or filling, of the stream upstream from the mouth. The wide variation of material load carried, between flood and normal stages, must be kept in mind.

Wave action is the primary source of energy at the shore. Its importance is overriding everywhere on the open coast and is not seriously diminished even in the immediate vicinity of large tidal entrances. The waves of concern are the characteristic progressive oscillatory surface waves so familiar to all, the long-period swell not usually so apparent, and the very long "tidal" waves, or tsunamis, of infrequent occurrence.

The surface waves and swell result from atmospheric disturbances over the ocean and are propagated over long distances with very slowly diminishing intensity. As a general rule these waves are always present on the ocean to a greater or lesser degree, from the days of dead calm when only very long and low remnants of swell from far-distant storms exist, to the days of violence when waves are being generated by a storm on or near the coast. Their action is unceasing and of an almost unbelievable magnitude. It is believed at present that waves derive their energy entirely from the wind systems to which they are subject, and that this energy is dissipated, except for a negligible fraction lost to internal friction, on the shores ultimately attacked by the waves. "Tidal" waves or tsunamis constitute a special case; because of their infrequent occurrence in highly violent form, they will not be discussed here, although they can by no means be neglected.

Theoretical studies, confirmed by laboratory experimentation and some observations in nature, show that surface waves involve motion of the water only to a depth equal to about one-half the length of the wave from crest to crest. The water moves in circular or elliptical orbits or paths which do not quite close; that is, there is a small excess forward motion associated with the passage of each wave, and this motion has been called the mass transport of the wave. Its magnitude is only a few percent of the actual orbital velocities, which range from 0 to a frequently attained maximum in excess of 6 feet per second. The velocity at the bottom varies with the depth of water, the wave length or period, and the wave height; it is the velocity most effective in causing erosion and transportation of material. As a wave moves into shallow water, the bottom velocities due to the wave increase by reason of decreasing depth but the wave energy is not greatly changed. Finally the wave breaks in surf, dissipating most of its energy in a highly turbulent, violently churning area, with the remainder dissipated during the rush of the wave up the beach. It is commonly held that the most active erosion and transportation zone is the area involved in the break of the wave and its uprush on the beach.

A fair evaluation of the energy available from the waves can be made if knowledge of the wave periods and heights is available, because it is known that the wave energy is proportional to the product of the period squared (T^2) and the height squared (H^2):

$$E \sim T^2 H^2$$

If in addition information on the direction of approach to the shore of the waves is available, it is possible to formulate a time distribution of wave energy with respect to direction of application of the energy.

From the engineer's point of view the adequate solution of shore-control problems is contingent on accurate knowledge of the following factors: (1) the sources and character of the beach material; (2) the rates of supply and loss of material to and from the problem area; (3) the manner of movement of material from the source to the beach and from the beach to other areas.

It is immediately apparent that, if answers to these questions can be obtained, the physical conditions that cause the problem situation will be defined. This is the definition of the problem. It has been the experience of investigators in this field that there are usually unique definitions of the problems; each problem is separate and distinct, thus making the solution of such problems by analogy an unsatisfactory method. There has been a tendency to practice solution by analogy, which still continues to some extent but which is on the decline as more and more knowledge of the problem is developed.

The physiographic unit is of value in determining the source and character of the beach material, particularly if the surficial geology of the area has been adequately studied. The material-energy balance of the unit, and specifically of that portion of the unit requiring control, greatly facilitates the consideration of the last two factors.

Frequently the definition of the physical conditions at the site by consideration and study of the above factors will suggest the type, and sometimes the detail, of the control to be employed. However, to arrive at a rigorous engineering solution the problem is analyzed further.

The analysis reduces to consideration of the following factors: (1) the feasible methods of modifying rates of supply and loss of material to achieve the desired results, and the effects of such modification; (2) the design requirements of the feasible methods of modifying supply and loss rates; (3) the economic cost of each of the feasible methods of modifying rates of supply and loss.

COLLABORATION BETWEEN GEOLOGISTS AND ENGINEERS

Geology can help engineering in connection with all but the last factor, but it must, to be of maximum value, focus attention on the dynamic aspects and significance of its findings. The average engineer who needs the information has little, if any, means of translating classical geologic information into a form he can use. A geologist informing an engineer that a beach deposit is recent in the geological sense, and derived from Quaternary deposits, is not helping the engineer. He will have made a major contribution if instead he can say that the beach deposit was probably made in the last half-century and that the material was derived from sandstone in such and such locality by erosion and stream transportation to the beach. The possibilities of interpretation of geologic information with a view to defining the dynamics of the environment have been realized but slightly. In this field they are particularly rich.

The geologist is eminently qualified to furnish basic data needed for shore control. Particularly is this true with respect to the definition of physiographic units and the determination of the source and character of littoral material. The engineer must contribute to the geologist knowledge of the forces acting and the nature of their distribution in time and space. Close collaboration between the two professions in the study of these problems will produce results of high economic significance with a minimum of effort. Almost all that is required is a wedding of the static and dynamic analyses now available.

Another area of collaboration that promises useful results quickly lies in study of the sources and transportation phenomena of material movement. Geologists already are informed about many of the details of the derivation of beach material from source rocks and the resulting characters of the material. Some of the gross details of transportation effects on the materials, for example, selective transportation and rounding, have been explored, chiefly by geologists, but much remains that can be done. The engineer can provide knowledge of some of the dynamic situations to which the material has been subjected, for example, the mechanisms of bed-load movement and the rate of mass movement of material in various types of watercourses. Here again the study of cause and effect with particular reference to the dynamic processes involved should produce immediately useful relations of value to both the geologist and engineer. The establishment of verified relations between material characteristics (including both discrete particles and aggregations of particles) and the dynamic processes in which the material has participated should open possibilities for a much better understanding of the details of transportation and deposition of sediments.

Engineers and geologists could well consider collaboration with a view to detailed study of the energy-material relations controlling physiographic changes at the shore. Engineers can supply much information on wave energy, its distribution along the shore and modification with changes in the shore conditions, which, coupled with geological knowledge of the materials present to be acted upon, might well lead to quantitative descriptions of physiographic processes.

Collaboration can probably be effected best by liaison between individuals and collaborative research. The field is too narrow and the specialization required too high to permit a more general approach. Some instances of combined application of engineering and geological knowledge, notably by W. C. Krumbein in the study of the details of littoral movement, have led to highly useful concepts that explain formerly complex situations.

FUTURE RESEARCH

In our present state of knowledge needed future research can be specified only in general terms. It can be stated that wide opportunities exist for research, and any discussion of needed research will certainly encompass only a part of the wide variety of problems that may exist.

One of the first problems to be studied as a concurrent geological-engineering problem is that of the definition of physiographic units for shore-study purposes. The problem involves, among others, determination of material sources and the transportation sequence from the source to the shore, with considerable attention to be devoted to the volume and time scales of the phenomena; determination of the limits of littoral transportation, both eolian and submarine; and consideration of the energy situation.

A second problem of prime importance in the study of many shore processes is the relation between shore-material characteristics and the dynamic processes of transportation to which the material has been subjected. Shore studies are concerned with the behavior of a population of particles in which the behavior of the population is conditioned by, and may be chiefly dependent on, the characteristics of the individual particles. The possibility exists that modification of one small fraction of the particles may change appreciably the behavior characteristics of the population. For example, it is known that the admixture of small amounts of clay in sand greatly increases the velocity required to erode the sand. More adequate definition of the erodibility and transportability of beach materials in terms of their particle characteristics is required. Determination of the erosion susceptibility of beach materials when subjected to the oscillatory velocity fields of wave action is a virgin territory, as is the problem of transportation on the bottom or in suspension under the same conditions.

There is some justification for thorough study of the details of selective transportation of shore material, with particular reference to the disposition of material. Closely related to this problem is the question of thickness of beach and near-shore deposits and their stratification as indicators of the past history of the beach.

In these suggested research problems as well as in others, the usefulness of tracers having the same dynamic characteristics as the materials under study is apparent. There is some evidence that adequate knowledge of the petrography of shore and source materials would make possible the use of some fractions of the native materials as tracers, thus greatly simplifying some of the technical phases of the studies.

It must be apparent that there are great deficiencies in our knowledge of even some of the less complicated phenomena of shore-line behavior. It is believed that future developments in shore study and control are contingent principally on fundamental research in materials, the transportation of materials, and methods of modifying the material-energy balance of an area. In the field we are concerned with, as in many other technical fields, there is presently a serious, almost critical shortage of basic information. Without such information, the technical man finds it impossible to advance applied research. In this field engineers have progressed as far as they have mainly on fundamental research in other fields, notably geology, hydraulics, mathematics, and physics. Our greatest need is for work on shore materials and the transportation of materials under maritime conditions. Available knowledge of unidirectional transportation of material in streams is of only partial value to us in our problem of multidirectional flow under periodically varying forces. We must know more about the details of material derivation from source rocks, the degradational processes of derivation and transportation, the possibilities of metamorphosis of constituent particles and populations of particles. It is essential to know the details of littoral movement of materials, the relationships of energy to material movement, the features of energy transfer and dissipation in the coastal environment, the requirements for modification of the material-energy balance, and even how to compute an accurate, quantitative material-energy balance.

REFERENCES

Bagnold, R. A. (1946). Motion of waves in shallow water—interaction between waves and sand bottoms: *Proc. Roy Soc. London,* ser. A, vol. 187, pp. 1–18, Oct. 1946.

* Beach Erosion Board (1942a). A summary of the theory of oscillatory waves: *Tech. Rept.* 2.

* ——— (1942b). Abrasion of beach sand: *Tech. Memo.* 2.

——— (1944a). Shore processes and beach characteristics: *Tech. Memo.* 3.

——— (1944b). Shore currents and sand movement on a model beach: *Tech. Memo.* 7.

——— (1948). An experimental study of submarine sand bars: *Tech. Rept.* 3.

Brown, E. I. (1940). Beach erosion studies, *Trans. Amer. Soc. Civ. Engrs.*, vol. 105.

Carey, A. E., and Oliver, F. W. (1918). *Tidal lands, a study of shore problems:* Blackie and Son, Ltd., London, 284 pages.

Cornish, Vaughan (1910). *Waves of the sea and other water waves:* George Allen and Unwin, London, 374 pages.

* Gilbert, G. K. (1883–84). Topographic features of lake shores: *U. S. Geol. Survey,* 5th Ann. Rept., pp. 69–123.

* Gulliver, F. P. (1899). Shoreline Topography: *Proc. Amer. Acad. Arts Sciences,* vol. XXXIV, pp. 149–258.

* Johnson, D. W. (1919). *Shore processes and shoreline development:* John Wiley & Sons, New York (out of print), 584 pages.

* Krumbein, W. C., and Pettijohn, F. J. (1938). *Manual of sedimentary petrography:* D. Appleton-Century Co., New York, 566 pages.

* Lobeck, A. K. (1939). *Geomorphology:* McGraw-Hill Book Co., New York, 731 pages.

Mason, M. A. (1942). Related problems in shore processes and oceanography: *Trans. Amer. Geophys. Union,* Pt. 2, pp. 325–326.

——— (1947, 1948). Lectures on beach erosion and shore protection: *Bull. Beach Erosion Board,* vol. 1, Nos. 2, 3; vol. 2, No. 1.

* Matthews, Ernest R. (1934). *Coast erosion and protection:* Charles Griffin and Co., Ltd., London, 228 pages.

Munk, W. H., and Traylor, M. A. (1947). Refraction of ocean waves: *Jour. Geol.,* vol. 55, No. 1, pp. 1–26, Jan. 1947.

Nevin, Charles (1946). Competency of moving water to transport debris: *Bull. Geol. Soc. Amer.,* vol. 57, pp. 651–674, July 1946.

* Schou, Axel (1945). *Det Marine Forland:* H. Hagerups, Copenhagen, 236 pages.

Shepard, F. P. (1946). Beaches and wave action: *Scripps Inst. Oceanography,* Wave Project Rpt. 56.

* Trask, P. D., *et al.* (1939). *Recent marine sediments:* American Association of Petroleum Geologists, Tulsa, Okla., 736 pages.

* Twenhofel, W. H. (1926). *Treatise on sedimentation:* Williams and Wilkins Co., Baltimore, Md., 661 pages.

* Important references.

16

Reprinted from *Internat. Navig. Congr., 23rd, Ottawa, 1973*, Sec 2, Subj. 2, 25 pp.

MEANS OF CONTROLLING LITTORAL DRIFT TO PROTECT BEACHES, DUNES, ESTUARIES AND HARBOUR ENTRANCES

George M. WATTS,
U.S. Army Coastal Engineering Research Center, Washington,

Limberios VALLIANOS,
U.S. Army Engineer, District Wilmington,

and

Robert A. JACHOWSKI,
U.S. Army Coastal Engineering Research Center, Washington.

INTRODUCTION.

The importance of controlling littoral drift for purposes of beach stabilization and maintaining desired entrances to harbors and estuaries has long been recognized by those having an interest in the coastal zone. However, the importance of this subject has greatly increased in recent years in the United States. The primary reason can be related to the rapid growth of population in the coastal states and associated development and demand for residential, commercial, and recreational usage of the coastal zone. For example, over the past 10 years, 90 % of the national population growth was in the coastal states. Also, 75 % of the nation's population now reside in the 30 coastal and Great Lakes states.

This increased interest and use of the coastal zone emphasizes the need to better understand shore processes and establish improved guidelines for planning, designing, constructing, and maintaining coastal works for their intended purpose. This need for an improved understanding of littoral drift and related shore processes along the United States coastal zone is also influenced by the increased concern regarding environmental effects of constructed coastal works. An example of this environmental concern involves the excavation of sand from bays or lagoons for all purposes including beach restoration or periodic beach nourishment. It is now necessary to evaluate the environmental impact this operation will have on the borrow site and adjacent areas. Since bays, lagoons, and similar bodies of water are important to the maritime life, the constraints on utilizing these bottom deposits for beach-fill purposes have increased. This has necessitated exploration of the subbottom in offshore coastal waters for beach fill material. However, environmental considerations must also be recognized in borrowing from these offshore deposits.

Perhaps the greatest impact has been the focus of attention on the value of the sand within the littoral zone, and conservation of that sand. If a beach is eroding due to a natural deficiency of supply, the economically feasible borrow sources to effect artificial nourishment to make up this deficiency are becoming fewer. If groins are considered to reduce the littoral drift movement (longshore littoral material transport) out of the shore area to be stabilized, an accurate prediction of the functional and structural behavior of the groins, including effects to adjacent shores, is becoming of greater importance in order to evaluate justification of the groins.

If backshore or upland construction is contemplated, many states now have statutory regulations concerning preservation of the foredune formations immediately behind the beach zone. If works are to be constructed at entrances to harbors or estuaries, and these will cause impoundment of the longshore moving littoral drift, a plan must also be incorporated for appropriate transfer of these impounded littoral materials back into the littoral stream (in most cases transfer to downdrift shores) so the adjacent shores will not be adversely affected. Thus, due to the increased importance of the coastal zone it is now necessary to evaluate a much broader spectrum of beneficial and detrimental effects when new coastal works or modification of existing works are considered.

This report presents the present techniques employed in the United States for controlling littoral drift for purposes of beach and dune stabilization, and stabilization of entrances to harbors and estuaries. There are few cases in the United States in which a beach was mechanically constructed in a reasonably exposed coastal sector where no beach previously existed. There are numerous examples of artificial beaches constructed on inland natural or man-made reservoirs or lakes, and similar waterways. Also, recreational beaches have been constructed in some harbor complexes where no beach area previously existed. It should be pointed out that in the United States the term beach restoration or beach fill normally refers to the mechanical placement of suitable sand to built or widen an existing beach to a desired dimension to serve single or multiple purposes, such as, to accommodate recreation usage or to protect structures located on the backshore. The term artificial nourishment normally refers to the placement of suitable sand (periodically) in a proper location in the beach zone so wave-induced currents will transport the sand to and along the shores desired to be nourished and stabilized. The placement of sand to serve as a feeder beach for the benefit of downdrift shores is considered to be a form of artificial nourishment. If the local condition involves stabilitzation through beach restoration or periodic artificial nourishment techniques, the subject of controlling littoral drift is actually confined to a supply and demand relationship of littoral material.

ARTIFICIAL FILLS AND STRUCTURAL CONTROLS FOR BEACH AND DUNE SYSTEMS.

Shore erosion is considered that condition where the rate of loss of sediment from a given reach of shore exceeds the rate of sediment supply to the shore. This condition prevails along a significant proportion of the oceanic, estuarine, and Great Lakes shores of the United States which, in combination, comprise a total distance of 84,200 statute miles (135,000 km).

A recent study conducted by the U.S. Army Corps of Engineers [1] * defined the magnitude of erosion along the U.S. coastal margin. This study revealed that 20,500 miles (33,100 km) of shore are experiencing significant erosion. Excluding Alaska, which has 47,300 miles (76,500 km) of shore, approximately 15,400 miles (24,200 km) of coastline or 43 % of the U.S. shore outside of Alaska are undergoing significant erosion. However, only 2,700 miles (4,400 km) of shore, classified as having " critical erosion, " are so located that economic, industrial, recreational, agricultural, navigational, demographic, ecological and other relevant factors indicate that engineering works to halt erosion may be justified. The remaining areas experiencing significant erosion, but " non-critical erosion " appear amenable to land-use controls and other management techniques rather than measures to halt erosion.

Erosion along the U.S. coasts may be a result of human activities as well as natural forces. Some causes of erosion are long term and some short term. These are briefly discussed below.

Many sections of the U.S. coastline are experiencing a rise in sea level which may be the result of an absolute increase of the elevation of the ocean levels or subsidence of the land masses that form the coastal margins. For example, Marmer [2] established a sea level rise of 0.55 foot (16.76 cm) at Charleston, South Carolina over the period 1930 to 1950, and 0.25 foot (7.62 cm) at San Francisco, California over the period 1900 to 1950 Assuming no change in wave climate, a rise in sea level would be attendant with readjustment and landward movement of the beach profile. A study conducted by Bruun [3] concludes that sea level rise is a major long-term cause of beach erosion along the approximately 1,000 miles (1,600 km) of sandy shore in Florida.

Littoral materials in transport along U.S. shorelines encounter numerous obstructions and entrapments. An example of the severity of littoral drift entrapment is provided by Caldwell [4] who determined that 7 inlets along 85 miles (140 km) of the southern New Jersey coast are responsible for the permanent entrapment of 1,760,000 cubic yards (1,346,000 m^3) per year. Upland development and control that influence sediment discharge to the coast can affect shore stability. This is a problem for some sections of the California coast. A study of the Ventura and Santa Clara rivers, in California, by Watts [5] showed that prior to 1948 these rivers contributed approximately 1,702,000 cubic yards (1,302,000 m^3) of sediment annually to the littoral zone. By 1963, controls had reduced the quantity of sediments supplied to the coast to about 716,000 cubic yards (547,700 m^3) per year.

A common short-term natural erosion phenomenon is associated with the frequent storms experienced along coastal areas of the U.S. The combination of storm tides and waves move sediment offshore, thereby creating erosion. Normally, following storm wave action, long-period waves or swells return some sediment to the shore. Shorefront developments located too close to the shoreline to allow these natural short-term beach fluctuations, are subject to structural damage. The amount of shoreline retreat resulting from storms varies. Based on field measurements at various locations, Saville and Watts [6] concluded that major storms can cause erosion of about 100 feet (30.5 m) on many U.S. beaches. Natural long-

* Number denotes references listed at end of paper.

and short-term shore erosion can also be created by the concentration of wave energy due to refraction. Wave refraction analyses combined with field observations conducted by Munk and Taylor [7] in the vicinity of La Jolla, California, demonstrated the influence of wave energy variability on a shoreline with respect to sediment transport and the associated alternating erosion and accretion zones.

Human activities along the U.S. coastal margin are often instrumental in creating erosion conditions. An example of manmade obstructions to longshore movement of littoral drift and resultant downdrift erosion are jetties at navigation entrances. This particular problem and techniques used to eliminate it are discussed later in this report. Manmade groins, intended to prevent erosion, sometimes create new erosion problems or relocate an existing problem. Even with judicious employment of groins, accurate prediction of the response of downdrift shores and possible property damage is usually most difficult. Lillevang [8] cautions "... the engineer who undertakes projects including seacoast groins needs legal counsel, special engineering knowledge and experience, and a healthy respect for the continuing acceptability of the client's seacoast groins to neighboring proprietors." The rapid growth in many sections of the U.S. coast also, has often been accompanied by unwise developmental practices, particularly in the destruction of coastal dunes and vegetation.

The causes of erosion, mentioned above, by no means include all possibilities. Though cause and effect may be generalized, most erosion problems have unique aspects. Following is a discussion of the general concepts employed in the United States for the planning and design of artificial fills and structural controls for dune and beach systems.

ARTIFICIAL FILLS.

General. For the past half century, artificial fills have been employed extensively along the U.S. coastline for the purpose of restoration or stabilization of beaches. Normally, artificial beaches fills serve the dual functions of providing an area for water oriented recreation, and upland protection. In providing coastal protection, artificial fills may be combined with one or more additional protective features such as sea walls, revetments, groins, or dunes. Frontal dunes, which often comprise a part of the cross-section area of artificial fills, are so dependent on the behavior of the beaches which front them that this report treats dunes and beaches as an integrated topic.

In 1953, Hall [9] reported on 60 major artificial-fill projects in the United States. Since 1953, many other artificial fills have been constructed. At present, the U.S. Government [1] has under construction or planned a total of 58 coastal protection projects, involving approximately 468 miles (760 km) of shore, which include artificial fills to restore or stabilize beaches. Basic in the planning and design of artificial fills are the following elements:

a. *Determination of Longshore Transport of Material.* Information and data concerning the direction and magnitude of longshore sediment transport as well as the deficiency of natural supply in the vicinity of a proposed artificial fill are important to evaluation of fill operations, and particularly important for planning frequency and magnitude of fill replenishment. Precise determination of the longshore transport rate of littoral material is usually very difficult. Often adequate quantitative data are not available, and considerable engineering judgement is required in the analyses. Several procedures to establish initial and periodic fill requirements are employed in the United States. Comparison of repetitive surveys of material impoundment areas or eroding areas which cover a long period of time is considered the best means of estimating the net rate of littoral transport or deficiency of material supply. When surveys suitable for volume measurements are not available for the area of interest, alternatives of proven reliability are very limited. One approach being given attention is if nearshore wave characteristics are available, refraction analysis can be used to develop longshore current velocity. This velocity can be computed by various methods, as for example, the method developed by Eagleson [10]. The computed longshore current can then be applied to a bedload equation such as proposed by Bijker [11] and a suspended-load computational procedure such as developed by Einstein [12] to estimate total longshore transport. It has been found that wave refraction analysis and computations of longshore transport are highly desirable procedures to be explored even if there are sufficient surveys available to determine net drift in a given area; particularly, if the shore in question is exposed to littoral drift reversals. This information provides insight into the cause and effect relationships of the existing shore regime. For example, a judgement concerning the use of artificial fills may assume a stockpile of material at some updrift location will rapidly influence a downdrift coastal segment along which it is desired to stabilize the beach. Wave refraction analysis and sediment transport computations could show that the material transport capacity along a possible updrift stockpile location is not great enough to transport the fill material downdrift at a rate that would produce the intended shore stabilization.

To use tracers to study littoral drift, there are, as reported by Noda [13], numerous techniques including the use of; (a) natural tracers - heavy minerals and naturally radioactive minerals, and (b) artificial tracers - sand tagged with radioisotopes and sand coated with fluorescent dyes. There are various methods of monitoring tracer movement after injection into the littoral zone. Duane [14] demonstrates that radioactive tracers provide a rapid means of defining time distribution and direction of littoral drift movement. Problems associated with the use of tracers include; (a) inability of present methods and techniques to quantify the long term volume rate of littoral transport, (b) operational difficulties of making measurements in a high wave energy environment, and (c) the possible costly and time-consuming delays which may occur when it is necessary to conduct the tracer studies under a specific set of conditions. Notwithstanding these problems, there are many complex practical situations as well as research investigations which can effectively use tracer techniques in the littoral environment.

To develop a first approximation of the net volume change in the littoral zone under study, it is common to use the relationship that 1 square foot (929 cm^2) of change in beach surface area is produced by the addition or subtraction of 1 cubic yard (0.765 m^3) of material from the active beach profile. The active beach profile is defined as extending from the landward limits of the beach berm to an offshore point beyond which there is little or no discernible profile adjustment under normal wave activity. Along the U.S. coastline exposed to the open ocean, the seaward limits of the active profiles are normally found at depths of 20 to 40 feet (7 to 12 m) as referred to mean-low-water datum. In determining the amount of beach change when there is a paucity of survey data it is common practice to use controlled aerial photographs, as some aerial photographic coverage is usually available for most sections of the coast. However, it has been found that caution must be exercised in using a sequence of aerial photographs to measure changes in beach areas, as each photographic coverage, in all probability, has captured a unique combination of tide level, wave runup and beach condition established during the existing or previous tide cycle.

b. *Selection of Artificial Fill Material.* Three important factors must be considered in the selection of artificial fill materials; (a) location of the material borrow area with respect to environmental side effects resulting from changes in the borrow area attendant with material withdrawal, (b) the quality of the fill material as compared to existing littoral material which comprises the active beach profile, and (c) location of the material borrow area with respect to the method which will be utilized to excavate, transport, and place the fill. These factors are discussed in the paragraphs that follow.

Material borrow areas employed for artificial fills in the United States have been located in the upland area, estuarine waters, and in the nearshore oceanic zone. The use of upland areas has been infrequent, since use of earth-moving equipment to convey materials which are usually not near the shore is rarely economically feasible or acceptable. Moreover, as a result of population growth in the coastal zone, land use and increase in property value usually preclude the possibility of acquiring lands for use as fill borrow areas. In addition, denudation of vegetation and land cuts or pits associated with upland material borrow areas can, if located near population centers, create undesirable landscape features. The relatively small utilization of upland borrow areas for artificial fills in the United States is reflected in the listing of projects given in reference [9] in which only 3 out of 60 projects involved upland borrow.

The most frequently used artificial fill borrow areas have been located in estuarine waters. Excavation, transportation, and placement is accomplished by cutter-suction dredges. To select an estuarine borrow area, careful consideration must be given of environmental side effects. As noted by Cronin et al. [15], the estuarine areas are of importance as nursery and production grounds for many species of fish, crabs, shrimp, oysters, clams, and other marine forms. Damage to the existing ecological system can affect both local production and the catch of the species over a wide area. Determination of ecological impact before dredging in estuarine areas is required in the United States. Precise evaluation of the impact is frequently difficult. For example, the creation of borrow holes in shallow lagoons can provide a new kind of productive habitat for certain marine life. On the other hand, a similar hole in another estuarine site could be undesirable because of certain pollution accumulation.

Artificial fill borrow areas in the nearshore ocean zone are being sought with increasing frequency in present planning efforts in the United States. There is a twofold reason for this development. First, there are often, as mentioned above, undesirable ecological side effects associated with borrow areas located in estuarine waters. Second, ocean bottom sediment along many sections of the U.S. coast are generally of acceptable quality as artificial fills. The trend toward employment of ocean borrow areas, Duane [16], is evidenced by the continuing program of the U.S. Army Coastal Engineering Research Center in exploring and delineating ocean deposits of sand suitable for beach restoration and periodic nourishment. Many beach fills have been constructed in the United States where the borrow was offshore, and a conventional cutter-suction dredge was used. However, most of these projects are located in semi-sheltered waters such as the Connecticut and New York coasts in Long Island Sound. Some recent beach-fill projects using modified cutter-suction dredges have been conducted in exposed ocean locations at Redondo Beach, California, Fisher [17] and at Pompano Beach, Treasure Island, and Fort Pierce, Florida. Also, an experimental artificial fill operation was performed by the Corps of Engineers [18] in the North Atlantic off New Jersey, utilizing a hopper dredge equipped for direct pump-out to shore.

Although modified conventional dredging equipment can operate in moderate wave action, high costs are presently associated with such projects and there are many operational difficulties. The recently completed projects in California and Florida clearly indicate the progress in plant development and techniques to transfer offshore deposits to the beach zone. It is reasonable to assume this progress will reduce operational costs. Several large U.S. dredging equipment manufacturers are showing great interest in developing high-production dredging plants capable of placing artificial fill on a shore under heavy sea conditions.

In evaluating the compatibility of a given source of artificial fill, emphasis is focused on comparing particle size distributions of the sediment in potential borrow area and the sediment on the surface of the active beach profile in or near the placement location. It is reasonable to assume that if material, with average characteristics of the active-profile material, is placed on the shore, the prevailing hydraulic forces will, in time, redistribute and sort the material into a similar distribution. However, if characteristics of the fill material are not similar to those of the beach sediment, a part of the fill will be transported away from the normal limits of the active profile. The average size characteristics of the beach and fill material can be determined by procedures developed by Krumbein [19] which is based on a large amount of data demonstrating that particle size distributions of beach sediments approach log normalcy. The basic parameters describing the characteristics of the beach and borrow material are the mean particle size and the standard deviation. The work reported in reference [19] was developed further by Krumbein and James [20], resulting in a method of determining the amount of fill material required to be equally as effective as material composing the active beach profile. The ratio of total amount of fill required to the fill remaining, after sorting action by waves, is referred to as the "critical ratio" ($R\phi crit$), where:

$$R\phi crit = \left[\frac{S\phi_b}{S\phi_n} \right] \left[e^{-\frac{(M_n - M\phi_b)^2}{2(S\phi_n^2 - S\phi_b^2)}} \right]$$

$S\phi_b$ = standard deviation of borrow material, in phi units;

$S\phi_n$ = standard deviation of native material, in phi units;

$M\phi_b$ = phi mean of borrow material;

$M\phi_n$ = phi mean of native material.

This expression was tested by Vallianos [21] and found to give good results in connection with the behavior of a large artificial fill placed on the coast of North Carolina. The above procedure is applicable only to the evaluation of potential fill materials with particle size distributions having standard deviations exceeding that of the active beach profile.

c. *Dimensions and Configurations of Artificial Fills.* The dimensions and configurations of fills are determined on the basis of the intended purpose. If the purpose is to restore an eroded beach to protect backshore improvements from damage by major storms, the width may be determined as the protection width which has been lost during storms of record. Edelman [22] describes computational procedures which can be used to determine beach and dune retreat during storms if no actual records are available. Where the beach is to be used for recreation, the width of the fill may be governed by the area required to satisfy the anticipated recreational demands. The current U.S. Government practice is to provide a minimum area of 75 square feet (7 m^2) per beach user. If the purpose of the artificial fill is to act as a feeder beach to stabilize an existing or restored beach, the stockpile is made wide enough to provide for expected recession during the intervals between replenishment operations.

Dunes are commonly a feature of artificial fills placed in the United States. Their purpose being to protect the upland areas from the effects of major storm tides and wave action. When material is placed to form a composite section comprised of a dune fronted by a beach area, the dune part of the artificial fill vegetation planting is included to stabilize against erosion by wind. Figures 1 and 2 show a developed beach front in North

Fig. 1. — Carolina Beach, N.C. — Before fill placement.

Fig. 2. — Carolina Beach, N.C. — After fill placement.

Carolina prior to and after the hydraulic placement of an artificial dune and beach fill. It is also common practice in the United States to artificially induce the creation and stabilization of coastal dunes by use of sand fences and dune grasses. Large scale prototype experiments conducted by Savage and Woodhouse [23] demonstrate the effectiveness of this procedure to create and stabilize dunes behind existing beaches on barrier islands.

It is common practice to build the berm elevation of the artificial fill, to an elevation equal to or greater than that determined for he natural berm. If constructed to a lower elevation the berm will be subject to frequent wave overtopping. This will cause a ridge to form along the crest, and may cause ponding or temporary flooding of the backshore.

High quality fill materials will have a particle size distribution with larger standard deviations than the composite material comprising the existing active profile. Accordingly, on exposure to wave sorting action, the fill will attain a particle size distribution similar to the natural beach and material will reestablish the active profile configuration in a position seaward of the initial profile location. Therefore, if it is desired to restore a shoreline to a given position, enough fill must be placed to advance every point on the active profile by the same seaward distance.

Difficulties arise at the transition zones required to connect an artificial fill with adjacent shore if the transition is too abrupt. Inman and Bagnold [24] demonstrated with wave-refraction concepts that if the transition was too abrupt the ends of the fill zone would be subject to relatively high erosion. Groins can be employed to make the transition, but this is rarely done in the United States. The transition is accomplished by a gentle closure, or taper, of less than 5° with the normal shore orientation.

STRUCTURES TO CONTROL LITTORAL DRIFT.

General. Groins are the most widely used structures in the United States to control longshore movement of littoral drift. There are few cases in the United States where offshore structures have been constructed, and maintained for a number of years, for the primary purpose of beach stabilization. Experimental works have been installed in the offshore zone for this purpose, but few have been well documented concerning their behavior or effectiveness. The paragraphs that follow discuss the use of groins, offshore structures, and selected experimental works to control littoral drift and stabilize beaches.

Groins. There are two general types of groins - impermeable and permeable. The basic difference is that an impermeable groin will not allow passage of littoral drift through its structural components. Figure 3 shows an impermeable groin constructed of rubble or stone. Figure 4 shows a permeable groin constructed of precast concrete members. Although there are many permeable groin installations in the United States, the most common installation is the impermeable type. Berg and Watts [25] [26] show many examples of variations in design of these two types of groins in the United States. They report on 15 impermeable and 10 permeable groin design variations, and indicate these designs illustrate only approaches, for there are many more actual variations. There have been many articles published relating to groins. Balsillie and Bruno [27] recently compiled an annotated bibliography on groins which cites 463 references. Only a limited number of published documents, however, provide specific guidelines on the functional and structural design of groins. It is believed that groin design guidelines contained in the U.S. Army Coastal Engineering Research Center's Technical Report No. 4 [28] are widely utilized by coastal engineers in the United States for the design of impermeable groins. Neither this report nor any other known published document provide specific guidelines on the functional and structural design of permeable type groins. The primary purpose of permeability in a groin is to avoid the abrupt offset in shore alignment which normally occurs at impermeable

Fig. 3. — Typical rubble groins.

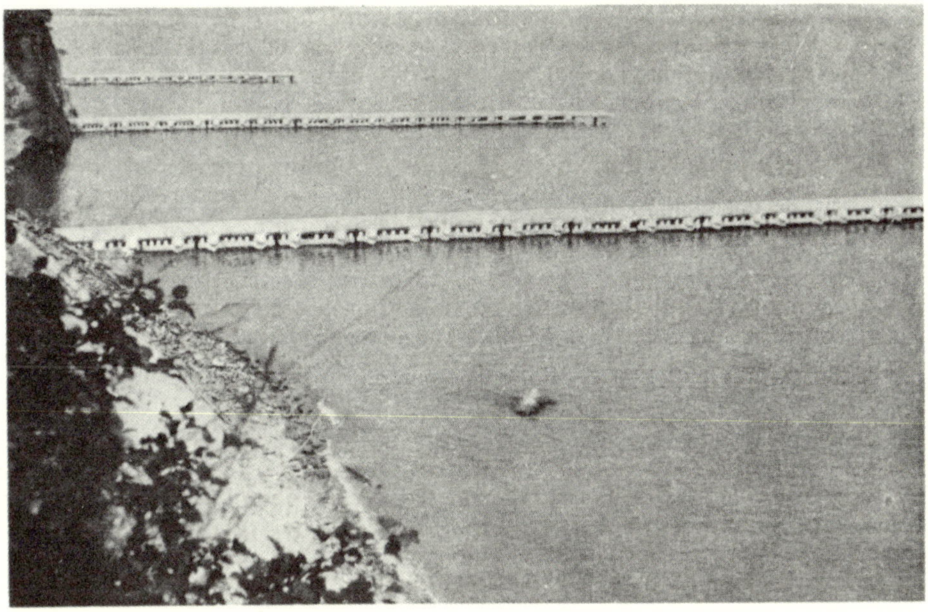

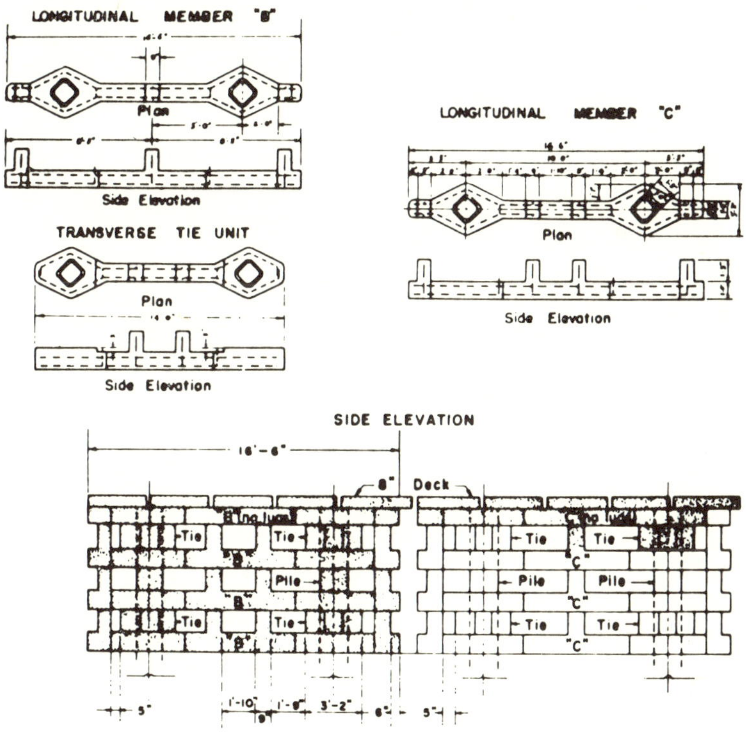

Fig. 4. — Typical permeable groin.

groins by permitting some of the littoral forces and materials to pass through the structure, inducing material deposition on both sides. The degree of permeability above the ground line affects the pattern and magnitude of material deposition within the limits of the groin's influence. Insufficient empirical data have been compiled to establish quantitative relationships between the littoral forces, groin permeability, and resulting shore behavior. Until such data are collected and used to develop criteria for functional design — similar to that available for impermeable groins — engineering evaluation and design will be less than exact when considering permeable groins.

Groins are used in the United States to control littoral drift in conjunction with or without a restoration beach fill. It is generally accepted by coastal engineers that a groin or groin system can cause erosion of downdrift shores if the system is not artificially filled with suitable sand concurrent with construction. Since the basic purpose of a groin or groin system is to increase the width of the beach zone, an economic analysis is generally made to develop comparative costs for the groin plan verses appropriate artificial fill and periodic nourishment to attain desired beach width without the groins. The estimated annual cost of the groin system, will be an indicator of how much the system will need reduce annual or periodic artificial nourishment of suitable sand to justify inclusion of groins in the plan of improvement. Employment of groins to impound longshore moving littoral drift has provided a broad spectrum of success and failure in the United States. This has motivated much research and development, and although progress is slow due to the complicated movement of littoral material at a groin structure, a significant amount of technical information is presently being developed. In this respect the U.S. Army Coastal Engineering Research Center has recently constructed an experimental groin facility at Pt. Mugu, California. Berg and Watts [26] describe this facility including the various groin configurations that will be tested.

Offshore Structures. The use of offshore structures to control littoral drift for beach stabilization in the United States is very limited. The principal reason is the high cost of offshore structures to achieve shore stabilization compared to the cost of alternate plans using beach fill and periodic nourishment, with or without groins. This type of structure, properly designed and positioned along any segment of coast, will be effective in impounding longshore movement of littoral drift and protect the shores within its influence, but unless appropriate plans for periodic sand bypassing are incorporated, downdrift shores will erode. The effect of a series of offshore breakwaters on a shoreline is shown in figure 5. This series was constructed in 1931-33, for shore protection

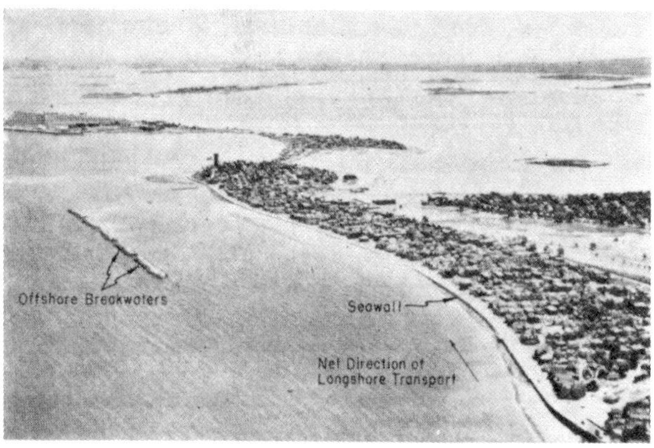

Fig. 5. — Offshore breakwater, Winthrop Beach, Mass.

purposes. Accretion at the shoreline in the lee of the breakwater is evident as well as some erosion to downdrift shore zone.

A submerged breakwater was constructed offshore in Lake Michigan near Chicago Illinois [29] and although this breakwater has been in place for over 30 years, the effect on shore stability within its influence is not known. A relatively small offshore rubble breakwater was constructed at Venice, California [30] in 1905 and eventually a tomobolo formed from the shore to the structure, which created a groin effect. This littoral drift barrier also induced erosion of the shores downdrift from the breakwater.

The relationship between wave energy transmitted to the shore and the rate of longshore transport of littoral drift has not been well established. It is therefore difficult to predict shore responses to an offshore structure which only partially dissipates the incident wave energy. Concentrated research and development work involving the relationship of wave energy and longshore material transport is presently underway in the United States. Das [31] recently presented all readily available data on this subject and illustrates the need to refine this relationship.

Experimentation relating to offshore works to control littoral drift includes some recent work on the concept of a " perched beach ". This involves construction of an offshore submerged sill or mound, and then artificially filling the zone with suitable sand between the sill and some point on or near the shore. Dunham [32] discusses this concept in regard to an offshore causeway across a portion of Santa Monica Bay, California. The concept includes a beach zone to serve recreation and shore protection, a land strip on which a highway system could be formed, and creation of a harbor embayment for small craft between the road system and the original shore. The concept is shown in figure 6. The " perched beach " part of the system has recently been studied in laboratory tests, and preliminary analysis indicate the concept and approach show promise.

Another example of experimentation of offshore structures to dissipate wave energy for purposes of shore stabilization is the installation of artificial seaweed. Wicker [33] and Brashears and Dartnell [34] describe tests carried out in New Jersey and at Wallops Island, Virginia in which a matrix of synthetic materials to simulate natural seaweed were placed to dissipate wave energy and induce deposition of littoral drift. Results of these tests revealed anchoring problems for the clusters of synthetic material, and benefits to shore stabilization were inconclusive.

As mentioned, there are many undocumented experiments involving placement of objects or materials offshore to reduce wave energy reaching the shore and effecting some control of longshore movement of littoral drift. This is indicative of the need to improve on the techniques of controlling littoral drift with offshore structures including the establishment of practical alternatives for use in developing the most economical and effecive plan of improvement.

CONTROL OF LITTORAL DRIFT AT HARBOR AND ESTUARY ENTRANCES.

The control of littoral drift at harbor and estuary entrances in the United States is predominantly in the interest of navigation. There are instances of coastal inlet creation and stabilization of an existing inlet for other purposes such as water exchange to reduce stagnation, creation

(Dunham, 1968)

Fig. 6. — Proposed Santa Monica causeway and cross–section of perched beach.

of a pass or waterway for fish and marine life, and reduction or control of flooding due to surges created by hurricanes or other severe storms. However, in nearly all cases where inlet improvement is for purposes other than navigation, the interests of navigation are recognized and incorporated into the plan of improvement. The reason for this is the rapid increase in recreational and commercial vessels in the United States and the desire to have additional suitable waterway systems connecting the ocean, gulf, or lake with inland bodies of water and harbor complexes.

As in many other countries there is also a wide variation of harbor and estuary entrance conditions and resultant littoral drift shoaling patterns in the United States. There are harbors along the ocean coast, and particularly in the Great Lakes, wherein tidal flow through the entrance is so slight that, if appropriate corrective measures are not taken, littoral drift shoaling will be concentrated at one location in the entrance channel. At harbor entrances where tidal flow is sufficient to move littoral materials transported into the inlet complex, the littoral drift shoaling is concentrated in the classic patterns of an ocean bar created by tidal ebb flow and a middleground shoal associated with the flood flow of the tide. Average annual fresh water discharge at some tidal harbor and estuary entrances is sufficient to negate the deposition of littoral drift to create a middleground shoal, and only an outer or ocean bar shoal pattern prevails.

Wave induced currents are the dominant factor in longshore littoral drift movement to harbor entrances. Tidal currents are generally only important in littoral material transport within the influence of the harbor entrance. The degree of importance is related to many factors such as entrance geometry, range of tide, and manmade entrance works. The rate

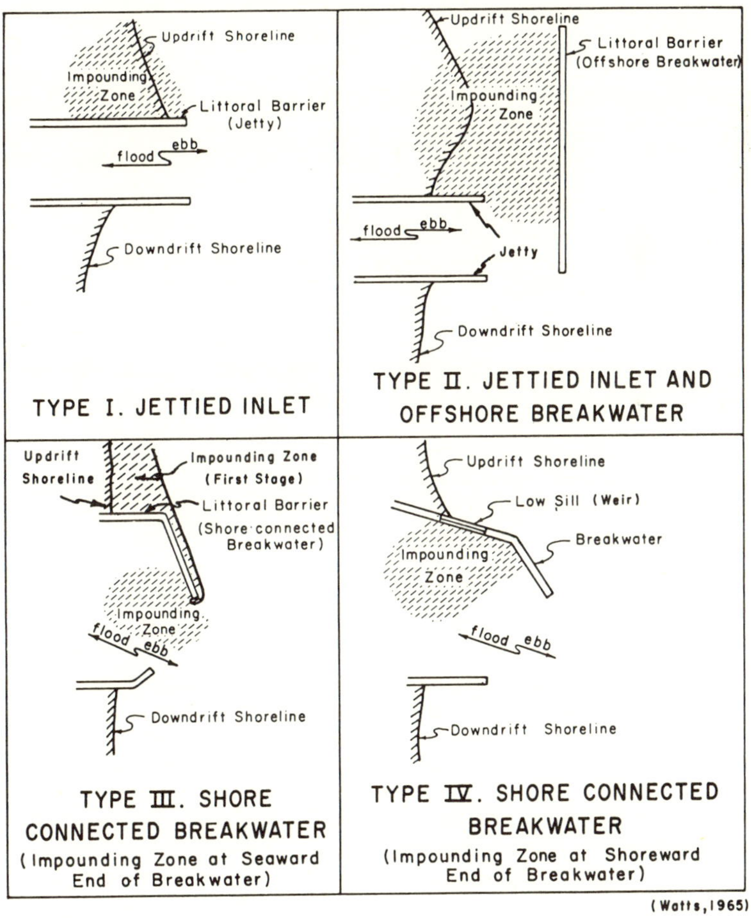

FIGURE 7. TYPES OF LITTORAL BARRIES WHERE SAND TRANSFER SYSTEMS HAVE BEEN EMPLOYED

that littoral drift is transported to a harbor entrance is extremely important in planning stabilization of the entrance. Waves vary in energy and direction of approach to the shore over a specified period of time therefore the longshore transport rate in either direction along the coast varies accordingly. The rate of longshore movement most commonly used in the United States is based on an average year, and the rate is delineated as net or gross. The net rate is the quantity of sand passing a particular point in only one direction during the course of an average year. The gross rate is the total quantity moving past a particular point in both directions during the average year. Coastal engineers have long recognized the importance of both the net and gross longshore transport rates in the design of harbor entrance stabilization works and channels to serve navigation. If the entrance complex is considered to be a complete littoral barrier, the littoral drift transported to it should be mechanically transferred to the appropriate side of the entrance. In the United States this transfer operation of the impounded littoral drift is called sand bypassing or sand transfer. Several alternative systems are presently employed to accomplish sand transfer.

The types of entrance structures or littoral barriers which have been generally employed in the United States in connection with inlet and harbor improvements are shown in figure 7. Where littoral drift movement predominates in one direction, any of these types would cause accretion to the updrift shore and erosion of the downdrift shore, unless provision is made for sand bypassing.

At a jettied inlet shown as Type I in figure 7, the sand bypassing operation can be performed by a land-based dredging plant or land vehicles. A floating plant can be used only when the impounding zone is subject to periods of light wave action, or by creating an access channel into the landward part of the impoundment and dredging behind the beach berm thus leaving a protective barrier for the dredge. This latter type operation was performed at Port Hueneme, California in 1953 and described by Savage [35]. The degree to which littoral drift passes around the updrift jetty is directly related to the size and maintenance of the sand reservoir created on the updrift side. Angas [36] describes a sand transfer operation at Shark River Inlet, New Jersey which is a jettied inlet. Trucks were used to transfer 250,000 cubic yards from the updrift to the downdrift side of the inlet. The excavated area on the updrift side provided a reservoir which was effective in impounding littoral drift for a short period of time, but as soon as the reservoir filled, material moved around the updrift jetty into the entrance channel. This operation is illustrated in figure 8. Fixed bypassing plants have been employed at jettied entrances, but have been less than 50 % effective; on an annual basis at least one-half of the littoral drift moves around the updrift jetty into the entrance channel. The fixed bypassing plant is essentially a stationary hydraulic suction dredge located near the seaward end of the updrift jetty. The swing of the suction intake creates a crescent-shaped reservoir of limited capacity. When littoral drift movement is great, the filling rate into the reservoir exceeds the plant capacity; when drift movement is low the plant may be idle due to the lack of sand accumulation in the reservoir. Fixed bypassing plants are located at Lake Worth Inlet, Florida

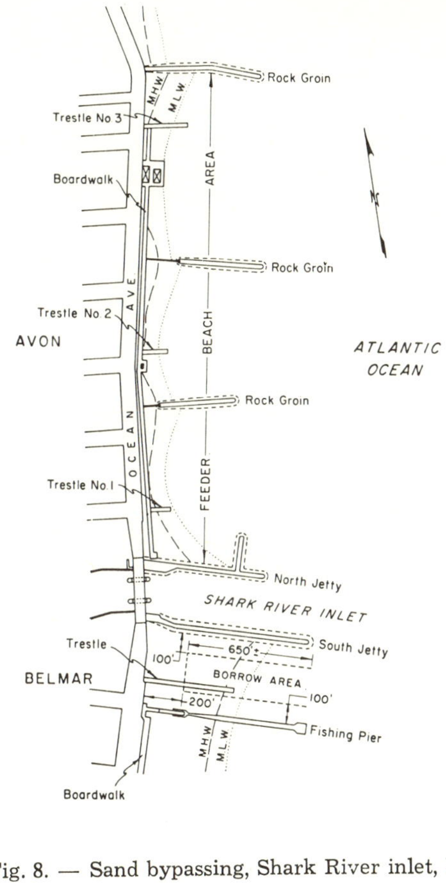

Fig. 8. — Sand bypassing, Shark River inlet, N.J.

and at South Lake Worth, Florida, and are described by Middleton [37], and Watts [38] respectively. The installation at Lake Worth Inlet is illustrated in figure 9.

Harbor entrance structures shown as Type II in figure 7 consist of a jettied inlet with an offshore breakwater. The offshore breakwater functions to protect the channel entrance area of the jetties for navigation purposes, and it impounds all the littoral drift moving longshore toward the harbor within the zone between the breakwater and the shore. If the sand trap or reservoir area is maintained, no entrance channel shoaling from that side of the harbor will result. A Cutter-Suction dredge is normally employed to transfer the impounded drift to downdrift shores. The protection provided by the breakwater permits the dredge to operate in all but the most severe wave conditions. This jettied inlet and offshore breakwater arrangement is considered the most effective type of improvement for both navigation and sand bypassing. However, the initial or first cost of this system is high due to the cost of the offshore breakwater. Although this type of harbor entrance works is nearing completion at Ventura Harbor, California, the only completed installation is at Channel Islands Harbor, California, shown in figure 10. Herron [39] has provided details of this harbor complex. There are a number of interesting design features of this type of entrance works. For example, the offshore breakwater must be positioned so it will provide the best protection for vessels entering and leaving the jettied entrance area; it must be positioned seaward of the active littoral zone, but function to create a complete barrier to longshore littoral drift movement; and it must establish a sand trap area in its lee. The characteristics of this trap area should permit low cost periodic dredging and placement of the dredged material on the feeder beach down-drift of the harbor entrance.

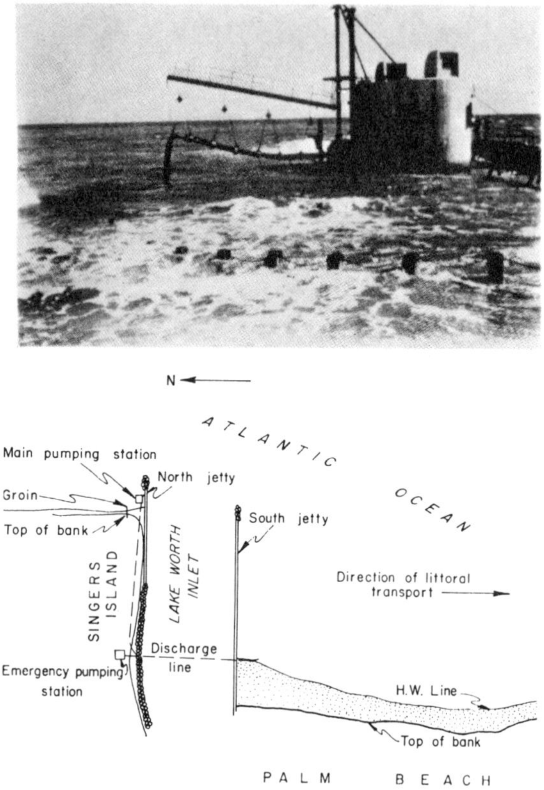

Fig. 9. — Fixed bypassing plant, lake Worth Inlet, Florida.

A shore-connected breakwater with impoundment at its seaward end, shown as Type III in figure 7, has been used effectively. Bypassing is performed by a floating dredge plant, although high waves could cause delays when the outer part of the impoundment is being dredged. Some shoaling of the navigation channel is likely between dredging operations. Although some experience has been obtained for this type of harbor

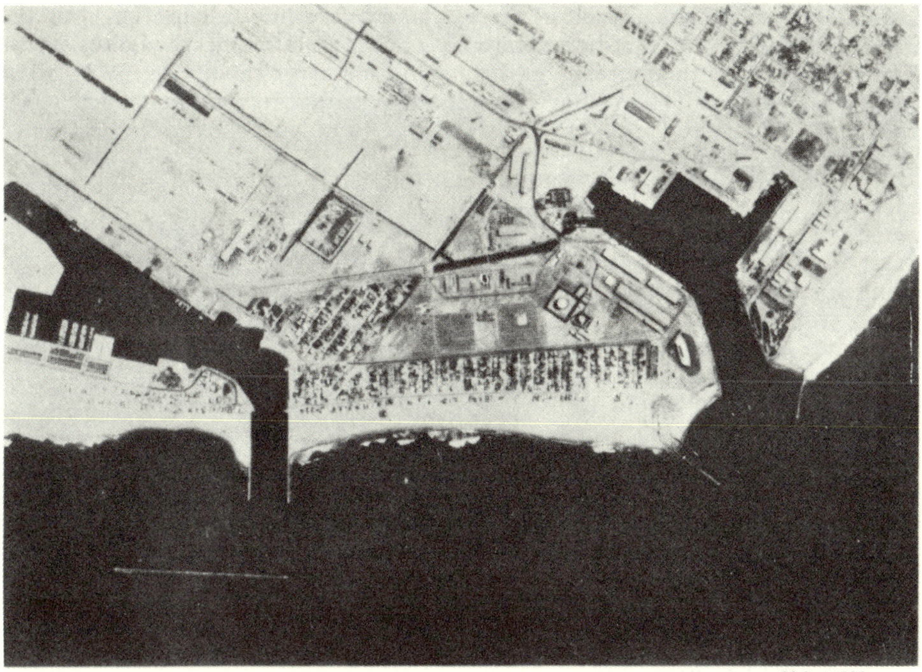

20 Sept 1965

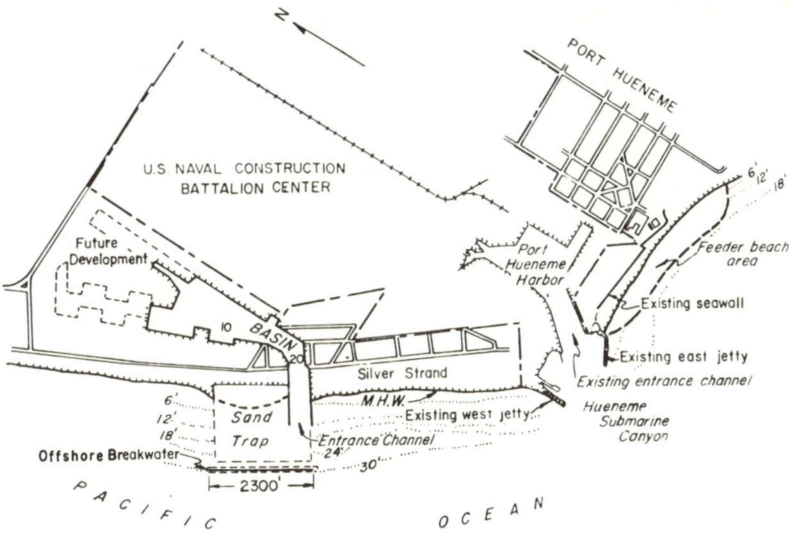

Fig. 10. — Sand bypassing, channel islands harbor, Calif.

entrance structure at Fire Island Inlet, New York, and Oceanside Harbor, California, the most documented case in the United States is the entrance structures at Santa Barbara Harbor, California. Wiegel [40] provides details of this harbor complex and the bypassing operation. This harbor layout and the bypassing arrangement are shown in figure 11. This type of harbor entrance bypassing of littoral drift from the impounding zone to the downdrift shores can be accomplished on a periodic or continuous basis. At Santa Barbara, experience has shown that by dredging on the harbor side of the impounding zone and leaving part of the shoaled material to provide protection for the dredge, continuous operation is not possible. Even with the protective shoal, wave and weather conditions limit actual dredging operations to about 70 % of the time as referred to a normal 8-hour working day. Periodic dredging of the entire impounded littoral drift must be coordinated with the more favorable periods of waves and weather during the year.

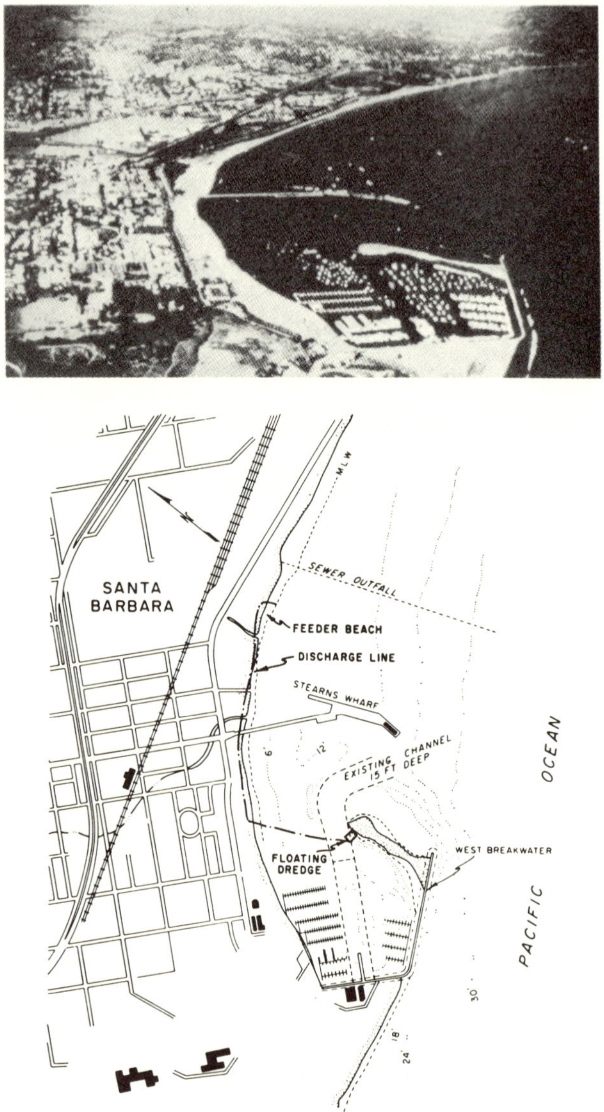

Fig. 11. — Sand bypassing, Santa Barbara, Calif.

A modification of the Santa Barbara Harbor entrance structure is shown as Type IV in figure 7. The updrift structure is a shore-connected breakwater, but a low-crest weir or sill is positioned at the shoreward end of the breakwater. The low-crest weir allows littoral drift to pass over that section and deposit in a deposition basin, in the lee of the outer segment of the breakwater system. The impounded sand is transferred to the downdrift shore by a conventional pipeline dredge. This general concept was first tested in 1955, at Hillsboro Inlet, Florida and described by Hodges [41]. The operation has continued since that date. At this location a natural reef formation exists with a top elevation of about mean sea level and oriented so that the longshore moving littoral drift is transported over the crest of the reef by waves and tides and deposited in the inlet area just in the lee of the reef. A small hydraulic dredge excavates the shoaled sand and pumps it to the downdrift shore. The natural features at Hillsboro Inlet were utilized for guidance in the design and construction of a similar system at Masonboro Inlet, North Carolina. This was the first artificial system of this kind constructed in the United States. Details of the project are provided by Magnuson [42].

The general plan of he project at Masonboro Inlet is shown in figure 12. The updrift shore-connected weir and breakwater structure is frequently referred to as " weir-jetty ". In this project, the 1,000 foot-long weir section

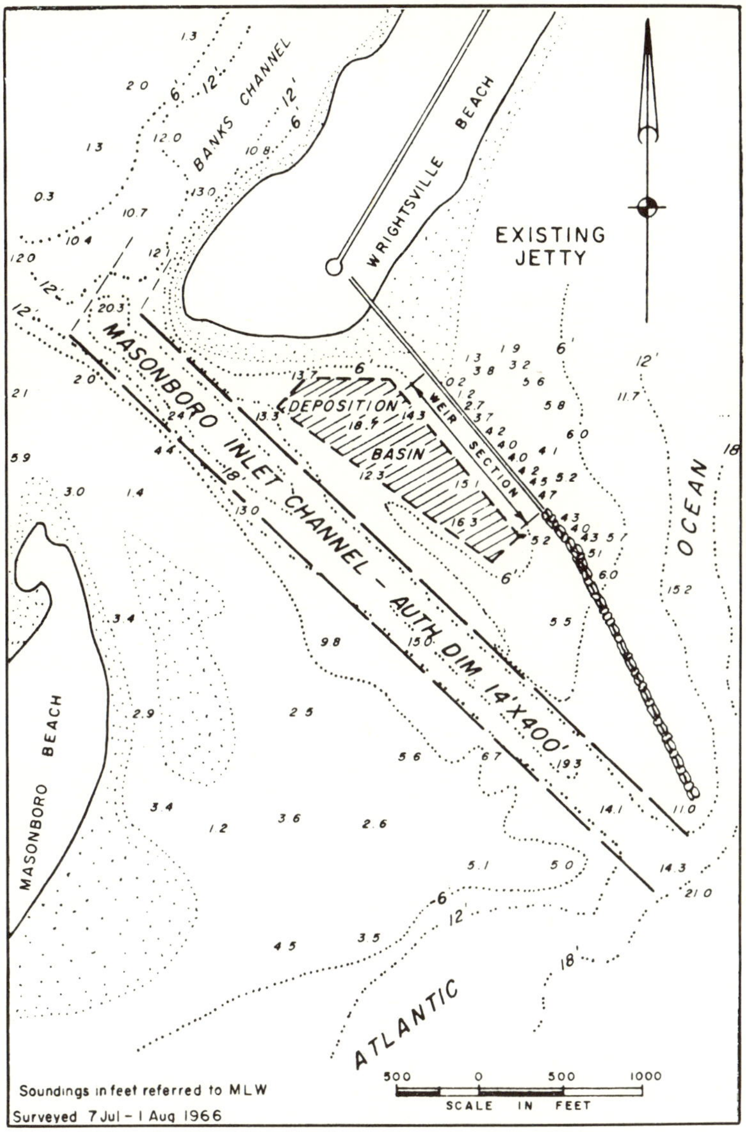

FIGURE 12. WEIR JETTY, MASONBORO INLET, N.C.

was constructed of precast, prestressed concrete sheet piles. The crest of the weir was positioned at mean sea level. The mean tide range is about 4 feet. The outer arm of the breakwater is a typical rubble or rock structure 1,900 feet long with a top elevation 5 feet above mean low water. This outer arm dampens incident wave energy from a large directional arc, and provides protection to a pipeline dredge plant when it is positioned in the deposition basin to dredge the basin and transfer the material to adjacent shores. The weir section provides some protection to the dredge in the basin, and it ensures that a channel will not form between the shoreline and the outer rubble breakwater. The capacity of the deposition basin is about 380,000 cubic yards. Redredging of the basin is required about every 2 years. The Masonboro Inlet weir-jetty system has been in operation since July 1966. The advantages of this system relative to the Type III entrance structures have yet to be completely proven, but due to the favorable behavior of the Masonboro Inlet project, three other projects employing this same approach have been constructed. These are located at East Pass, Florida, Perdido Pass, Alabama, and Ponce de Leon Inlet, Florida. Performance and behavior of these projects are presently being studied. Data developed from these studies will provide valuable guidance on refinement of the functional and structural design criteria for this Type IV entrance to harbors.

At unimproved inlets, including entrances to large estuaries, along the coastline of the United States control of littoral drift is varied. There are many unimproved tidal inlets, particularly along the middle and south Atlantic and the Gulf of Mexico, where the predicted benefits that would result from constructing of appropriate inlet stabilization works to control the littoral drift would not equal costs involved. Techniques of controlling littoral drift without incorporating stabilizing structures are under investigation. Watts [43] describes an experiment presently in progress at Carolina Beach Inlet, North Carolina. A deposition basin was initially dredged in the throat of that inlet, and is periodically redredged after it is filled by littoral drift transported into the basin by flood tide currents. The dredged materials are pumped to downdrift shores. The natural channel system in the ocean bar complex is monitored to determine changes that can be attributed to the denial of a part of the littoral drift transported to the ocean bar during ebb tide flow out of the inlet.

Littoral drift transported into the entrance channels of large estuaries such as the entrance to Raritan Bay (New York Harbor), Delaware Bay, Chesapeake Bay, Tampa Bay, and San Francisco Bay, has been, and is being generally handled by conventional type large hopper dredges. The apparent reason for employing only dredging at these estuary entrances is the length and associated costs that would be required for structures to control the littoral drift. In light of present evaluations being made to provide entrance channels and terminals to accommodate vessels of 60 to 90 feet of draft at these estuary entrances as well as other large inlets it is apparent the use of entrance structures to control littoral drift is being given greater attention.

It is beyond the scope of this report for detail coverage but the importance of control of littoral drift at the entrance to tidal inlets as a function of sedimentation in the interior reaches of an estuary should be mentioned. Simmons [44] and others have reported on sedimentation and channel shoaling associated with the position change of the interface of the salt water wedge with fresh water discharge in a tidal estuary. The control of littoral drift at the entrance to the estuary can be influential on the volume of tidal exchange in the estuary, thus, it can be very important to the position where the saline and fresh water meet in the estuary with resultant floculation and precipitation of suspended sediment to the bottom. The problems associated with this migratory shoaling pattern in the upper reaches of the estuary attributable to littoral drift entering the estuary entrance can be of critical importance in some instances. Considerable progress has been made in the United States on hydraulic and mathematical modeling of salinity intrusion in estuaries, but, the fact that much remains to be understood about this phenomenon cannot be overlooked.

If the littoral drift movement at a tidal inlet and the art of modeling this type inlet were more fully understood, many of the problems of controlling littoral drift to serve multipurposes could be solved. Research is presently being conducted in the United States on these factors. Until these two factors are more fully understood, it is believed progress toward improving techniques to control littoral drift at tidal inlets will be slow.

SUMMARY AND CONCLUSION.

This paper has presented a summary of techniques utilized in the United States for controlling littoral drift to protect or stabilize beaches and dunes and to protect entrances to harbors and estuaries. It describes present practices followed in the design of artificial fills and periodic nourishment operations, including environmental factors that must be considered in the borrow of suitable material for beach fills. It points out that although groins are widely used, a better understanding of the functional and structural design of these structures is needed.

Only general type conclusions can be developed from material presented in the paper. It can be concluded that population growth in the coastal sectors has caused this zone to greatly increase in importance and use, and, the impact of this is transmitted to almost all facets of coastal engineering. It has focused attention on the necessity of evaluating the environmental impact with each alternative plan to control littoral drift for whatever purpose that control is intended to serve. It can be concluded that, although progress is being made toward development of improved guidelines for use in the design of coastal works related to the control of littoral drift, it is apparent much research and development work still needs to be done.

ACKNOWLEDGEMENTS.

This paper was prepared under a portion of programs conducted by the Coastal Engineering Research Center and other U.S. Army Corps of Engineer offices. Permission of the Chief of Engineers for publication is appreciated.

REFERENCES.

[1] Department of the Army, Corps of Engineers, " Report on the National Shoreline Study ", Washington, D.C., August, 1971.

[2] MARMER, H.A. : *Changes in Sea Level Determined from Tide Observations.* " Proceedings of the 2nd Conference on Coastal Engineering, Council on Wave Research ", pp. 62-67, 1951.

[3] BRUUN, Per : *Sea-Level Rise as a Cause of Shore Erosion.* " Journal of the Waterways and Harbors Division, Proceedings of the American Society of Civil Engineers ", Vol. 88, No. WWI, February, 1962.

[4] CALDWELL, J.M.: *Coastal Processes and Beach Erosion.* "Journal of the Boston Society of Civil Engineers", Vol. 53, No. 2, April, 1966, pp. 142-157.

[5] WATTS, G.M.: *Sediment Discharge to the Coast as Related to Shore Processes.* Federal Interagency Sedimentation Conference, Jackson, Mississippi, 1963.

[6] SAVILLE, T., Jr.; WATTS, G.M.: *Coastal Regime, Recent U.S. Experience* "Papers 22nd International Navigation Congress, PIANC", Paris, 1969, Section II, Subject 4, pp. 249-269.

[7] MUNK, W.H. and TAYLOR, M.A.: *Refraction of Ocean Waves: A Process Linking Underwater Topography to Beach Erosion.* "The Journal of Geology", Vol. IV, No. 1, January, 1947.

[8] LILLEVANG, O.J.: *Groins and Effects - Minimizing Liabilities.* "Proceedings of the Coastal Engineering Santa Barbara Specialty Conference", October, 1965, pp. 749-754, American Society of Civil Engineers, 345, 47th Street, New York, N.Y., 10017.

[9] HALL, J.V., Jr.: *Artificially Nourished and Constructed Beaches.* "Proceedings 3rd Conference on Coastal Engineering", Cambridge, Massachusetts, October, 1952, Council on Wave Research, pp. 119-136.

[10] EAGLESON, P.S.: *Theoretical Study of Longshore Currents on a Plane Beach.* "Department of Civil Engineering Hydrodynamics Laboratory Report", No. 82, 1965, Massachussets Institute of Technology.

[11] BIJKER, E.W.: *Littoral Drift as a Function of Waves and Current.* "Proceedings of 11th Conference on Coastal Engineering", London, England, September, 1968, American Society of Civil Engineers, 345 East 47th Street, New York, N.Y., 10017.

[12] EINSTEIN, H.A.: *The Bed-Load Function for Sediment Transportation in Open Channel Flow.* "U.S. Department of Agriculture, Technical Bulletin", No. 1026, September, 1950.

[13] NODA, E.K.: *State-of-the-Art of Littoral Drift Measurements.* "Shore and Beach", Vol. 39, April, 1971, No. 1; "Journal of the American Shore and Beach Preservation Association", pp. 41.

[14] DUANE, D.B.: *Synoptic Observations of Sand Movement.* "Proceedings of the 12th Coastal Engineering Conference", September, 1970, Washington, D.C., American Society of Civil Engineers, 345 East 47th Street, New York, 10017, pp. 799-813.

[15] CRONIN, E.L., et al.: *Effects of Engineering Activities on Coastal Ecology.* Report by the Office of the Chief of Engineers, Corps of Engineers, U.S. Army, August, 1971.

[16] DUANE, D.B.: *Sand Inventory Program - A Study of New Jersey and Northern New England Coastal Waters.* "Shore and Beach", October, 1969; "Journal of the American Shore and Beach Preservation Association".

[17] FISHER, C.H.: *Mining the Ocean for Beach Sand.* "Proceedings of Ocean Engineering Conference", Miami Florida, Dec., 1969, American Society of Civil Engineering.

[18] MAURIELLO, L.J.: *Rehabilitation of Beaches with the Hopper Dredge.* "Shore and Beach", Vol. 34, No. 2, 1966.

[19] KRUMBEIN, W.C.: *A Method for Specification of Sand for Beach Fills.* Technical Memorandum No. 102 BEB, U.S. Army Coastal Engineering Research Center, Washington, D.C., October, 1957.

[20] KRUMBEIN, W.C. and JAMES, W.R.: *A Lognormal Size Distribution Model for Estimating Stability of Beach Fill Material.* Technical Memorandum No. 16, U.S. Army Coastal Engineering Research Center, November, 1965.

[21] VALLIANOS, L.: *Recent History of Erosion at Carolina Beach, N.C.* "Proceedings of the 12th Coastal Engineering Conference", September, 1970, Washington, D.C., American Society of Civil Engineers, 345 East 47th Street, N.Y., N.Y., 10017, pp. 1223-1242.

[22] EDELMAN, T.: *Dune Erosion During Storm Conditions.* "Proceedings of the 11th Conference on Coastal Engineering", London, England, September, 1968, Vol. 1, pp. 719-722, American Society of Civil Engineers.

[23] SAVAGE, R.P. and WOODHOUSE, W.W.: *Creation and Stabilization of Coastal Barrier Dunes.* 11th Conference on Coastal Engineering, London, England, September, 1968, Vol. 1, pp. 671-685.

[24] INMAN, D.L. and BAGNOLD, R.A.: *Littoral Processes.* "The Sea", Vol. 3, pp. 529-553, General Editor: M.M. Hill, Interscience Publishers, 1962.

[25] BERG, D.W. and WATTS, G.M.: *Variation in Groin Design.* "Journal of the Waterways and Harbors Division, ASCE", Vol. 93, No. WW2, Proc. Paper 5241, May 1967, pp. 179-100.

[26] BERG, D.W. and WATTS, G.M.: *Groins and Groin Systems: A review of Research.* "Shore and Beach", Journal of the American Shore and Beach Preservation Assoc., Vol. 39, No. 2, October, 1971.

[27] BALSILLIE, J.H. and BRUNO, R.O.: *Groins; An Annotated Bibliography.* "Misc. Paper", 1-72, 1972, U.S. Army Coastal Engineering Research Center.

[28] U.S. Army Coastal Engineering Research Center: *Shore Protection, Planning and Design.* TR No. 4, 3rd Edition, 1966, U.S. Government Printing Office, Wash., D.C.

[29] U.S. Army Corps of Engineers (1953), Illinois: *Shore of Lake Michigan, Beach Erosion Control Study.* House Document No. 28, Public Law 504, 83rd Congress, 1st Session.

[30] U.S. Army Corps of Engineers (1954) Appendix II, Coast of California, Pt. Mugu to San Pedro Breakwater, Beach Erosion Control Study. House Document No. 277, Public Law 153, 83rd Congress, 2nd Session.

[31] DAS, M.M.: *Longshore Sediment Transport Rates: A Compilation of Data.* U.S. Army Coastal Engineering Research Center, Wash., D.C. 20016, Misc. Paper 1-71, September, 1971.

[32] DUNHAM, James W.: *Proposed Santa Monica Causeway Project.* "Journal of the Waterways and Harbors Division, ASCE", Vol. 94, No. WW4, Proc. Paper 6219, November, 1968, pp. 425-436.

[33] WICKER, Clarence F.: *Report on Artificial Seaweed.* "Shore and Beach", No. 2, published by the American Shore and Beach Preservation Association, Vol. 34, October, 1966.

[34] BRASHEARS, Robert L. and DARTNELL, John S.: *Development of the Artificial Seaweed Concept.* "Shore and Beach", No. 2, published by the American Shore and Beach Preservation Assoc., Vol. 35, October, 1967.

[35] SAVAGE, R.P.: *Sand By-Passing at Port Hueneme, California.* Beach Erosion Board TM No. 92, March, 1957, Coastal Engineering Research Center, Wash., D.C.

[36] ANGAS, W.M.: *Sand By-Passing Project for Shark River Inlet.* "Proceedings, American Society of Civil Engineers", Vol. 86, No. WW3, September, 1960.

[37] MIDDLETON, S.R.: *Installation and Operation of Fixed By-Passing Plant at Lake Worth Inlet, Florida.* "Shore and Beach", Vol. 27, No. 1, June, 1959.

[38] WATTS, George M.: *A Study of Sand Movement at South Lake Worth Inlet, Florida.* Beach Erosion Board TM No. 42, October, 1953, Coastal Engineering Research Center, Wash., D.C.

[39] HERRON, W.J.: *Beach Erosion Control and Small Craft Harbor Development at Point Hueneme.* "Shore and Beach", Vol. 28, No. 2, 1960.

[40] WIEGEL, R.L.: *Sand By-Passing at Santa Barbara, Calif.* "Proc. American Society of Civil Engineers", Vol. 85, No. WW2, June, 1959.

[41] HODGES, T.K.: *Sand By-Passing at Hillsboro Inlet, Florida.* "Beach Erosion Board Bulletin", Vol. 9, No. 2, April, 1955, Coastal Engineering Research Center, Washington, D.C.

[42] MAGNUSON, Nels C. : *Planning and Design of a Low-Weir Section Jetty at Mansonboro Inlet, North Carolina.* " Proceedings of Speciality Conference on Coastal Engineering ", Santa Barbara, California, American Society of Civil Engineers, 1965, Chap. 36, pp. 807-820.

[43] WATTS, George M. : *A Study of Ocean Bar Response to Dredging in the Throat of an Inlet.* " Proceedings ", IV, World Dredging Conference, December, 1971.

[44] SIMMONS, Henry B. : *Tidal and Salinity Model Practices* and *Field Experience in Estuaries.* In " Estuary and Coastal Hydrodynamics ", Mass. Institute of Technology, Cambridge, Mass., 1963.

RESUME.

Le rapport ci-joint donne un aperçu des techniques et des méthodes utilisées aux Etats-Unis pour le contrôle du cheminement du sable littoral afin de réaliser la stabilisation des plages et des dunes et pour protéger les accès aux ports et aux estuaires.

Une analyse détaillée est présentée concernant l'application des remblais artificiels pour remettre une plage aux dimensions désirées et pour la mise en place périodique du sable adéquat pour la nourrir à ces dimensions. Les méthodes utilisées pour déterminer la dimension des remblais et pour l'évaluation des quantités requises pour l'entretien, sont comprises dans cette discussion. On passe en revue quelques facteurs qui influencent sensiblement les projets, la conception, la construction et l'entretien des remblais artificiels des plages. Ces facteurs comprennent l'accroissement de population dans les régions côtières et associent l'augmentation de l'importance des régions côtières, les besoins pour l'évaluation des effets qu'aura sur le milieu le projet d'ensemble des remblais des plages et les contraintes d'utilisation des matériaux du fond de la baie et des régions de lagunes, lesquels requièrent de dépendre plus largement des dépôts de sable du large et de développer des techniques plus efficaces pour le transport de ces dépôts de sable vers la zone côtière.

On a également passé en revue le rôle des épis et des ouvrages construits au large pour contrôler le cheminement littoral longitudinal. On a souligné que les épis sont largement utilisés, mais les nombreux cas où les épis n'ont pas atteint les objectifs désirés reflètent la nécessité de recherches scientifiques supplémentaires pour mieux comprendre le comportement structurel et fonctionnel des épis.

Les ouvrages d'accès utilisés aux Etats-Unis pour contrôler le cheminement littoral dans les ports sont examinés. Cet examen comprend la description des techniques et des dispositifs utilisés pour chaque type d'ouvrage d'entrée pour transporter le sable littoral retenu vers les plages contiguës. Le dispositif le plus effectif pour contrôler le cheminement littoral de sable transporté à l'entrée d'un port comporte un brise-lame au large du côté de l'apport de sable en amont du port.

17

Reprinted from *Shore and Beach* **27**(2):8–13 (1959), journal of the American Shore and Beach Preservation Association

SOME EXAMPLES OF LARGE SCALE SHORE PROTECTION PROJECTS

by

R. O. EATON

Chief Technical Advisor
The Beach Erosion Board

Thank you, Mr. Thompson. General Holle, friends of The American Shore and Beach Preservation Association and as well of the newly formed Florida Shore and Beach Preservation Association, it is always a very great pleasure to me to be invited to address this group. It is an opportunity to present thoughts or ideas which seem to me to warrant your consideration.

I have chosen for my theme today a topic which I hope may aid in furtherance of the basic objective of this American Shore and Beach Preservation Association, which was to promote the mutual effort and mutual understanding and mutual attack upon problems of shore protection from governmental agencies at all levels and on down to the individual; to gather people together so that we could understand and attack their problems mutually.

I wish that I could qualify for the capabilities that Mr. Thompson credited me with; that is, being acquainted with all the grains of sand on our shores and how they behave. Only God has that information and he hasn't imparted it to me. At this stage, however, we are getting better acquainted with the behavior of these little sand grains and better equipped to cope with them and properly put them in their place, not necessarily to keep them there, but to see that they keep coming when needed.

I have chosen today as the locality that I am going to use as an illustration an area a long way from Florida. This was done with malice aforethought. I thought perhaps if I talked about a place not too many of you would be familiar with you wouldn't know just how accurate the things I may say are: but also to lend credence or faith in what I have to say, I have chosen to illustrate my talk with a number of photographs that have not been retouched. You will thus be able to see evidence of the work which has been done; and I hope I will be able to describe the reasons why and how the local groups have gotten together on the problems.

It may be sacrilegious to talk about California in Florida, although I think there are so many people visiting both states today that the competitive attitude from a recreational standpoint is perhaps becoming a thing of the past. In any event, the area shown on this map is 3,000 miles away, so perhaps you haven't all seen it.

The segment of shore you see on the map is approximately 20 miles in length, the Santa Monica mountains to the north and the Palo Verdes Headlands on the south end. The center of the area is 12 miles west of the City of Los Angeles, so it is an area of rather dense population. In fact, along this twenty-mile stretch it has been estimated that one and a half million people have visited the beaches in a single day. Some of our Beach Erosion Board members who have toured this shore on Sunday in the summertime I am sure will not question those figures.

The actual beginning of works that might be said to have affected the shore processes in this area started before the turn of the century. Henry Huntington, when he was bringing his railroad across the country, was bound to bring it directly to the sea. After reaching the then modest village of Los Angeles, he kept proceeding westward and reached the Pacific Ocean at the northern end of what is now Santa Monica, at about this location. The impetus developed in building the railroad was so great he couldn't stop it immediately and it projected about three thousand feet into the ocean before they could shut it off.

The ocean extension was built on a pier. They had been crossing all types of terrain and had done it successfully, but they had not encountered the sea before and they had a lot of trouble with piling. So in the customary railroad manner they dumped a lot of stone in there and pretty soon had a stone jetty about one thousand feet long to protect the pier piling in surf area. The immediate effect was a progressive buildup on one side and erosion on the other side.

No one worried much about the beach in those days so no measurements were made of the behavior of the shore. Mr. Huntington at the turn of the century lost his battle to establish this as the site of Los Angeles Harbor and removed the structure. So the lessons which might have been learned were lost to science except in the memory of a few of the people who were around at that time.

A good many years later—in 1933—a second structure, which was to have an important effect on the shore, was built in the Santa Monica region. By the way, there are approximately nine political boundaries in this shore area. The northernmost section here and six miles in the center are in the City of Los Angeles. The rest of the frontage is divided between

Santa Monica in 1933 shortly after initiation of breakwater construction. Note beach width in vicinity of the pier. (Photo by Fairchild Aerial Surveys)

the five smaller municipalities and two segments which are not incorporated are under the jurisdiction of Los Angeles County. This makes seven local political groups that each had something to say as to what went on on these shores, plus the State and Federal Governments in their respective interests.

Santa Monica's citizens in 1933 decided they should have a harbor and to create it by building a breakwater two thousand feet long and two thousand feet from the shore. They wanted a place to moor small craft to add to the recreational attraction of Santa Monica. They gave little thought as to how the shore might be affected and no erosion problems were anticipated. This is the way this segment of shore looked in 1921. This point at the bottom of the slide is the Santa Monica pier.

This picture of the same area was taken in 1933 and it shows the beginning of construction of the Santa Monica breakwater. The first rock is poking its head out of the water. The core had been completed and the first of the cap stone is showing up. About eight months later the breakwater was completed. I show you this to call to your attention the alignment of the shore at that particular time.

This is the same area in 1935, two year later, with the breakwater finished; and you can see the beginning of a substantial accretion in the lee of the breakwater. There had been a minor bulge in the shore due to the groin effect of the pier, but I think the accretion that is now starting is quite conspicuous as a new development. Note, however, that to the north the beach is still relatively narrow beyond the immediate protection of the breakwater.

This is a view in 1938 looking in the other direction. I think this illustrates quite conspicuously the erosion zone on the down drift side. Also, you can see some evidence of the width of beach that is being created backing up to the north from the accumulation back of the breakwater. I wish the picture were clear enough for you to see it better but in this region, which is in the center of the Los Angeles City segment, the beach has practically receded to the point where the waves are reaching the bluffs. The segment of the shore you can see in this picture is approximately twelve miles long.

This is a view in 1949, once again of the accretion against the Santa Monica breakwater. The accretion has by this time extended a mile and a half to the

Santa Monica in 1935, less than a year after completion of the breakwater. Accretion of the beach in the lee of the breakwater is already becoming conspicuous. (Photo by Fairchild Aerial Surveys)

Santa Monica in 1938. Accretion behind the breakwater has reached substantial proportions and erosion of downdrift shores (upper part of picture) is apparent. (Photo by Pacific Photographic Service)

north. The shore down here to the south of the breakwater is not as bad as it was in the last picture due to the fact that the City of Santa Monica had been engaged in digging out accreted material in the breakwater lee and pumping it down to the shore to the south. They used a home-made dredge and the costs were high so they soon stopped this attempt at by-passing and substituted periodic dredging with a large and powerful dredge.

This is Santa Monica again in 1957, just two years ago. This photograph, incidentally, was taken just prior to the second major by-passing operation. In 1950 the State and Cities of Los Angeles and Santa Monica engaged in a joint project with a large dredge and pumped a million cubic yards from the accreting zone behind the breakwater and deposited it on the shore to the south. Those beaches in the distance beyond Venice were built from another source which I will show you in later photographs as we go along. I am concentrating on events in the Santa Monica area at the moment.

In 1958 the State of California, independently as far as finances are concerned, dredged an additional quantity of material in this area; and it was similarly pumped to the shore to the south. This photo shows the State operation in progress in 1958. You can see that the operation is rather messy in that it

Santa Monica in 1957. Note great expanse of accreted beach in foreground even though 1,000,000 cubic yards was dredged from this area and pumped to downdrift beaches in 1950. (Photo by Pacific Air Industries)

strews the beach up with pipe and equipment. The dredge is operating in this corner. For possibly two or three months the area is disrupted for recreational use while this method of by-passing is being accomplished.

Progressing southward now to the center of the Santa Monica Bay area, this is called the Playa delRey region. This picture was taken again in 1938, 21 years ago, and the bluffs that you see back here are all sand dunes. The region in this segment of the shore had been benefited by the presence of this groin in part and by some of the material that had been taken from the dunes in the years past. You note particularly that there is no beach beyond that groin to the south and that the bluffs and dunes are subject to direct attack from waves with no protection whatever.

This next photo shows the beginning in 1947 of a major project to restore the beaches in this area. In this operation fourteen million cubic yards were pumped from the dunes to the shore, using a sluicing operation to move sand from the dunes to a central sump where powerful pumps distributed it along the shore for a distance of approximately five miles to the north toward the Santa Monica area. Later we will see pictures of the results of this development.

This operation, which was in effect a dual purpose project, was planned and conceived by the City Engineer of Los Angeles. He had the problem of building a large sewage disposal plant in this region and had a choice of building it back of the dunes and piping the effluent through them to the sea or siting it in the middle of the dunes and thereby requiring the excavation of a large volume of sand. Selection of the latter site, in addition to several other advantages, made possible rebuilding of the depleted beaches in this central segment of Santa Monica Bay and as we shall see in later pictures, improving as well the bay shores to the south.

This photograph is of Ballona Creek, which is approximately two miles updrift (north) from the area

Santa Monica in 1958 showing by-passing operation in progress. Method is to be improved when planned harbor improvements are accomplished. Note excellent appearance of beach in background, restored by 14,000,000 cubic yard fill pumped from dunes and maintained by nourishment from by-passing at Santa Monica. (Photo by Pacific Air Industries)

Santa Monica Bay Shores looking South from Playa del Rey (16 miles south of Santa Monica) in 1930. Note narrow beaches along dune bluffs. (Photo by Fairchild Aerial Surveys)

we just looked at. This is a flood control outlet and the largest one at present that drains into Santa Monica Bay. This shows it just after the completion of the jettied outlet by the Federal government in 1938.

This is a picture of the same area in 1947 just after completion of the beach fill nine years later. The jetties at Ballona Creek are in the foreground and have been lengthened 500 feet since the previous picture. The shores have been widened by the pumping from the dunes at Hyperion in the earlier picture you saw. Note that Santa Monica pier in this view is the last structure you can see. Note now the appearance of this shore along this entire reach and the protection that has been afforded.

This 1949 photo is taken from considerably below Ballona Creek. Ballona Creek is at the upper part of the picture. You can see there the broad expanse of beach where there had been previously no beach at all and where the waves were reaching the bluff area.

This is the same area at the same time looking downdrift to the south toward Redondo Beach in the background. This was one year after the completion of the beach restoration operation. This next view shows the completed sewage treatment plant that was built in the area excavated from the dunes.

Now we progress to Redondo Beach, which is near the southern extremity of this lengthy shore segment I have been discussing. This particular photograph shows the condition of Redondo Beach in 1921, which was before the major erosion problems that received nationwide publicity had occurred. The structure near the center is called Monstad's Pier. I should say at this point that in the knowledge that has been gained of littoral processes in the whole area I have been talking about, we have learned that the predominant drift is southerly throughout, but it diminishes as we progress southward towards Redondo. The rate of drift at Redondo is probably between thirty and fifty thousand cubic yards a year, whereas, it is about three hundred thousand cubic yards at Santa Monica and approximately one hundred sixty thousand cubic yards at Ballona Creek.

El Segundo (8 miles south of Santa Monica) at start of 14,000,000 cubic yard beach fill pumped from dunes to provide sewage treatment plant site at left center of picture. Beaches in background were widened by nourishment received from 1,800,000 cubic yards fill from same source in 1938. (Photo by Pacific Air Industries)

This is a picture in 1941 with Monstad's Pier in the foreground. The hooked breakwater shown here which was the initial construction feature of a planned harbor. It was stopped at this point because there was not enough money to finish, and it was left in this manner from 1939 until 1958. This accretion south of the breakwater which is in the opposite direction of normal drift, occurred because the southward component of wave action was blocked by the breakwater. The northward component, effective

Same area as shown in preceding picture but taken two years later in 1949. Note broad expanse of beach in foreground and sewage treatment plant in excavated area at left center. (Photo by Fairchild Aerial Surveys)

Looking north (toward Santa Monica) at completed beach fill. Ballona Creek outlet at upper left. (1949 Photo by Fairchild Aerial Surveys)

about five months of the year, moved sand in the lee of this structure. There is actually also an accretion starting here to the north which is not very apparent in this picture. Bear in mind this was only three years after this structure was completed and the net drift rate is relatively low.

This next view shows Redondo in 1950 with Monstad's Pier in the background. You can see a seawall which was jointly built by the State and the City of Redondo Beach in an effort to save the improved property along the shore. This was built after two rows of buildings, mostly apartment houses, had been destroyed.

This is a view of the area in 1954, giving a very good indication of the slow but persistent accretion north of the breakwater. This is 15 years after the construction of the Redondo breakwater, which you see near the center.

Looking north from Ballona Creek in 1949. Fill ended just beyond first structure beyond Ballona Creek jetties. Last pier in background is at Santa Monica. (Photo by Fairchild Aerial Surveys)

This is Redondo again in 1958. Note the very nice appearance of the beach which now extends north of the breakwater. The Federal Government has finally stepped into the breach and adopted a project to complete this harbor, which project has the dual purpose of protecting the shore and providing the harbor that Redondo Beach tried to build to begin with and found it couldn't finance. This work was completed in November of this past year and plans are now ready and work in progress to build interior facilities for quite a sizable harbor, to accommodate some fourteen hundred boats.

In summing up this photograph display, the point I wish to make is that the sand moves along this Santa Monica Bay shore without much regard to political boundaries. The evidence of what happened to Santa Monica and adjoining Los Angeles shores during the period they were left unattended demonstrates quite clearly what might be expected. The City of Los Angeles was wealthy enough to step into the breach, to make a fourteen million cubic yard beach fill, and the State has aided in by-passing at Santa Monica and in emergency measure at Redondo. Now the beach is excellent from Redondo to Santa Monica. It represents a unified effort which in the last twelve years has been guided to an increasing extent by the State of California in the expanding interest it is now taking in coordinating shore projects of this type. The combined interests and mutual efforts of the State, County and Municipalities concerned, with an occasional assist by the Federal Government, has now made possible what I regard as a well coordinated segment of shore.

I have not mentioned up to now the prospective future developments which will have an extremely important effect on this area. Obviously, small craft harbors are still in great demand. You saw the one

Ballona Creek in 1940 prior to beach fill. Jetties were lengthened 500 feet when beach fill was made. Playa del Rey Harbor site in background marshes. (Photo by Fairchild Aerial Surveys)

Santa Monica Bay in 1950 looking north from Redondo Beach, in 1950. Widest beach at upper part of picture is between El Segundo and Ballona Creek. Santa Monica is in extreme background. Badly eroded Redondo Beach shore in immediate foreground is now protected (1959) by extension of breakwater to provide inclosed harbor. Nine political boundaries divide the shores shown in this picture. (Photo by Pacific Air Industries)

being built at Redondo. Construction is in progress on a huge marina at Playa del Rey, with the entrance adjoining Ballona Creek. Santa Monica has plans for a nine and a half million dollar enlargement and improvement of its harbor. The problem, as usual, is that at every one of these sites the littoral drift will be completely blocked.

The plans for the enlargement of Santa Monica harbor are being designed to provide a most efficient and non-disrupting method of by-passing. It may be possible to regulate sand supply from this point so that additional by-passing will not be required at Playa del Rey. The plans are not finalized yet, and I am sure they will be brought before the Beach Erosion Board before the plan is finally adopted. I understand that they are presently about to be submitted to the State of California.

That completes my presentation of this somewhat involved shore protection problem. I hope that you have been able to see the pictures. I also hope that I have been able to make myself clear in stressing the necessity for cooperative and coordinated effort in such problems and, to my mind, the great desirability of State participation in effecting this coordinated effort. Where the States, such as New York, California, Massachusetts, Ohio and others, have taken a leading hand in effecting coordination I think the most outstanding results have been secured. We in the Corps of Engineers stand ready as always to help municipalities, counties or any higher level of government. We think the best results occur when everyone does his bit in solving the problem.

MR. THOMPSON: Thank you, Mr. Eaton, for giving a broader picture in solving these problems. Too often we are too concerned with the smaller aspects so we don't realize the bigger problems. So it is good to hear about it.

ADMIRAL COLBERT: I would like to bring out the point that you are operating on a more modern concept. You are working between two or more "hard points" and considering a stretch of our coastline without regard to local subdivisions. In other words, you are attacking the problem in a scientific manner.

MR. THOMPSON: Mr. Paul E. Stephenson is taking the place of Mr. Walter G. Schulz and will deliver the next paper on the morning schedule.

SARASOTA, FLORIDA—LIDO KEY—(Erosion and Accretion)

United States Coast and Geodetic Survey maps of the 1870's show Lido Key and Longboat Key to be a continuous strip with a noticable depression between the two. A name meaning "Dry Month" for this section between the two keys was used by the Spanish and Indians.

Twice during the early years of this century, heavy floods and small hurricanes hit this area and as a result the section between the Keys was made deeper. During the 1920's some dredging was done to provide an adequate passage for privately owned boats and some fill also for construction work was accomplished. Up until that time the shoreline was almcst straight from the North end of Lido to a point a few yards South of the Lido Beach Casino. However, since then, and quite apparent in the picture, the center portion of the key shoreline (adjacent to and extending to the North and South of the Casino), is the major area of erosion and at the extremities are evidences of accretion. A study has been completed under the direction of Dr. Per Bruun, Coastal Engineering Laboratory, University of Florida, Gainesville, Florida, regarding recommendations for controlling the erosive activity. Four more rock groins and artificial nourishment were recommended. That briefly, is the story of erosion and accretion on Lido Key.

18

Reprinted from *Shore and Beach* **40**(2):32–35 (1972), Journal of the American Shore and Beach Preservation Association

A CASE HISTORY OF SHORELINE EFFECTS OF JETTIES AND CHANNEL IMPROVEMENTS AT THE MOUTH OF ST. JOHNS RIVER

By Oscar G. Rawls
City Engineer, Jacksonville, Florida

INTRODUCTION

The last time I was called on to do a case history of St. Johns River was at the Third Conference on Coastal Engineering at Massachusetts Institute of Technology in October, 1952. Unfortunately, the emphasis in that paper was enough different from today's subject so that I could not just dust it off and use it again, but had to write a new one.

I have, however, borrowed the opening statement which runs like this: "An early book on Jacksonville begins by paraphrasing a statement attributed to the ancient Greek historian, Herodatus, in which he states in summarizing the economy of an empire, 'Egypt is a gift of the Nile.' The paraphrase continues: 'In a somewhat similar sense, Jacksonville is a gift of the commerce-bearing St. Johns.' In a true sense it may be said that Jacksonville owes its existence, its early development, and much of its present commercial status to the river which passes through the city and provides its outlet to the Atlantic Ocean." I cite this merely to emphasize how important this river is to the economy of Jacksonville and its service area.

With the full knowledge that I shall be telling most of you things you already know about beach processes, I shall nevertheless do so. This is because it seems necessary to do in order to train your thoughts toward what is really involved in a case history of the ocean entrance to the mouth of a river which serves as the navigation access to a major deep-water port. The maintenance of the channel at the bar, the jetties, and the effects on adjacent beaches are all parts of this drama.

The enigma—We start off by facing a rather serious enigma that runs about like this:

1. The mouth of the river in its natural state migrated and became shallow. The channel shifted and was not dependable for use by any but small craft.
2. So, we dug a channel for ships, but without maintenance it quickly filled in. The cost of keeping it open was terrific. A cheaper solution had tc be found.
3. So, to reduce maintenance, large jetties were built on both sides extending across the bar and out into the ocean.
4. The predominant littoral drift is southward at Jacksonville, as is generally true for the whole South Atlantic Coast. The jetties trapped the southward migrating sand and held it north of the north jetty. Sand accreted there. That accretion meant, however, that there was almost equal starvation of the beaches south of the jetties. Thus, the natural southern flow of this along-shore "river of sand" has been interrupted and we have an eroding beach to the south of the jetties extending for several miles.
5. Then, we may seek to halt the erosion by building a sand-transfer plant, or bringing in sand from somewhere to periodically replenish the eroding beaches—like retreading a tire. A sand transfer plant has not been provided at St. Johns River, but there are several at other Florida East Coast sites.
6. The nice quartz sand coming down the coastal streams from Appalachia has also been reduced in amounts as farmers adopted contour plowing and other anti-erosion measures for their lands. Also, dams have been built on those streams and their tributaries. The total supply of sand to the coast has been reduced, resulting in further erosion by curtailing the source of natural replenishment.
7. The net result is that we have a fine harbor channel but have also acquired a rather erosion-prone beach.

My point, in case you are in doubt, is that it is not bad to do contour plowing on farms and to build check dams on tributaries to prevent erosion. Neither is it bad to build bigger dams on the main streams for flood control, power, and other multiple purposes. It is also not bad to provide navigation projects from the coast to inland harbors to serve commerce. No, those things are not bad, they generally are good. But, there is an innocent victim of all three. It is obviously the beach—particularly the several miles south of the mouth of the river. Further, my point is that when we start out to do any of the several good things mentioned above, we must realize that part of the cost of doing those things is also to provide protection for the victimized beaches. Such beach protective costs need to be justified as part of the cost of the harbor development, or flood control development, since they help cause the beach erosion. If the benefits are not sufficient to do the basic project *plus* the compensating beach protection, I cannot see that we have an economically-justified basic project.

I believe that I can claim some credit for getting a policy to implement the above thought established—at least for navigation projects—while I was with the Corps of Engineers. My people submitted the first report that I know of where the costs of compensating beach nourishment became part of the costs of the navigation improvement. This apparently fell on fertile ground and is now, I believe, the usual practice.

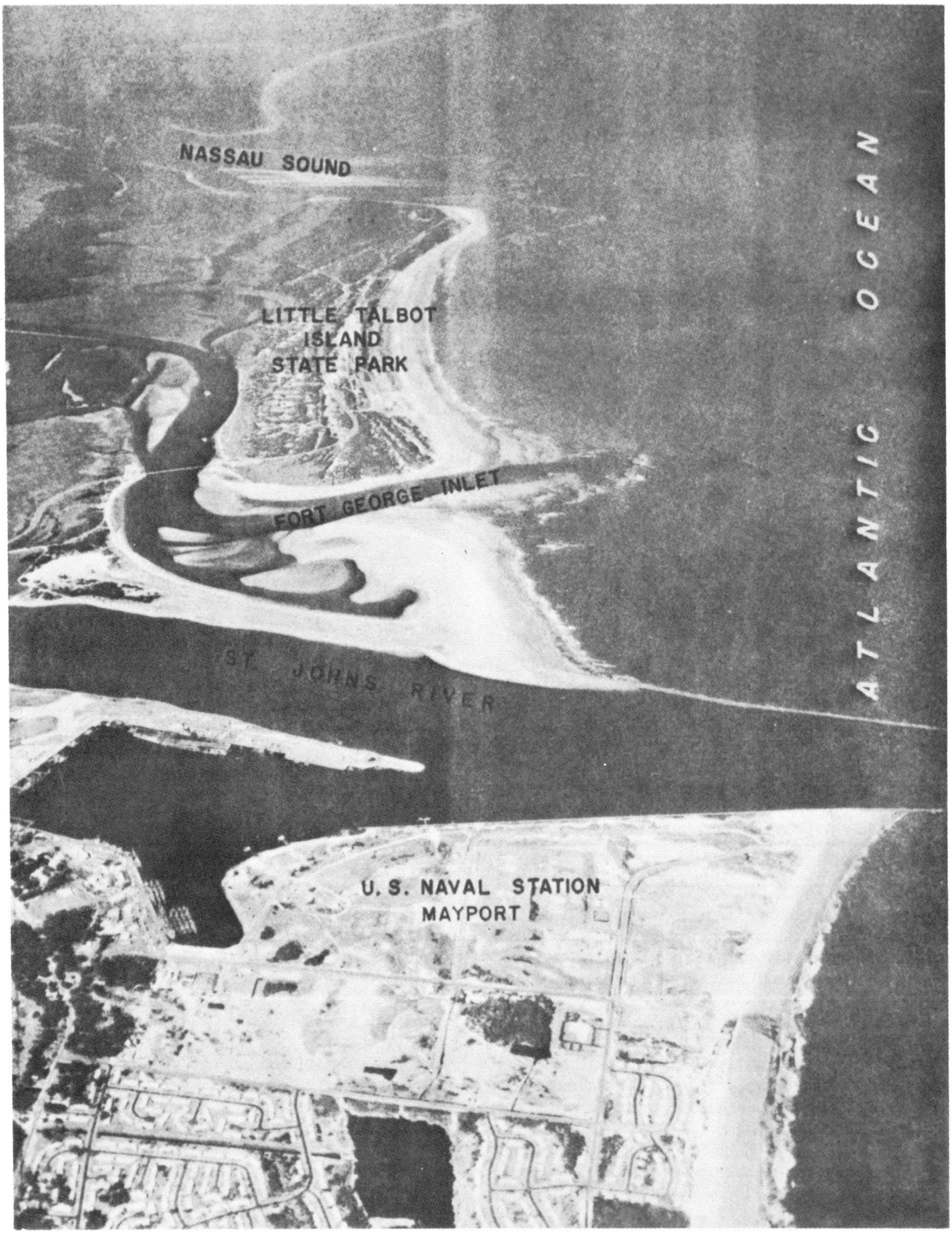

Fig. 1. View of the coast from the St. Johns River to Nassau Sound.

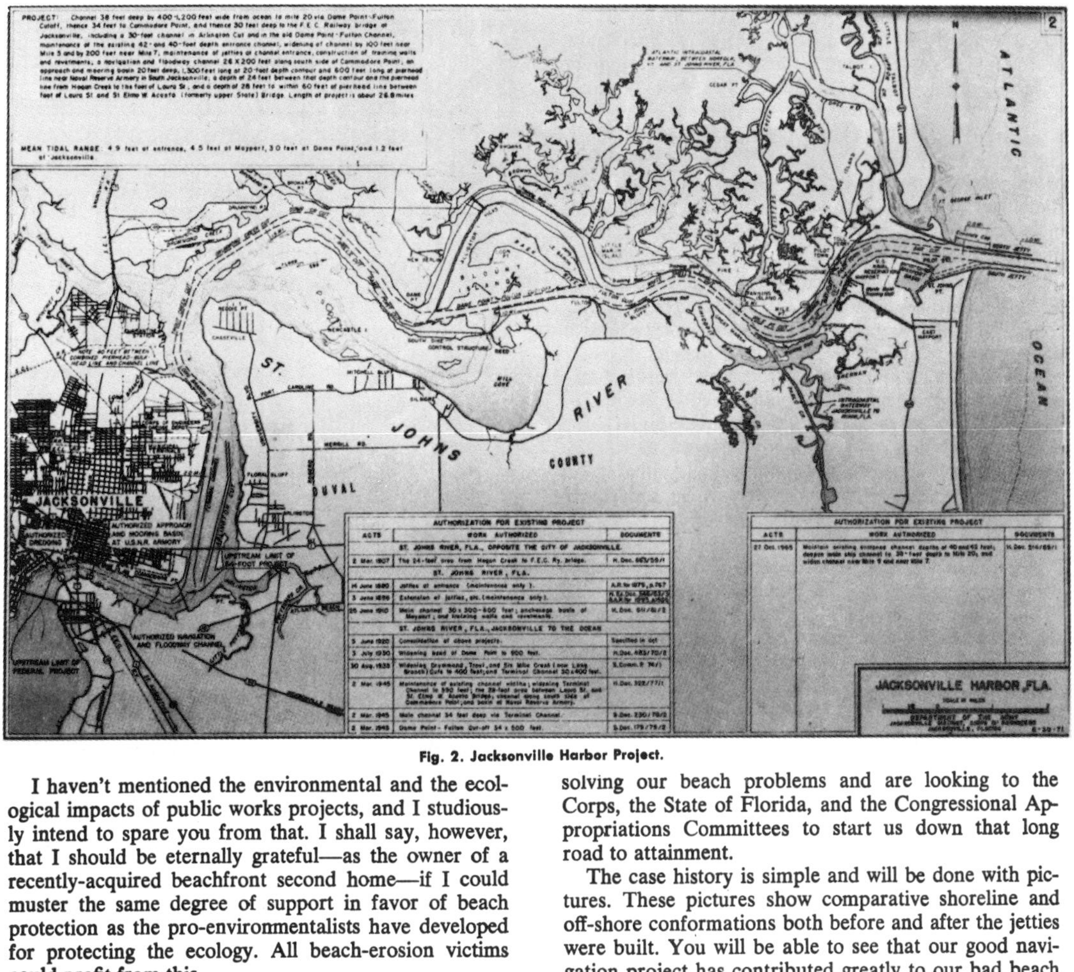

Fig. 2. Jacksonville Harbor Project.

I haven't mentioned the environmental and the ecological impacts of public works projects, and I studiously intend to spare you from that. I shall say, however, that I should be eternally grateful—as the owner of a recently-acquired beachfront second home—if I could muster the same degree of support in favor of beach protection as the pro-environmentalists have developed for protecting the ecology. All beach-erosion victims could profit from this.

Case History—Now for the case history of Jacksonville Harbor and how its improvement has influenced down-drift beaches. Please note that I start off by saying: we like what the Corps of Engineers has done to improve our harbor. We like it so much that we have fought for each step of further improvement. We are delighted that it is right now being even further improved, and we expect to be advocating even more future improvement when the present enlargement work is finished. The harbor is a vital factor in our Jacksonville economy. We love the harbor project, but we do not love the erosion it has caused and want tangible relief. I keep saying "we" because since retiring from the Corps in July, 1970, I have been City Engineer for Jacksonville, which under our consolidation of city and county governments means all of the 844 square miles of Duval County.

In all fairness, let me say that the Corps has an authorized project for restoration of all the beaches in Duval County south of the St. Johns River jetties—some 10 miles of them. We hope to get pre-construction planning funds for the Corps to get this project started in the fiscal year 1973 Federal budget. We need help in solving our beach problems and are looking to the Corps, the State of Florida, and the Congressional Appropriations Committees to start us down that long road to attainment.

The case history is simple and will be done with pictures. These pictures show comparative shoreline and off-shore conformations both before and after the jetties were built. You will be able to see that our good navigation project has contributed greatly to our bad beach erosion problem. As is to be expected, it has followed the classic script that I have already outlined. Here, as expected, there is accretion north of the jetties and erosion to the south.

ILLUSTRATIONS

Figure 1. This is an fairly modern photograph of the mouth of St. Johns River looking from south to north. The Navy's Mayport carrier basin is at mid-left. The river and the jetties cross the middle horizontally. A somewhat unusual condition is shown, since the south, or near shore, extends farther to sea than the north shore. I believe this is accounted for by the eddy effect in the lee of the jetties. The more severe erosion takes place farther south. Also, we have the unusual bit of topography that brings Fort George River into the ocean just north of the north jetty. This causes local effects that are unusual, but the broad picture follows the classic situation.

Figure 2. This is a drawing of Jacksonville Harbor from the mouth of St. Johns River on the right, upstream some 20 miles to downtown Jacksonville on the left. The jetties are shown, as is the course and inflow of Fort George River. The latter is a tidal estuary fingering out to the north into rather vast sawgrass marshes.

Figure 3. This also shows the upper reaches of Ft. George River. Principally, however, the beach project of the Corps of Engineers is shown. Note that the shoreline would be initially restored and periodically maintained for the entire 10 miles of frontage from the mouth of St. Johns River to the St. Johns County line.

Figure 4. Shows the after-effects of hurricane Dora at Atlantic Beach (September 1966). Shortly thereafter the Corps of Engineers with emergency funds placed a rock revetment and added sand nourishment throughout the more populous frontage reaches. It was supposed to provide only about 10-year protection. Its depreciation rate is on or ahead of schedule. With at least a few years before we can hope to start construction and a few more to completion, we shall probably be needing this project when it finally is accomplished. We hope for no more Doras until then.

As we would expect the history of this inlet since before the jetties were constructed (1879-1887) has shown accretion on the north side and erosion to the south. This is very evident for a comparison of the shorelines and the 6, 12 and 18 foot contours.

For the 30-foot contour, there is no great difference between the before and after photos except in the excavated throat of the jetties channel. Thus, it seems that if we could move all of our human coastal activities out to the 30-foot depth contour our erosion problems would all be solved and we could spare ourselves the need for listening to dully presented case histories.

Jean Ribault, the French explorer, was perhaps the first visitor to the mouth of the river who needed a deep and stable channel. In 1565 (55 years before Plymouth Rock) he was forced to anchor his large ships off shore at the bar and do his upstream exploring in the smaller ones. The present year also marks the 120th anniversay of the first efforts toward real improvements at the bar. Local citizens persuaded Congress to appropriate $10,000 in 1852 for the first study to try to decide how to keep the mouth of the river open.

My own association with the jetties, which were built between 1879 and 1897, and with the beaches to the south, goes back more than half a century. When I was a teen-age boy I could stand on the beach and easily sit on top of the seawall. Now I would not be able to touch the top with a ten-foot pole. I also remember a beach that must have been about 400 feet wide at low tide and as smooth as velvet and as hard as concrete. Automobile races on an elongated oval speedway were an absolute must for all Jaxons on Labor Day and the Fourth of July.

The Corps of Engineers in its November 1964 project document for the beach protection project showed that the mean high water shoreline has receded about 1500 feet inland for the mile of beach south of the jetties since the jetties were constructed. Also, between 1923 and 1963 (40 years) the average annual net losses from St. Johns River to the St. Johns County line (10 miles) were 191,000 cubic yards landward of the 18-foot depth contour and 47,000 cubic yards seaward of the 18-foot depth contour, or a total of 238,-000 cubic yards of erosion loss a year. Adjustments for material artificially added to the beach raises the true loss rate to 260,000 cubic yards annually for the 10 miles. At that rate, it is easy to see that the emergency sand nourishments provided by the Corps in 1962 and 1964 are nearing the end of their useful lives.

In summary, the St. Johns River harbor development has had the expected effect on the eroding beaches eo the south. My abiding hope is that we shall not stop harbor improvements, but that we shall recognize the facts and see that in making the harbors better we also must take compensating protective measures to insure that we can also keep our wonderful, delightful, enjoyable, and splendid beaches.

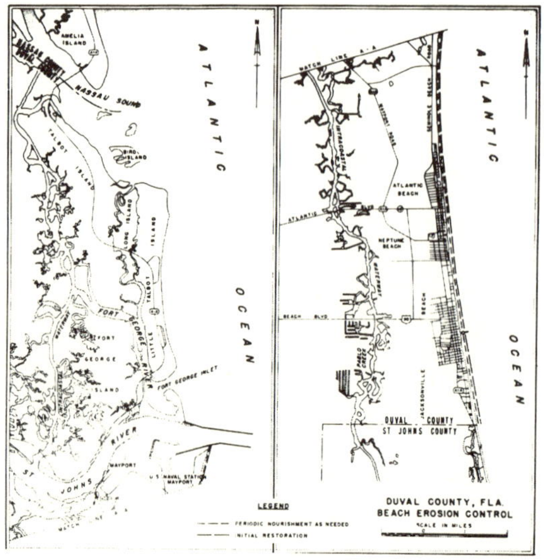

Fig. 3. Erosion Control Project, Duval County.

Fig. 4. Effects of Hurricane Donna at Atlantic Beach.

19

Reprinted from *Coastal Eng. Conf., 13th, Vancouver, 1972, Proc.*, pp. 801–821

CASE HISTORY OF MISSION BAY INLET, SAN DIEGO, CALIFORNIA

William J. Herron, Jr.

Moffatt & Nichol Engineers, Long Beach, California

ABSTRACT

The Mission Bay Inlet was designed as a "Non-Scouring" entrance channel by the Los Angeles District, Corps of Engineers, in 1946. Construction of the inlet was completed in 1959 and the entire project was completed in 1963. A channel with over twice the cross-sectional area required by the "O'Brien" equation was developed to reduce the average cross-sectional tidal currents to less than two feet per second. The design depth of -20 feet MLLW eliminated bottom movement induced by wave action- except during the most severe storms. The jetties are sealed to the +4 foot elevation and extend to the -25 foot depth almost entirely eliminating the intrusion of littoral drift. The channel has shoaled at a rate of less than 20,000 cubic yards per year since final dredging in 1959, indicating the soundness of this concept. This case history was prepared under contract to the Coastal Engineering Research Center, U. S. Army, Corps of Engineers, and project data and aerial photographs were obtained from the Los Angeles District, U. S. Army, Corps of Engineers, and the City of San Diego.

HISTORY

Mission Bay, prior to 1946, was a natural estuary of over 4,000 acres (Photo 1). The major drainage feature into the estuary was the San Diego River with a drainage area of 435 square miles. San Diego is a semi-arid area and the river is normally dry, but during severe floods it may flow in excess of 100,000 cubic feet per second and carry large quantities of silts, sands, cobbles and floating debris. There has not been a flood of this magnitude since 1927. The San Diego River originally discharge into either San Diego Bay or into the southeast corner of Mission Bay. The latter entrance to the ocean was a natural inlet through a sandy beach. While the thread of the inlet meandered constantly, it generally had a controlling depth of 6 feet mean lower low water. The tidal currents in the natural inlet could exceed 4 knots and were sufficient to counteract the force of waves and infringement of littoral drift and maintain a permanently open tidal inlet.

It was realized that if the San Diego River were allowed to continue to discharge into San Diego Bay, serious shoaling would result and would interfere with commercial shipping. In 1876 a permanent levee was constructed, protecting San Diego Harbor. From 1876 to 1946 the river was permitted to discharge its debris into the "useless" Mission Bay.

DESCRIPTION OF AREA

Geomorphology

Indian legend has it that Pt. Loma was originally an island. This is undoubtedly true. There is also little doubt that a few hundred years ago there was much more rainfall than at present and the San Diego River supplied a great deal more sediment to the coastal plain. However, the history of the river up to 1876 was alternate discharge into either San Diego Bay or Mission Bay, or False Bay as it was originally called.

Littoral Drift

Studies by the Beach Erosion Board of the Corps of Engineers in 1942(2) and 1949-50(3) indicate that the Mission Bay area is essentially a beach compartment extending from La Jolla on the north to Point Loma on the south. The San Diego River, originally discharging through Mission Bay, is the principal source of beach sand. It was observed that while there was some upcoast sand movement in the summer and downcoast in the winter the area was essentially in littoral balance. They also found in their 1950 studies that there was a very positive onshore-offshore movement from summer to winter.

Tides

At the entrance to Mission Bay the mean range of tide is 3.8 feet and the diurnal range is 5.4 feet. The extreme range varies from 2.5 feet below to 7.0 feet above mean lower low water. Mean sea level is 2.8 feet above MLLW. The tides are characterized by a diurnal inequality and, as the maximum run is from higher high tide to lower low tide, inlets in southern California generally have higher ebb-tide than flood-tide velocities. This infers an ability to move more sediments during periods of ebb tide than during flood tides.

Waves

Wave action is generally mild in this area having periods of 6 to 16 seconds and heights of less than 3 feet. However, waves 18 feet in height can occur at the entrance to Mission Bay and a significant wave height(H) of 16 feet was used for design of the jetty structures.

Also, based upon experience and consideration of the larger storm waves it was decided to use a navigation project depth of -20 feet MLLW to insure the occurrence of non-breaking waves in the entrance channel at all times except during the most severe storms.

Winds

Winds rarely exceed 30 knots, and their main import is how they affect the maneuverability of boats under sail.

Rainfall

Average annual rainfall in this semi-arid region ranges from 10 inches along the coast to 38 inches in the mountains. 80% falls from December to March inclusive and, in the lower reaches, the San Diego River is dry most of the time. The last major runoff was in 1927. During peak floods it may run as high as 100,000 cubic feet per second. Thus, while the river has almost no influence on the tidal characteristics of the bay and its ocean inlet, the sediments carried during periods of major runoff can be of large quantity and materially affect the bay and/or the adjacent shoreline.

ANALYSIS OF DESIGN

Project Authorization (Fig.1)

The first mention of plans for development of Mission Bay as a navigation area was in 1941 in studies by the City of San Diego and the Corps of Engineers to improve the lower San Diego River for purposes of flood control. As this study progressed it became evident that maximum benefits could be obtained by a combined flood control-navigation project.

The combined project was adopted by Congress in July 1946, as presented in House Document No. 760, 79th Congress, 2nd Session. Detailed design and cost estimates were presented by the Corps of Engineers in "Definate Project Report on San Diego River and Mission Bay", dated January 1949. The Federal Government was responsible for the main channel and its sideslopes, the dredging of the east and west basins, the dredging of the navigation entrance channel and construction and maintenance of the three jetties defining the navigation and flood control channels. Navigation depths of -20 feet MLLW were authorized. The Ventura Boulevard Bridge and everything inland of it was the responsibility of the City, and navigation areas were to be dredged to a depth of -8 MLLW. The Federal portion of this project was essentially completed in 1959 at a cost of about $10,000,000. Local public interests have expended about $19,000,000 to date and private interests over $18,000,000. Ultimate public recreational development is expected to exceed $50,000,000 (4).

Design Objectives

A well conceived harbor, among other considerations, represents a balanced sedimentation system. It is hopefully a large quiet body of water with a relatively narrow protected channel to the sea

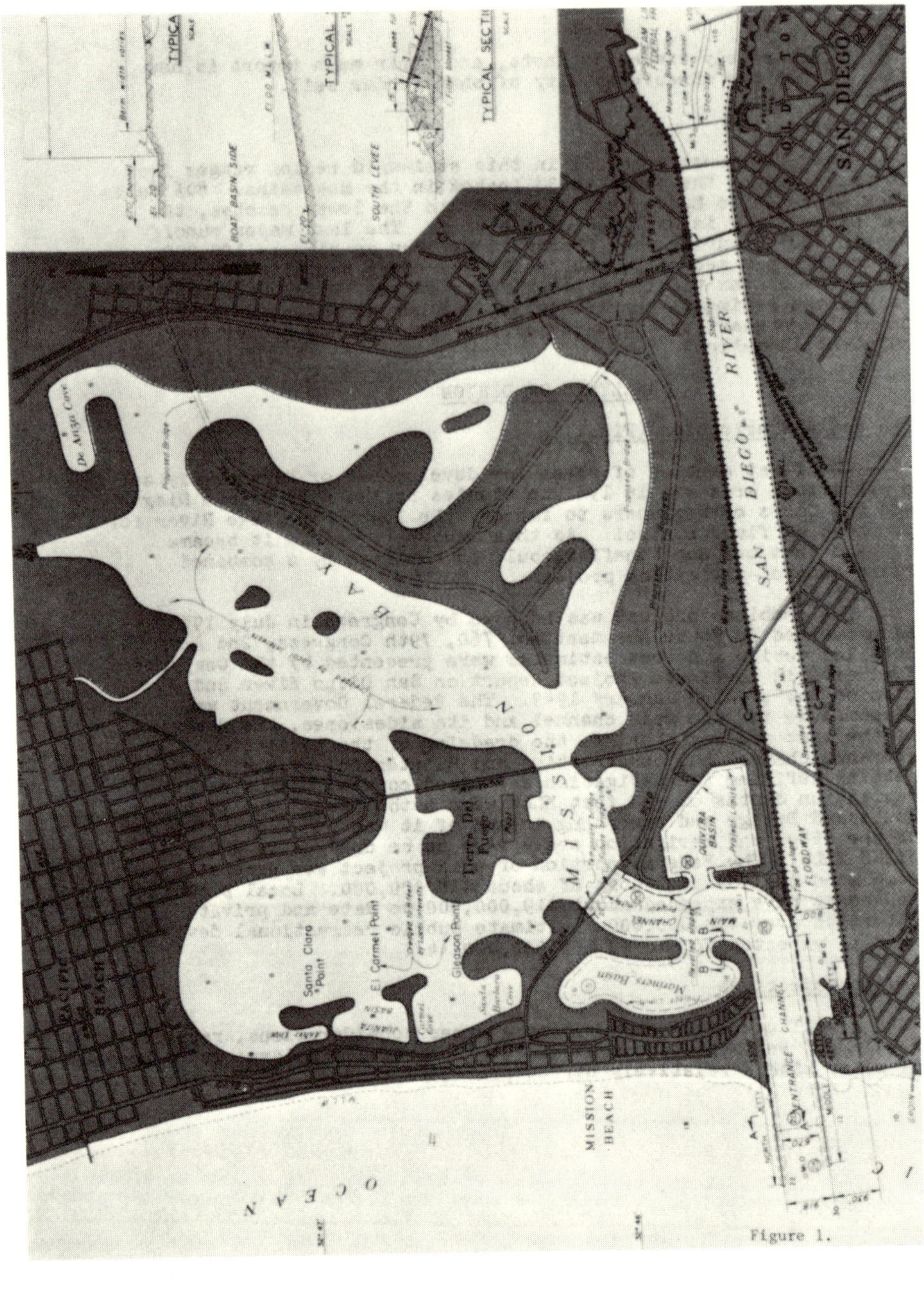
MISSION BAY
De Anza Cove
Tierra Del Fuego
Santa Clara Point
El Carmel Point
Gleason Point
Mariners Basin
QUIVIRA BASIN
SAN DIEGO RIVER
ENTRANCE CHANNEL
FLOODWAY
PACIFIC BEACH
MISSION BEACH
OCEAN
SAN DIEGO

Figure 1.

requiring minimal maintenance. Due to waves, tides or river discharge, currents of varying rates may be created. If these currents have sufficient velocity, they will either initiate movement and transport sediments or if sediments are already moving, continue them in motion. These currents and resultant sediment movement may cause internal changes in bottom depths or may cause a net gain or loss of sediments to the total harbor complex with resultant shoaling or scouring, either of which may be detrimental.

Engineers in Southern California learned as early as 1876 that if the rivers of this semi-arid country were diverted out of the harbor area and preferably discharged into the sea on the down-drift (littoral) side of the entrance a major source of harbor shoaling could be eliminated. This was demonstrated (successfully accomplished) in 1876 when the San Diego River was diverted out of San Diego Bay and in 1916 when the Los Angeles River was diverted out of Los Angeles-Long Beach Harbor.

The other major exterior source of shoaling is sand from the adjacent beach being carried into the entrance by tide and/or wave induced littoral current, creating shoals either in the entrance or in the quiet water area of the harbor. If the entrance jetties are sand tight, of sufficient elevation, and are extended seaward of the area of littoral drift, most of this type shoaling is eliminated. However, in areas of predominantly uni-directional littoral drift, a sand-bypass system will ultimately have to be established to prevent the impounded sand from being carried over or around the seaward end of the intercepting jetty and thus shoaling the entrance. Jetties extended to the 25 to 40 foot depth contour at Newport Beach in 1936 and Pt. Hueneme in 1941 had demonstrated this concept.

Even though there are no external sources of sediments to cause shoaling of a harbor or its entrance channel and the discharge from the uplands has been diverted, tidal or other currents still are a liability. They may either cause internal movement of sediments (with adverse shoaling or scouring effects) or interfere with the safe and efficient operation of boats using the harbor.

These experiences resulted in a concept known as a "non-scouring" tidal channel, i.e., enlarge the inlet cross section to a point where tidal current velocities are reduced below their capability to move bottom material. A secondary benefit in such reduced velocities is the elimination of hazards to boats. Such a maximized cross-sectional area requires a balanced design between width and depth. Increased depth permits deeper draft boats to use the harbor, prevents a breaking wave, or reduces the steepness of waves during the storms, making for a safer entrance. Wave induced bottom movement can also be eliminated, or greatly reduced. The principal effects from increased width of channel is less boat traffic congestion and greater maneuver area for sail boats. On the minus side are increased dredging costs, sometimes more costly jetties, increased land acquisition costs and the introduction of more wave energy into the inner harbor to the possible detriment of berthed boats and interior land perimeters.

BASIS OF INLET DESIGN 1946 to 1949

1. The design relationship between the dimensions of a harbor or bay and its channel connection to the ocean at this time was largely an emperical one. The most acceptable work in this field had been done by Lt. Col. T.M. Robins of the Corps of Engineers(6), and Morrough P. O'Brien of the University of California at Berkeley (5). They had developed the following relationships based strictly on observations of existing harbors, bays and estuaries.

In 1931 O'Brien developed the equation $A = 1{,}000\ T^{0.85}$ where A = area of the entrance section below mid-tide in square feet. T = volume of the tidal prism in square-mile feet between MLLW and MHHW, and has a value ranging between 7.0 and 3000. In 1933 Robins published the relationship developed by C.I. Grimm of his staff as A = T where A = Area of channel cross-section at mid-tide in square feet and T = The Tidal prism in acre-feet between MLLW and MHHW.

2. Mission Bay in its natural state reportedly had a tidal prism of about 10,400 acre feet (Photo 1). The minimal cross-sectional area of the entrance channel as surveyed in 1947 was 6000 square feet (Fig.2a). Compared to this field data, the Robins' equation shows a computed equilibrium entrance area of 10,400 sq.ft. while the O'Brien equation shows an area of 9,100 sq. ft. The reported 10,400 acre-feet of tidal prism may be in error or, if correct, the very flat slopes of the marshlands and the constructions of tidal flow at the two Ingraham Boulevard bridges may have reduced the resultant required channel cross-section from the O'Brien derivation of 9100 sq. ft. to the actual 6000 sq. ft.

3. The designers of Mission Bay increased the cross-sectional area of the planned entrance channel to 19,800 square feet below mean tide level and even though creation of the Aquatic Park resulted in some 30,000,000 cubic yards of dredging, the net result was a tidal prism of about 9,200 acre feet. However, the flat slopes of the marshland areas were eliminated and there is very little frictional or time loss in disposing of the tidal prism. Thus, using the O'Brien equation, the jettied entrance channel computes at an equilibrium flow area of 8900 square feet as compared to the design area of 19,800 sq. ft. Treating the tidal current produced from a spring tidal drop of 5.4 feet in 6 hours, average channel velocity would be 0.97 ft. per second. It was felt that maximum velocity would not exceed 2 ft. per second except during spring tides, and scouring velocity along the bottom of the channel should be even less.

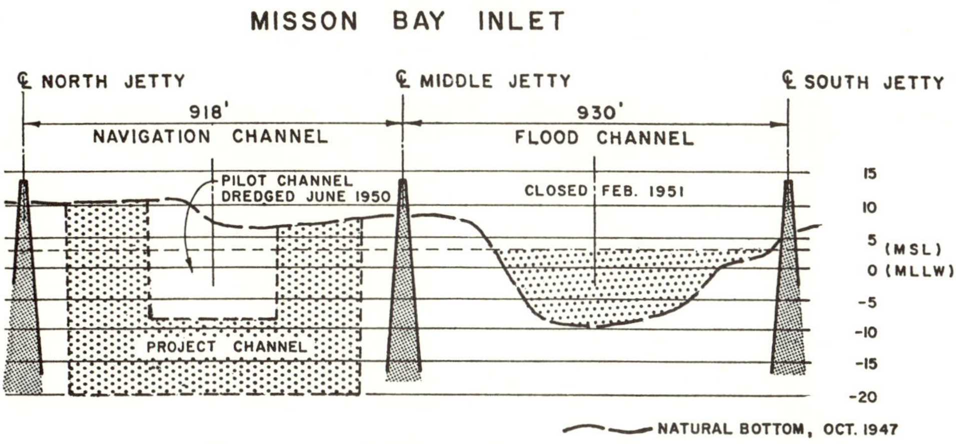

Station 40+00
FIGURE 2a

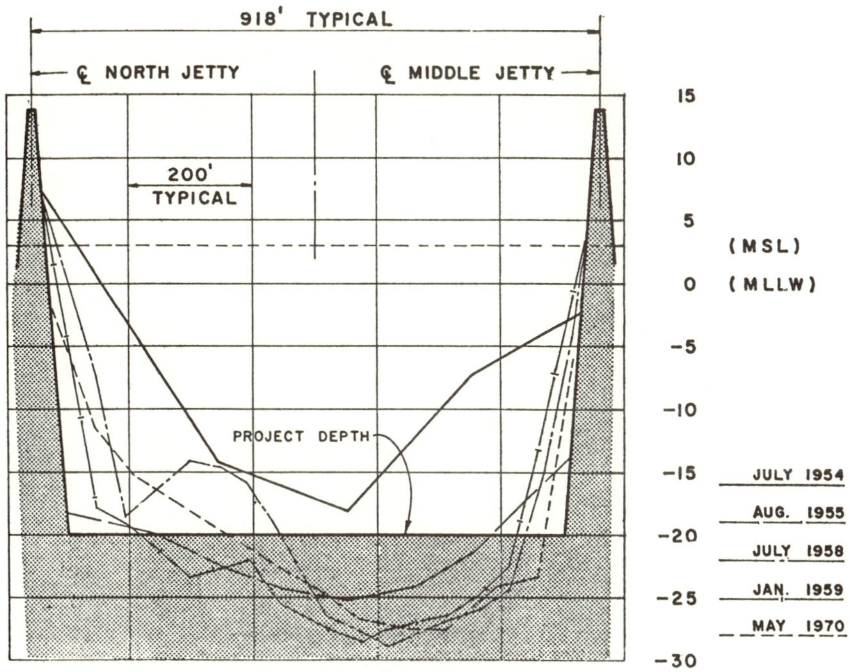

Station 48+00
FIGURE 2b

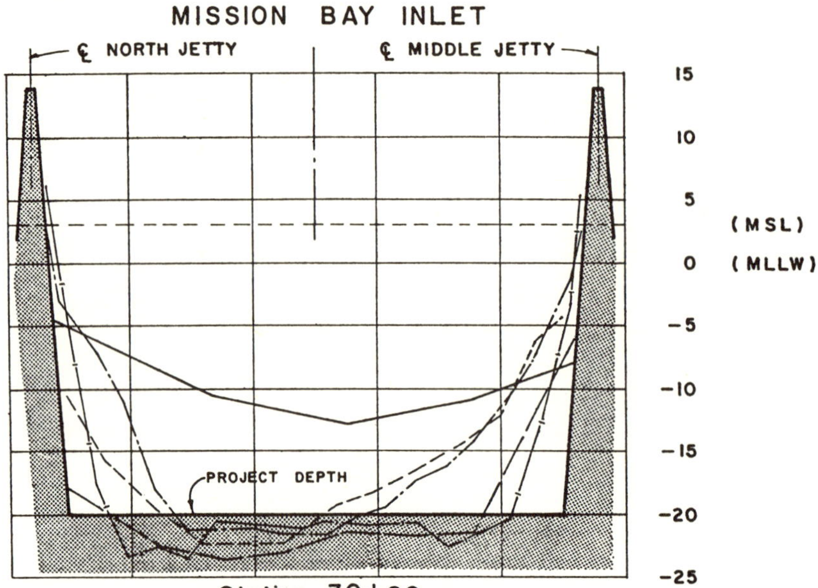

Station 30+00
FIGURE 2c

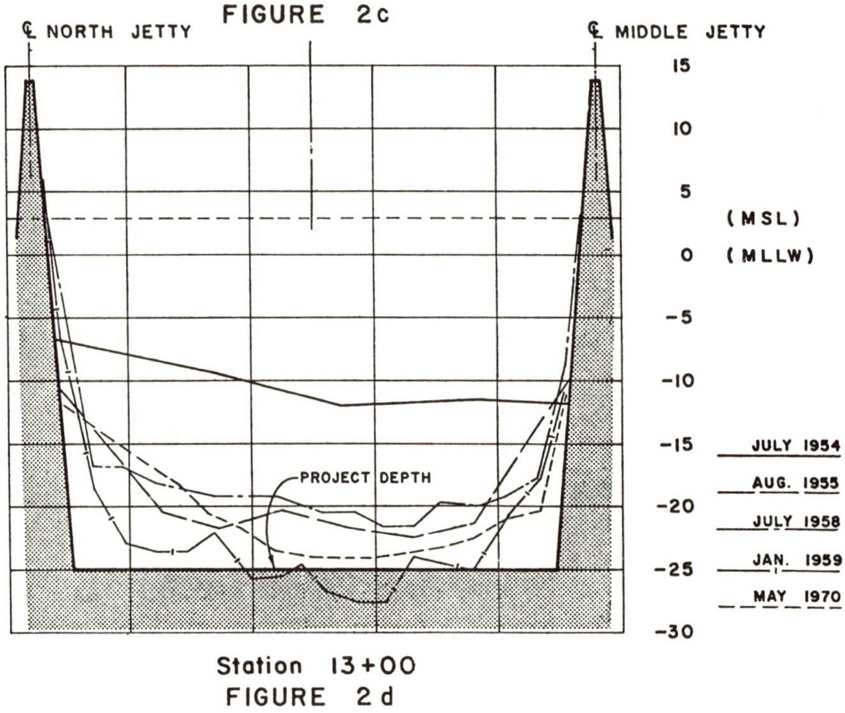

Station 13+00
FIGURE 2d

BEFORE CONSTRUCTION
GENERAL CONFIGURATION OF BAY AT HIGH TIDE
1947 NO. 1

START INTERIOR CONSTRUCTION
NOTE CONFIGURATION OF NATURAL INLET
FEB. 1948 NO. 2

START ENTRANCE CONSTRUCTION
MIDDLE AND SOUTH JETTIES
JULY, 1948 NO. 3

COMPLETION OF FLOOD CONTROL CHANNEL OUTLET
MIDDLE AND SOUTH JETTIES
FEB. 1949 NO. 4

NAVIGATION AND FLOOD CONTROL CHANNEL JETTIES
COMPLETION OF NORTH JETTY
NOV. 1950 NO. 5

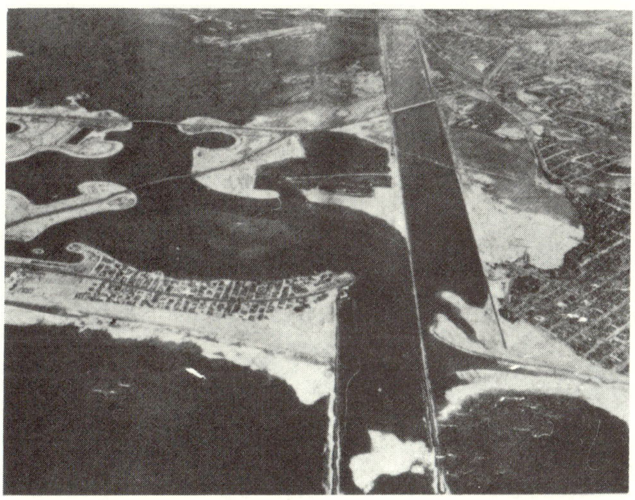

NAVIGATION AND FLOOD CONTROL CHANNELS
MIDDLE JETTY CLOSED, CREATING TWO CHANNELS
SEPT. 1951 NO. 6

DEVELOPMENT OF WEST MISSION BAY
NOTE CLOSURE OF FLOOD CHANNEL
JAN. 1953 NO. 7

START OF DEVELOPMENT OF EAST MISSION BAY
NOTE COMPLETED DREDGING OF ENTRANCE CHANNEL
JAN. 1958 NO. 8

COMPLETION OF MISSION BAY - SAN DIEGO RIVER PROJECT
1963 NO. 9

HYDRAULIC MODEL STUDY - MISSION BAY ENTRANCE
1969 NO. 10

CONSTRUCTION PROCEDURE

In any project of this magnitude, construction problems during an extended construction period (1946 to 1963) can create problems and misunderstandings. The time s quence for creation of the bay and the jettied entrance channel was as follows:

1946: Dredging and filling of interior portions oi the bay were initiated by the City. (Photo 2, 1948).
May 1948: The Corps of Engineers initiated construction of the south (flood control) jetty and the middle (combined flood control and navigation) jetty. (Photo 3, 1948); Photo 4, 1949).
June 1949: The Corps of Engineers initiated construction of the north (navigation) jetty.
November 1949: The south and middle jetties were completed.
September 1950: The north jetty was completed (Photo 5, 1950).
June 1950: A pilot channel (300 feet wide) was dredged to the -8 foot MLLW depth to divert tidal flow of the bay to between the navigation jetties. (Photo 5, 1950).
February 1951: The flood control channel was completed and the final section of the middle jetty was closed, separating the Flood Control Channel from Mission Bay. (Photo 6, 1951).
March 1951 to Nov. 1954: The Corps of Engineers project was shut down because of the Korean War. During this time period, the City essentially completed their dredge and fill program for the bay westerly of Ingraham Street (Photo 7, 1953).
December 1954: Dredging of the outer entrance channel by the Corps of Engineers was resumed.
May 1955: An experimental attempt was made to seal the jetties to prevent littoral sand from entering the navigation channel through voids in the jetty cap rock.
July 1955: The outer entrance channel was completed to full width and depth.
January 1957 to December 1957: The Corps of Engineers dredged a portion of the main channel and Quivera Basin to a depth of -20 feet MLLW. This relatively coarse sand was pumped to the eastern perimeter of the bay to stabilize the mud deposits along U. S. 101 Highway. This dredging essentially permitted full and unimpeded tidal flow through the West Bay, the main channel and the entrance channel. (Photo 8, 1958).
July 1958: The Corps of Engineers dredged the main channel and the west anchorage basin (Mariner's Basin) to a depth of -15 feet, MLLW. At the same time, the City initiated a 12 million cubic yard dredging contract to complete the development of the east bay. This project, completed in 1963, essentially completed the City's dredge and fill program, and while the natural tidal prism was reduced by about 10%, flow characteristics were greatly improved by elimination of the large marshy areas with flat slopes between high and low tide. (Photo 9, 1963).
August 1959: The middle and north jetties were sealed with concrete grout from mean lower low water to the +4 foot elevation through the littoral zone to prevent future shoaling of the navigation channel by intrusion of littoral sand through the large voids in the cap rock.

October 1966 to June 1969: Field studies, resulting in a hydraulic model study were made by the Corps of Engineers to develop a solution for occasional undesirable wave action in the two deepwater anchorage basins (Photo 10, Hydraulic Model Study).
1972: Maintenancedredging is planned to remove about 300,000 cubic yards of sand from the entrance channel.

DISCUSSION OF PROBLEMS ENCOUNTERED DURING CONSTRUCTION

The shutdown of this project as a result of the Korean War had a most adverse effect. While the flood channel was completed, only a minimum effort was made to open the bay to the sea by dredging a pilot channel through the barrier beach. The pilot channel was to be 300 feet wide to the -8 foot depth (Fig. 2a). In other words, looking at O'Brien's equilibrium equation the cross-sectional area of less than 4000 feet was completely inadequate and the channel shoaled and spread the full width between jetties to form a very dangerous bar. (Figs.2c and 2d). From an engineering point of view this was a very successful maneuver because in accord with the equilibrium relationship and because of the west coast tidal character of getting maximum tidal run from MHHW to MLLW, the minimal cross-sectional area of the entrance channel increased to about 11,000 ft.^{2} by July 1954, and over 250,000 cubic yards of sand were scoured from the channel and apparently carried out to sea. However, the boatmen were not adequately informed as to the hazards of this partial channel, and before its closure by City Ordinance in 1953, several boats were capsized and 11 lives lost.

Another major problem was the intrustion of littoral sand into the entrance channel. However, until 1954, in Southern California, it was the practice to only bring core rock, or "C" rock up to MLLW for jetties and then to continue construction through the wave zone with large cap rock. Thus both navigation jetties at Mission Bay consisted of 2 to 8 ton cap stone from MLLW to the +14.0 foot elevation. It was realized during the initial dredging of the entrance channel in early 1955 that a very appreciable amount of littoral sand was being carried through the voids of the cap rock and shoaling the navigation channel. In May 1955 a plan was conceived to introduce a concrete grout diaphram into the center of the cap rock section (7). After considerable testing, this was accomplished by drilling holes along the centerline of the structure and forcing a mixture of concrete and bentonite (driller's mud) through the holes and into the central voids of the cap rock. The intent was to create a diaphram with a minimum elevation of +4 feet. Some 2200 feet of the middle jetty and 600 feet of the north jetty were sealed in this manner.

During this same period of developing the entrance channel, it was observed that, all too frequently, waves were either breaking in the entrance channel or were so steep as to constitute a serious hazard to small boats. As a final move, in June 1960 the outer 1000 feet were dredged to the -25 foot depth. This appears to have greatly improved the navigability of this outer portion.

The remaining inlet problem at Mission Bay is excessive wave action inside the bay. It has been learned at Mission Bay, Alamitos Bay, Newport Harbor and Marina del Rey that while they are excellent harbors, so far as navigability and low maintenance are concerned, these excessively wide entrances do admit a great deal of wave energy which must be disposed of. Alamitos Bay is partially sheltered by the outer harbor breakwater and much of the interior perimeter is still in sand beaches which can absorb considerable wave energy. Much the same is true at Newport; the jetties are skewed some 20° to the south,and the north jetty acts, in part, as a breakwater. Newport Bay also consists mostly of sand beaches around its perimeter, and has had no interior wave problems. A very bitter lesson was learned at Marina del Rey, where subsequent to construction of jettied entrance, the interior walls were changed from rock revetment to vertical concrete bulkheads which almost totally reflected wave energy. An offshore breakwater had to be constructed across the entrance to keep this wave energy out of the boat basins. At Mission Bay, a wave problem exists in the two deepwater anchorages: Quivera Basin and Mariner's Basin. Very soon after installation of boat slips in 1959 and 1960 in Quivera Basin, there were complaints of waves as high as 2.5 feet at the berths with damage to both boats and floats. Studies were made in the field (8) and by hydraulic model at the Corps of Engineers' Waterways Experiment Station (9) and it was concluded that the main problem was excessive wave energy reaching the inshore end of the entrance channel. The inshore end of the entrance channel makes a 90° left into the main channel, and the shore is protected by a semi-circular rock revetted slope. This revetment not only tends to reflect too much wave energy, it also tends to focus this reflected energy towards the entrance to Mariner's Basin. Several corrective measures were tested and the most promising was to convert the revetted semi-circular end of the channel to a series of straight revetted sections in echelon that would tend to reflect this energy back out the entrance channel to sea (Photo 10). Estimated cost of this modification is about $1,000,000, and until there is further development of the two basins and the need is strongly demonstrated, no action is planned.

REVIEW OF DESIGN 1972

The design of the inlet for Mission Bay, as developed in 1946-1949, has since proved to be fundamentally correct. Dean O'Brien reviewed his work in 1969 (10) and arrived at the basic equation: $A = 4.69 \times 10^{-4} P^{0.85}$, expressing P in cubic feet rather than his original expression of square-mile feet. Thus the minimal cross-sectional area of the entrance channel for equilibrium flow would still be in the order of 9000 square feet. Since completion of the entrance channel dredging in 1959, the cross-sectional area has varied from 16,400 sq. ft. to the design area of 19,800 sq.ft. which is well above the minimal. The first maintenance dredging of the entrance is scheduled for the latter part of 1972. Making due allowance for sand entering the channel before completion of jetty sealing, the average maintenance dredging of the entrance between 1954 and 1972 amounts to less than 20,000 cubic yards per year.

In Feb. ,1972, Professor J. W. Johnson published a paper comparing the various tidal inlets of California, Oregon and Washington (11). His work confirmed the equilibrium curve developed by O'Brien. He realized, as did O'Brien, that factors other than tidal prism should have an influence on the equilibrium cross-sectional area of the inlet. In an attempt to prove this point, he compared the characteristics which he believed should have certain modifying effects on the equilibrium ratio. He gave identifying symbols to the plotting points for inlets with two,one or no jetties and with various degrees of exposure to wave action. He also made separate plots of his data adding such parameters as tide range and tide-cycle duration. Surprisingly, these factors and parameters showed no sifnigicant modifying trends. All inlets plotted within reasonable proximity of the equilibrium curve with the exception of Morro Bay, Marina del Rey, Alamitos Bay and Newport Bay, all of which showed entrance sections far larger than their tidal prisms should have allowed. Prof. Johnson stated that these radical departures from the norm appeared to be "primarily a function of wave climate and the limited movement of littoral drift as affected by long jetties and the possible initial over-dredging of the entrance channels". He concluded that since these inlets and a few others, including the Mission Bay Entrance,showed no evidence of shoaling, they were not in equilibrium in the usual sense for unimproved inlets because of the effects of limited wave action, low littoral drift or long jetties. It is apparent from Figure 3 of this study that, at least in the case of San Diego Bay, Mission Bay, Newport Bay, Marina del Rey and Alamitos Bay, the inlet has not been over-dredged by mistake, but the cross-section has been deliberately increased beyond "equilibrium" limits. This was done to reduce the tidal currents and to prevent appreciable sand transport in the entrance channel.

The various relationships between the tidal prism and the area of the inlet since start of construction in 1946 are shown in Figure 3 and Table 1. It is apparent that there has been a very major change in the relationship of tidal prism to inlet area with a resultant reduction in tidal velocities, yet no significant entrance shoaling has occurred as the equilibrium formula would have predicted.

During the original design of Mission Bay and its inlet, it was only possible to make a rough approximation of tidal current velocities through the critical sections. Since 1948, a great deal of study has been done on this subject and one of the most widely accepted methods for computing velocity of inlet currents was developed by Dr. G.H. Keulegan, as described in the U.S. Army Corps of Engineers' Committee on Tidal Hydraulics, Technical Bulletin No.14. His procedure takes into account the tidal oscillation in the ocean, the physical properties of the inlet channel, the physical properties of the basin and inflow from streams or rainfall, and it develops a mathematical relationship between the tidal range in the basin and the maximum velocity in the connecting channel.

TABLE 1
MISSION BAY INLET STUDY
RELATION TIDAL PRISM TO INLET

Date	Min.CrossSection-Area of Inlet A(Sq.Ft)	Tidal Prism Spring Tide Range P (Cu.Ft).	Ratio $P/Ax10^4$
1947	$7.2x10^3$ (1)	4.5×10^8	6.3
July 1954	$1.11x10^4$ (2)	4.4×10^8	4.0
Aug. 1955	$1.95x10^4$ (3)	4.4×10^8	2.1
July 1958	$1.55x10^4$	4.3×10^8	2.8
Jan. 1959	$1.86x10^4$	4.3×10^8	2.3
June 1963	$1.78x10^4$	4.0×10^8 (4)	2.3
May 1970	$1.64x10^4$	4.0×10^8	2.4
June 1971	$1.72x10^4$	4.0×10^8	2.0 (6)
Fall 1972	$1.98x10^4$ (5)	4.0×10^8	2.0 (6)

1. Natural Channel, including San Diego River
2. Adjusted pilot channel, dredged in June 1950
3. Entrance channel dredged to full dimensions
4. Tidal Prism completed to full dimension
5. Proposed Maintenance dredging of entrance channel
6. The O'Brien equilibrium equation for the relationship between the entrance channel with two jetties and the tidal prism for a tidal prism(P) of $4.0x10^8$ cu.ft. gives an equilibrium cross-sectional area(A) of $1.0x10^4$sq.ft. and P/A is $4.0x10^4$. This is twice the value of P/A for a "Non-Scouring" channel.

Mr. John M. Nichol, Moffatt & Nichol, Engineers, had occasion to use Dr. Keulegan's method in a study of currents at Channel Islands Harbor (13) in 1970. He developed the following maximum entrance current velocities for the "non-scouring" entrance channels of Southern California:

San Diego	2.1 ft/sec.
Mission Bay	1.9
Newport Bay	1.5
Alamitos Bay	1.1
Marina del Rey	0.4

Thus it is seen that "tidal scour" is a very minor or non-existant factor in design and maintenance of these particular harbors.

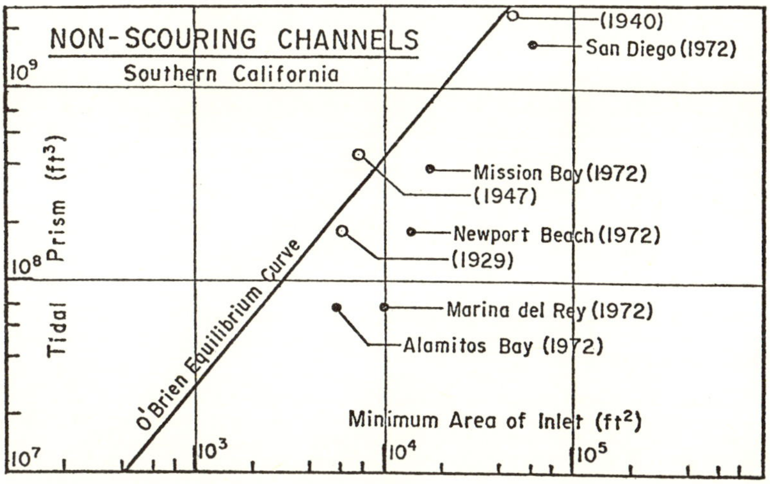

FIGURE 3

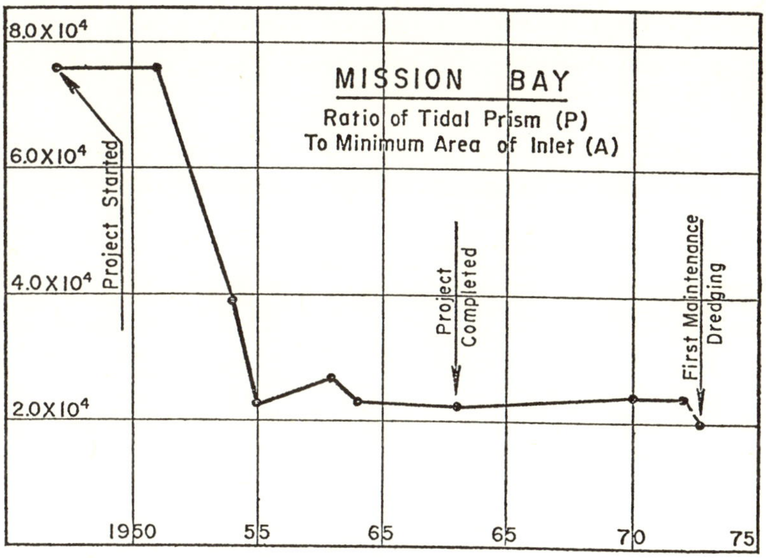

FIGURE 4

CONCLUSIONS

The principle of a "non-scouring-non-shoaling" channel, such as was developed for Mission Bay in 1946-49, is sound so long as the following criteria are satisfied: 1) there must be no exterior supply of shoaling material either from upland drainage or from littoral drift; 2) tidal velocities must be reduced to where there is no movement of bottom sediments within the bay or the entrance channel; 3) the entrance channel must be of sufficient depth to prevent appreciable movement of bottom sediments by wave induced currents.

Problems created by this type channel are: 1) possibility of such low tidal velocities as to fail to create adequate flushing action and cause pollution problems; 2) a wide entrance channel may permit excessive wave energy to be delivered to the inner basins. Measures must be taken to properly dispose of this energy by either absorption or reflection.

The most serious problem at Mission Bay resulted from the suspension of work on the project from 1951 to 1955 because of the Korean War, and the failure of the boating public to appreciate the hazards of using a partially dredged entrance channel. Completion of the entrance channel to full dimension eliminated this hazard.

A jetty-design deficiency had to be corrected because of the use of cap rock only in the tide and wave zone above mean lower low water level. The large voids (35 to 40%) inherent in caprock placement permitted an excessive amount of littoral sand to pass through the jetty into the channel. Sealing of portions of the jetties to the +4 foot elevation effectively reduced this shoaling factor.

Since completion of the entrance channel to full design dimension in 1959, there has been no serious shoaling or navigation problems. The first maintenance dredging scheduled for the latter part of 1972 indicates an average annual rate of shoaling of less than 20,000 cubic yards.

* * * *

REFERENCES

1. Westway Magazine(Oct. 1967). Southern California Auto Club.

2. Beach Erosion Board, U.S.Army, Corps of Engineers (Nov. 1942). Beach Erosion Study of Mission Beach, California.

3. Beach Erosion Board, U.S.Army, Corps of Engineers (1951). Unpublished Report by D. L. Forrest.

4. Patterson, John K. (Oct. 1965). Mission Bay, 1939-1965. City Engineering Dept., City of San Diego.

5. O'Brien, Morrough P. (May 1931). Estuary Tidal Prisms Related to Entrance Areas. Civil Engineering Vol. 1, No. 8, pp 738.

6. Robins, Thomas M. (Jan. 1933). Maintenance and Improvement of Entrance Channels to the Pacific Coast Ports. World Ports, Jan. 1933.

7. Los Angeles District, U.S. Army Engineers (March 1958). Addendum No. 1 to Definate Project Report, dated Jan. 1949, on San Diego River and Mission Bay.

8. Marine Advisers (Sept. 1961). A Study of Sea Swell and Seiches in Mission Bay. Wave Study at Mission Bay, Calif. (Sept. 1963).

9. U.S. Army Engineer Waterways Experiment Station, Vicksburg, Miss. (1970). Wave Action in Mission Bay, Technical Report No.________.

10. O'Brien, Morrough P. (Feb. 1969). Equilibrium Flow Areas of Inlets on Sandy Coasts. Journal of the Waterways and Harbor Division. Proceedings of the American Society of Civil Engineers Vol. 95, No. WW 1, Feb. 1969.

11. Johnson, J. W. (Feb. 1972). Tidal Inlets on the California, Oregon and Washington Coasts. University of California Hydraulic Engineering Laboratory. HEL-24-12, Berkeley,Calif.

12. Keulegan, G.H. (1967). Tidal Flow in Entrances. Technical Bulletin No. 14, Committee on Tidal Hydraulics, Corps of Engineers, U. S. Army.

13. Nichol, John M. (April 1970). A Study of the Effects of Waterway Expansion, Channel Island Harbor.

Part IV

COASTAL ZONE MANAGEMENT

Air photograph (1951) of Copano Bay and Fulton Beach Oil Field, Aransas County, Texas. (U.S.G.S. photo)

Editor's Comments on Papers 20 Through 24

> The [U.S.] Congress finds that the coastal zone is rich in a variety of natural, commercial, recreational, industrial, and esthetic resources of immediate and potential value to the present and future well-being of the nation. . . .

The above is an excerpt from the Coastal Zone Management Act of 1972. The Act was the result of an emerging concern for environmental quality and a growing interest in developing the resources of the oceans with minimal impact on the fragile coastal zone. The Act facilitates the development of policies, standards, and processes for coastal resources management by thrity-four United States coastal states and territories. It provides an opportunity for balancing competing uses which have a direct and significant impact on ocean and lake shorelines.

In view of recent developments involving proposed large scale exploration and exploitation of oil from the continental shelves, the U.S. House of Representatives passed a bill on March 11, 1976, to amend the Coastal Zone Management Act of 1972. The new bill, if it becomes law, would set up a new comprehensive coastal OCS-energy impact grant program, allowing coastal states more time to develop coastal management plans, and increase federal funding for state programs.

In Part III of this volume, human activities affecting sources of sediments and resulting in beach erosion problems were discussed and analyzed. However, man's activities affecting natural balance in the coastal zone are not limited only to those affecting sediment supply. Competing economic and industrial activities have resulted in pollution of coastal waters by human wastes, chemicals, pesticides, radioactive waste, oil spills, and thermal effluents. Dredge and fill activities have resulted in critical modification of the natural balance in the ecology of plants and animals in tidal areas, wetlands, and estuaries. A study on coastal area development conducted for the United Nations Economic and Social Council (1975) indicated that the coastal regions' value as fishing grounds is endangered by pollution and that oil spills threaten the world's busiest seaways. Concentration of coastal industry is increasing the risks of pollution in industrial countries like Japan and in the developing countries of North Africa. Although, only eighteen percent of the world's oil comes from offshore areas, the oil production and tanker traffic are beginning to harm fishing grounds. The study notes, in particular, the danger of oil spills in the Malacca Strait and the Persian Gulf.

In calling for a national policy for preservation and planning in the coastal areas of England and Wales, Steers (1944) stated that the narrow limits of the coastal belt and its great attraction are two of the principal factors involved in preservation problems and policies. He conducted a field survey of the coast, utilizing 1:25,000 maps on which all relevant details were directly inserted using a simple scheme of colors and notes. He evaluated the coastal scenery as exceptional, very good quality, or good quality; and showed natural coastal features as well as coastal developments. Such maps could easily be prepared for other coastal areas at the present time from air photographs and mosaics as well as from satellite imagery. Dolan and Vincent (1973) found that ERTS-1 (LANDSAT-1) imagery can provide a basis for land cover and land use mapping within the shoreline zone and that supportive data from underflights and field work should provide a planning base for future development of the sand beaches and barrier islands of the mid-Atlantic coast.

Steers (1944) stated: "The coastal areas must be enjoyed, and planning may suggest to some people limitation of access and control. It should certainly check access that leads to spoiling, but it should encourage any other form, including the right of walking around the coast. . . . There are, however, some areas where walkers should not be encouraged too freely. These include occasional dune areas where natural erosion is already serious and where wear and tear should not be made worse by man. . . . a great deal of harm can be done by walking along or through them and by making gaps which will be widened first by the wind and then by the sea."

Paper 21 by Professor Steers represents the last portion of his survey of the whole coastline of England and Wales. The paper covers mainly the southeast section of the coast as well as reassessment of his previous evaluation of the Dorset coast between Weymouth and Swanage. The major portion of the study was published in 1944 and is referred to above.

The paper by Professor Douglas Inman and Birchard Brush (Paper 20) presents an excellent summary and overview of the natural processes operating in the coastal zone and the extent of man's intervention. They state that man's intervention takes three interrelated forms: (1) the impact of increasing numbers of people competing in diverse and complex ways for portions of the coastal zone; (2) the pollution of the coastal zone; and (3) the critical modification of the natural balance in the ecology of plants and animals as well as in the sources of sediments that make the world's beaches.

A classic paper on the influences of man upon coastlines was published by Davis (1956). He classified man, in his relation to the coasts, sea, and land, into three ecological types, each of which affects the coastline processes and forms differently from the others. These are: (1) maritime men; (2) coastal men; and (3) inland men. *Maritime men* are similar to pelagic animals because they are dependent mainly upon the open sea for their livelihood. Their use of coastal lands has been mainly as ports. *Coastal men* are those who use the land near the sea and the shallow water of the sea. They engage in commerce and fishing near the shore. They cause the most significant effects on the coasts. They also engage in some agriculture, together with inland men crowded near the sea. These two groups have reclaimed large land areas from the sea. *Inland men* have been afraid of the sea but seek it for pleasure, recreation, commerce, and agriculture. They have built cottages and second homes along some coasts resting on dunes or in filled-in areas of the tidelands. Seawalls, groins, jetties, and other measures have been constructed to protect them from attack by the sea. In the Low Countries of Europe, great land reclamation projects have been undertaken by inland men for agricultural purposes. For thousands of years they have utilized the good soils of the deltas of the Tigris-Euphrates, Nile, Po, and Ganges rivers.

Davis also stated: "Man, in fact, is part of the total of natural forces, and he acts within a complexly integrated whole of natural processes. For this reason many of his activities indirectly affect numerous conditions along the coasts. Farming in the flood-plain and delta areas has had both direct and indirect effects, but how and how much is uncertain. We know little of the total or multiple effects of all these indirect activities and it will be difficult to predict some of the future influences."

Canals, dredging, and land reclamation and their relationship to land loss and environmental deterioration in deltaic coastal zones were examined by Gagliano (1973). He reported that within the 51,200-km^2 coastal zone of Louisiana, there are approximately 484 km^2 of man-made water bodies, more than 261 km^2 of spoil banks, and more than 192 km^2 of drained marsh. In addition to direct loss of habitat through dredging and spoil disposal, salt water intrusion, disruption of runoff patterns, accelerated erosion, and other secondary impacts of canals may result in severe environmental degradation. Urban encroachment into wetlands results in loss of valuable renewable resource areas.

Gagliano (1973) challenged the ideas of Wagret (1959) which claim that swamps and coastal lagoons are of little economic importance and when left in their original state they are worthless regions. Gagliano stated that "environmental scientists have learned that some of these 'unfavorable natural environments,' such as brackish marsh, are among the most productive places on earth. It has been established, in fact, that biologically, the brackish marsh is more productive than the 'rich, corn-growing land' into which it might be transformed. Further, as more and more swamps and marshes are miraculously transformed by the hand of man, the 'wretched inhabitants' come to realize that the natural landscape was very beautiful, and that which replaces it is often far less so. After they disappear, those things which were taken for granted are recognized as amenities which contributed to the quality of life of the areas' residents."

Gagliano and Van Beek (1975) reported that the Mississippi delta system has become the site of a serious conflict between optimum use and actual use of natural resources. Human activities have seriously altered the balance of the delta system and is largely responsible for the massive environmental degradation that has occurred during the past thirty to forty years. Most consequential has been man's efforts to confine the Mississippi river for navigation and flood protection. This eliminated overbank flooding which used to provide swamps and marshes with sediment and fresh water annually. Of equal importance are the channelization and dredging projects that resulted in deep, efficient channels, funneling sediment into deeper parts of the Gulf of Mexico, interrupting the natural process of delta building, and causing erosion of the beaches. In order to resolve such conflicts and prevent total collapse of the system, the authors proposed a multiuse management plan in which environmental management and landuse guidelines are closely integrated. Such a plan must protect the important ecological system and respond to the demands of urbanization and industrialization.

Other areas of high biological productivity along coasts, a prime source of food supply, are the estuaries. They represent areas of dynamic interaction, where fresh water from the rivers meets the salt water of the sea and where the tides play a major role in water circulation. Estuaries are also feeling the pressure of population growth and its intensified effects. The four largest cities in the world, Tokyo, New York, London, and Shanghai, are located around estuaries. The increasing pollution from human waste, oil, pesticides, industrial waste, and sediments from construction activities on land are seriously affecting the productivity of estuaries and in some cases have exceeded the natural capacity to absorb them. A good review on the role of man in estuarine processes was presented by Cronin (1967).

It is logical to conclude that a comprehensive coastal zone management plan should be based on careful analysis of the natural processes operating in an area, with an inventory of all natural features and resources, and an approach towards multiuse where developments can be compatible with biological productivity and geomorphic integrity.

The paper by Paul and Melinda Godfrey (Paper 22) is a good example of a study comparing the ecologic and geomorphic interactions between altered and unaltered barrier island systems in North Carolina. The Outer Banks is a chain of sandy barrier islands, fringed with salt marshes, and separated from the main land by lagoons. As a result of increased demands from recreation and other developments, combined with a program for stabilization of dunes and beaches, undesirable changes are taking place in the ecological and geomorphic processes which created and maintain the islands. The study is concerned with the natural system of the Cape Lookout National Seashore area. It paralleled the research done by Dolan (1972) on a developed section of the Outer Banks (Cape Hatteras National Seashore) and arrived at similar conclusions, independently and through different routes. The observations indicate that when a barrier system is not altered, its vegetation and geomorphic processes are closely interdependent; in contrast, an artificial system may separate them with sometimes dire results in that both vegetation and development may be very close to the ocean. They concluded that man, not the sea, is the worst enemy of barrier islands today, and that present undeveloped seashores should be preserved in their natural conditions.

Air photography and imagery from other remote sensors can provide valuable information for planners interested in coastal zone management. They have an added value in that they can be quantified and manipulated for automatic data readout. They provide a fast, relatively low-cost technique for determining water quality in estuaries and general vegetation characteristics in swamps and marshes. Jamison (1971) reported that with a color air photograph of the coastal zone of Puget Sound at a scale of 1:12,000, one can identify and plot the type of material that makes the shore, the slope of the beach, and some of the surface dwelling organisms such as seaweed, eel grass, and sand dollars. Kelly (1969) from a study of nearshore ocean biology using air photography concluded that four basic types of environmental information were available from study of remote photographs of the ocean bottom: (1) mapping of the distribution of biological cover on the bottom; (2) observation of the basic environmental relationships of the biological communities; (3) detection of unusual or previously unrecognized features; and (4) analysis of man's activities and their effects. From a study of high-altitude air photographs of the Oregon coast, Keene and Pearcy (1973) concluded that the photographs provided an exceptional "big picture" perspective of the ocean of nearly shipboard resolution. Together with space photographs, they show that oceanographic features related to productivity, ocean circulation, and pollution may be manifested as surface phenomena and can readily be detected from altitudes of eighteen kilometers, which enhances our understanding of the marine environment. A study on the impact of remote sensing in management of New Jersey's coastal zone by Feinberg and Mairs (1974) reported that repetitive high-altitude photography was quite useful in studying offshore waste disposal, ocean outfalls, and shore protection problems, while repetitive ERTS–1 photography has great value in detecting land use changes in coastal areas.

The paper by Reimold, Gallagher, and Thompson (Paper 23)

shows how it is possible to increase the accuracy of primary production estimates from large marsh areas by employing remote sensing techniques. Such techniques and measurements are important in planning industrial and recreational developments in the coastal zone as they provide information on the spatial distribution of species and primary production in the marsh.

The paper by Hunt (Paper 24) presents two recent applications of remote sensing in defining shoreline geology for engineering purposes. One area was in Barbados, where the project included a sewer system, treatment plant, and ocean outfall in an environment of coral reef formations. The other area involved an investigation of dredge spoil disposal sites along a fifty-four-kilometer stretch of Cape Fear River in North Carolina. The paper concluded that substantial information on the geologic environment of coastal regions was obtained from interpretation of air photography and that the use of conventional test borings and geophysical methods without the use of imagery can be extremely costly and time consuming.

20

Reprinted from *Science* **181**:20–32 (July 6, 1973)

The Coastal Challenge

Douglas L. Inman and Birchard M. Brush

Sand dune south of Pismo Beach, California. See page 20. [U.S. Navy]

Dr. Inman is professor of oceanography and Mr. Brush is a senior development engineer; both authors are affiliated with Scripps Institution of Oceanography, University of California, La Jolla 92037. This article is based on a talk entitled "Degradation of the coastal environment" that was presented 9 December 1970 at the San Francisco meeting of the American Geophysical Union.

The shoreline is the unique boundary separating the earth's three domains: the land, the sea, and the atmosphere. It is one of man's oldest frontiers and one of his most informative paleo-oceanographical tablets. The nearshore and estuarine waters exert a predominate influence in the everyday affairs of man. Although the continental shelves and nearshore waters comprise only about 5 percent of the area of the world, about two-thirds of the world's population lives near the coast. This imbalance has its roots in antiquity for mankind's first freeways were the seas which, once oars and sails appeared, became the means of bypassing the rigors of overland travel. Sailors traditionally have sought the comparative safety of the open sea in preference to the hostility of the shores' dangerous shoals, strong and unexpected currents, and destructive breaking waves.

To the casual observer, the coastlines of the world look craggy and indomitable, the rocky stretches seem tough, and the beaches appear to be permanent. But the sand on the beaches is easily moved and is in increasingly short supply. In reality, then, the beaches are fragile ribbons of sand that are frequently broken by acts of nature and man. The waters that bathe them have a limited flushing capacity, and yet they have become the most common depository for mankind's wastes (Fig. 1).

Fig. 1. Beauty died here.

Pictorially and esthetically diverse as they may be, the coastal and nearshore zones of the world all share common dynamic experiences (Fig. 2). Here, shore processes begin the mixing, sorting, and transportation of sediments and of runoff from land. Waves, winds, and currents mold the shorelines of the world, and their interaction with the land and its runoff determines the configuration of coastlines and the adjacent bathymetry.

Man's rapidly expanding use of the ocean and his increasing incursion into it commonly involve processes that take place in shallow water. These same shallow waters are experiencing the bulk of the impact of waste discharges, thermal and radioactive pollution, dredging, coastal construction, mining, and poaching.

The Coastal Zone

The oceans and seas have a profound effect on the continents, exerting the controlling influence on weather and climate. Indeed, the sea completely dominates the environmental aspects of the land bordering it. In this connection it is useful to describe coastal areas in terms of a "coastal zone" and a "shore zone" (Fig. 3).

The coastal zone may be defined in terms of the large-scale tectonic and erosional-depositional features which have lengths along the coastline of the order of 1000 kilometers and widths extending from the coastal plain out into the water of an order of magnitude less. The coastal zone is composed of the coastal plain, the continental shelf, and the waters that cover the shelf; it also includes other major features such as large bays, estuaries, lagoons, coastal dune fields, river estuaries, and deltas.

The shore zone is the sedimentary and solid surface associated directly with the interaction of waves and wave-induced currents on the land and on

Fig. 2. Aerial photograph of rip currents along a beach. The occurrence of rip currents is universal; the spacing between rips determines the longshore dimension of the nearshore circulation cell.

the runoff products from the land. The shore zone includes the beach, the surf zone, and the nearshore waters where wave action moves bottom sediments. The shore zone extends landward to the sea cliffs that border the backshore of a beach, and, where wave-deposited structures such as barrier islands and spits are narrow, across these features to the cliff or coastal plains bordering shallow lagoons. All bodies of water have "shore zones," the extent and configuration of which depend upon the length and height of the waves, the range of the tide, the degree of exposure to winds, and the size of the wave-deposited structures. As a practical concept, the waters involved in the coastal zone may be considered to include the shallow seas and waters covering all the continental shelves of the world which total an area of 29×10^6 km^2, or about 8 percent of the surface area of the world oceans. The nearshore waters are bounded on the landward side by shorelines that total about 440,000 km in length (*1*) and on the seaward side by the break in slope at the shelf edge, which marks the change from the relatively horizontal shelf to the steeper continental slope. The continental slope marks the topographical and structured edge of the continents and is the boundary between relatively shallow water covering the shelf and the great depths of the true oceans. The continental slopes, one of the striking geographic features of the earth, have a combined length of about 150,000 km (*1*).

Conventionally, the depth of the shelf edge is taken as about 200 meters (100 fathoms), although in some localities the depth may be as great as 400 m. On a worldwide basis, the average depth of the shelf edge is about 130 m. The width of the shelf ranges from nearly zero to more than 1300 km and averages about 74 km (*2*).

Because of the mutual interdepen-

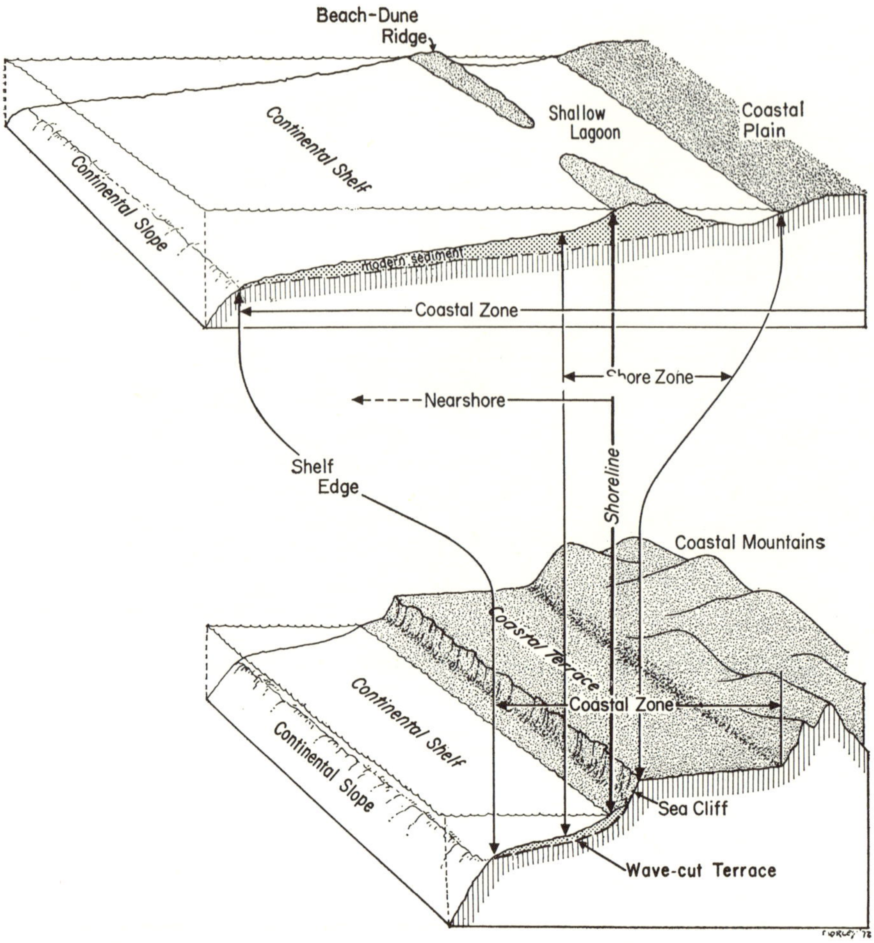

Fig. 3. Definition sketch for coastal zone nomenclature. The type of coast is related to its relative position on the moving plates of the tectosphere; wide-shelf plains coasts (upper part) and narrow-shelf mountainous coasts (lower part) are characteristic of the east coast (trailing edge) and west coast (collision edge) of the Americas, respectively. [After Inman and Nordstrom (*1*)] [Courtesy of the University of Chicago Press, Chicago, Illinois]

dence of processes in the coastal and nearshore zones, it is difficult to decide where to begin a discussion of the processes that affect the shore. One can break into this circle by asking: What are the important shore processes, and what coastal factors affect them? In a very general sense, the important factors are: (i) the degree of exposure to waves and currents, (ii) the supply of sediment and runoff to the coast, (iii) the topography of the continental shelf and the adjacent coast, (iv) the tidal range and intensity of the current, and (v) the coastal climate

Coastal climate is principally dependent upon latitude and the location of the major ocean current and wind systems; the topography of the shelves and coast are closely associated with the geologic setting of the coasts and the origin of the adjacent ocean basins (*1*). The potential of man's intervention becomes apparent when one compares the physical processes operative in nearshore waters and the natural balance of energy that drives them with man's requirements in terms of usage and waste disposal.

Physical Processes in Nearshore Waters

The important processes that operate in the nearshore waters of oceans, bays, and lakes are similar. They differ in intensity and scale, variables which are determined by the energy in the waves and in the physical dimensions of the surf zone. Moreover, it is becoming increasingly clear that processes in nearshore waters are driven by basic, interrelated forces that are systematic and essentially regular in form. These systematic driving forces lead in turn to the development of coherent processes such as the nearshore circulation cells (Fig. 2) and the longshore transport of sand that are basically similar the world over.

Most of the energy for these processes comes from the sea (Fig. 4). It is translated by the stress from winds which blow over the ocean and generate waves of many sizes and frequencies. It is produced by the gravitational attraction of the moon and sun acting on the mass of the ocean. It is transmitted by various, sometimes impulsive, disturbances at the boundaries of the ocean with the atmosphere and with the sea bottom, which generate additional waves of other sizes and frequencies. These waves travel across the ocean with little loss of energy, until the configuration of the landmass and of the adjacent shelves and the slope of the bottom direct and focus the waves for their final assault on the coast. Straight, gently sloping shores accept energy uniformly, whereas headlands, points, and peninsulas tend to attract and concentrate energy.

The energy available to nearshore waters is dissipated in various ways, including the reflection of waves, the generation of turbulence and currents, the transportation of sediment, and the formation of other kinds of waves.

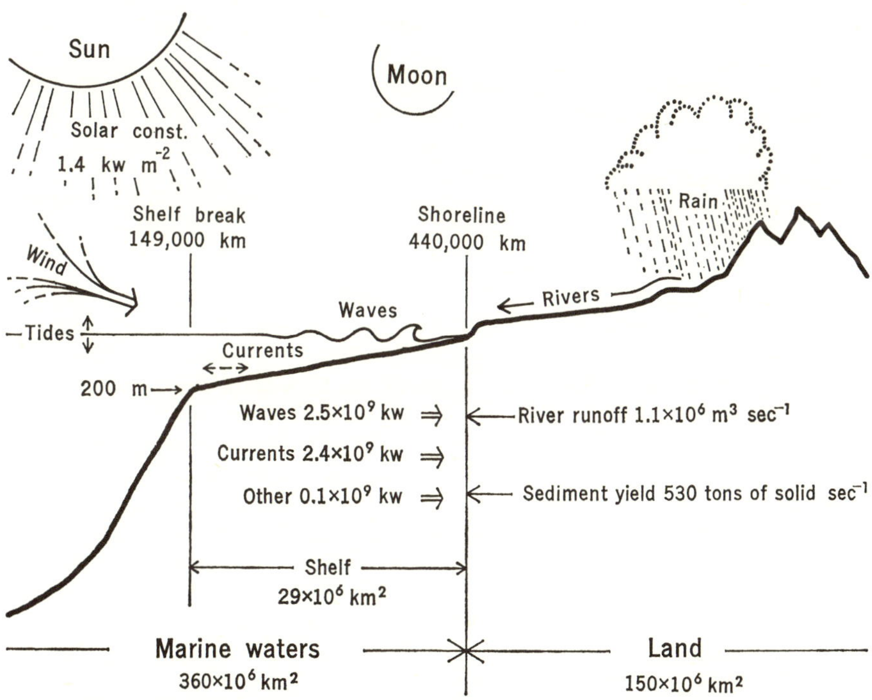

Fig. 4. Budget of energy and land runoff in the coastal zone. Most of the energy comes from the open sea.

Some energy appears to become "trapped" against the coast rather than reflected, thus leading to a higher energy level at the coast. The trapped energy may take the form of edge waves (*3*), which are special modes of surface waves that travel along the coast, or long-period resonant oscillations called "shelf seiche."

Circulation Cells and Mixing

The many forms of energy flux and the high rates of energy dissipation over the shelf and in the surf zone further complicate processes within the nearshore environment. Although in deep water the mutual interaction of waves is relatively weak, in shallow water the interaction of waves with other waves, of waves with currents, and of currents and waves with the bottom can produce strong interchanges of energy. For example, the waves in shallow water produce bottom-boundary currents flowing in the direction of wave propagation, and these currents play a significant role in the transport of sediments over the shelf (*4*). Pressure fields produced by waves traveling in shallow water change the average water level, reducing it near the breaker zone and increasing it where the waves run up the beach face (*5*).

The interaction of surface waves moving toward the beach with edge waves traveling along the shore produces alternate zones of high and low waves that determine the position of rip currents (Fig. 2). The pattern that results from this flow takes the form of a horizontal eddy or cell, called the nearshore circulation cell (*6, 7*). The rip currents are the "freeways" across the surf zone for experienced surfers and the greatest cause of drownings for inexperienced swimmers.

The nearshore circulation system produces a continuous interchange between the waters of the surf zone and the waters of the offshore zone, acting as a distributing mechanism for nutrients and as a dispersing mechanism for land runoff. Offshore water is transported into the surf zone by breaking waves, and particulate matter is filtered out on the sand of the beach face. Runoff from land and pollutants introduced into the surf zone are carried along the shore and mixed with the offshore waters by the rip currents (*8*).

Two well-defined mechanisms dominate mixing processes in the surf zone; each has distinctive length and time scales determined by the intensity of the waves and the dimensions of the surf zone. The first process is associated with the breaking wave and its bore, which produce rapid mixing in both the onshore and the offshore direction, alternatively. This mixing, when normalized and averaged over the surf zone width X_b, gives coefficients of eddy diffusivity of the order of $H_b X_b / T$, where H_b and T are, respectively, the breaker height and the period of the waves (Fig. 5). The second process is advective and is associated with the longshore and rip current systems in the nearshore circulation cell. A constant longshore discharge of water between cells Q_l gives a concentration N_n in the nth cell down-current (in the direction of current flow) from a continuously injected source of tracer of

$$N_n = N_0 (Q_l / Q_m)^n \qquad (1)$$

where N_0 is the concentration of tracer leaving the injection cell and Q_m is the maximum longshore discharge of tracer within a cell (Fig. 5). As an approximation, the concentration decreases exponentially with the distance y from the injection point when n is replaced by y/Y, where Y is the spacing between rip currents. This relation gives an apparent longshore eddy mixing coefficient of the order of $Y \cdot \bar{v}_l$, where $\bar{v}_l$ is the longshore current velocity. Along ocean beaches $H_b X_b / T$ and $Y \cdot \bar{v}_l$ are about 10 and 100 $m^2 \ sec^{-1}$, respectively.

Once the tracer front has passed through a particular cell, tracer is introduced into the adjacent offshore waters by the rip current. In the absence of coastal currents, the tracer remains in the offshore area where it is available for recirculation into the cell. Recirculation of tracer decreases the dilution capacity of the cell and causes the concentration to increase, as shown schematically in Fig. 6B. Thus the concentration becomes a function of both distance and time, and the simple equation for mixing (Eq. 1) no longer holds.

The tracer tends to remain in the secondary mixing zone (area, $X_r \cdot Y$) because the rip currents move slowly

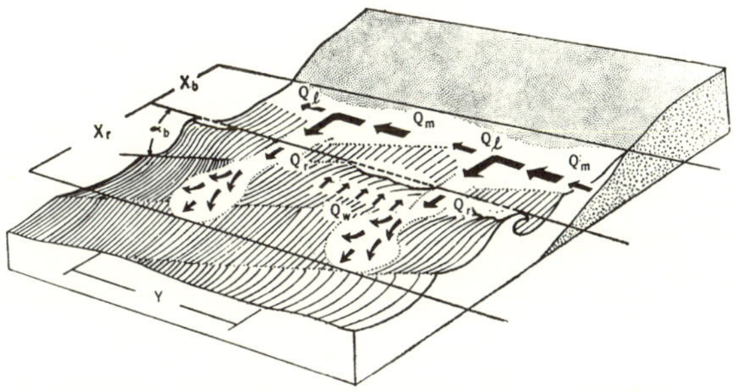

Fig. 5. Definition sketch of the model for mixing when waves break at angle α_b with the beach. The zone between rip currents constitutes a circulation cell; primary mixing occurring in the surf zone (area, $X_b \cdot Y$), and secondary mixing occurs in the offshore zone between the heads of the rip currents (area, $X_r \cdot Y$). The water discharges in volume per unit time are as follows: Q_l, from cell to cell; Q_m, maximum within a cell; Q_w, onshore transport associated with breaking waves; and Q_r, offshore flow in a rip current. [From Inman *et al.* (*8*)] [Courtesy of the *Journal of Geophysical Research*, Washington, D.C.]

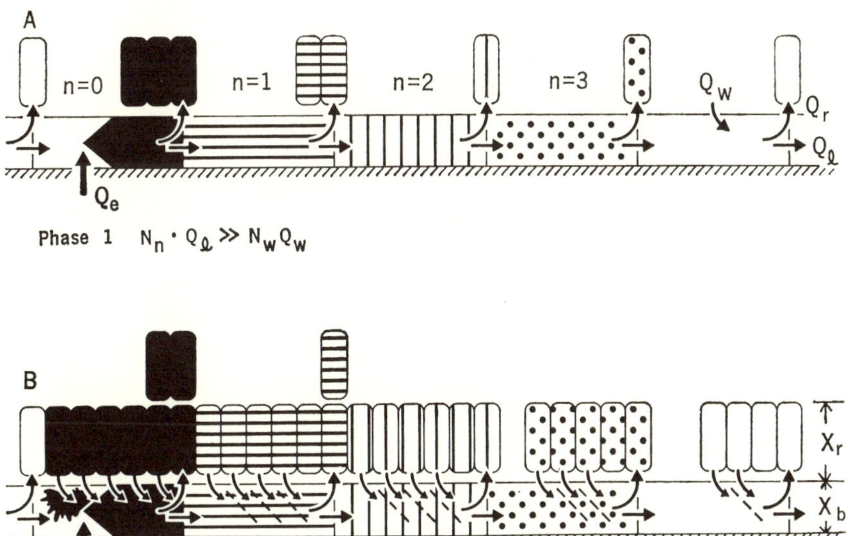

Fig. 6. Schematic diagram of the longshore mixing of tracer Q_e, continuously injected. (A) Phase 1 applies when recirculation from the secondary mixing zone is negligible; (B) phase 2 applies when the background concentration becomes significant and recirculation causes the system to become saturated. Symbols are defined in Fig. 5. [From Inman *et al.* (*8*)] [Courtesy of the *Journal of Geophysical Research*, Washington, D.C.]

down the coast with the longshore current. Thus the rip currents continue to flow seaward into tracer-free waters until the entire secondary mixing zone is filled with tracer of concentration N_n. A situation in which the secondary mixing zone is about half filled with tracer is shown by the extent of offshore discoloration in Fig. 7.

It is apparent from Fig. 6B that the presence of background concentrations of tracer in the offshore zone decreases the concentration gradient by recirculating material already there. This in turn leads to an exponential buildup of concentration until both the surf zone and the region offshore become saturated.

Circulation over the Shelf

Coastal circulation on a much larger scale but of reduced intensity occurs over the entire shelf. This circulation may take the form of eddies and countercurrents from the permanent ocean current systems that flow across the shelf; the circulation may be due to tidal currents, or it may be induced locally by the wind in the form of the upwelling of deep water and the horizontal flow of water along the coast.

Coastal circulation cells of large dimension are also associated with the submarine canyons that cut across the shallow shelves of the world. These canyons act as deep, narrow conduits connecting the shallow waters of the shelf with deeper water offshore. At times, strong seaward flows of water occur in the canyons, so that they resemble large-scale rip currents. The canyon currents produce circulation cells having dimensions of the shelf width and spacing comparable to that between the submarine canyons (Fig. 8). These strong currents in submarine canyons seem to be caused by a unique combination of air-sea-land interactions (*9*) consisting of the following: (i) a pileup of water along the shoreline caused by strong onshore winds; (ii) down-canyon pulses of water caused by the surf beat of the incident waves; (iii) a shelf seiche excited by the waves and by the pressure fluctuations in the wind field; and, finally, (iv) the formation of steady down-canyon currents as the weight of the sediment suspended by the currents overcomes the density stratification of the deeper water.

There is less pileup of water by the wind over the canyon than over the adjacent shelf, and, as a result, the shallower waters over the shelf flow

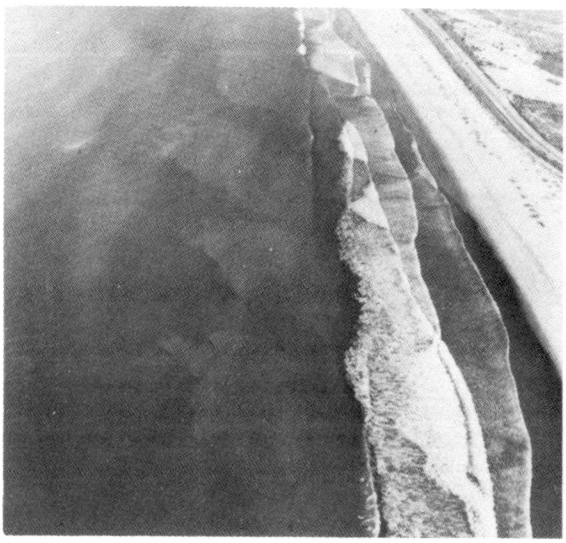

Fig. 7. Photograph showing phase 2 mixing and an increase in the offshore concentration by rip currents as shown by the plumes of discolored water. The discolored water is being recirculated, causing the mixing to approach saturation (see Fig. 6).

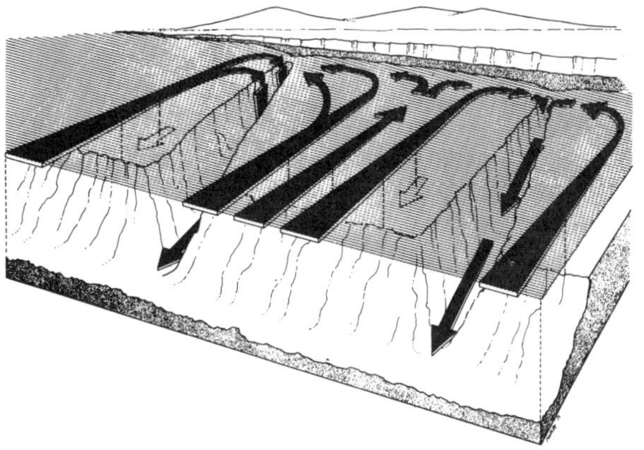

Fig. 8. Circulation over narrow shelves is markedly influenced by submarine canyons. Onshore winds produce an onshore flow of water at the surface with compensating seaward flow at the bottom. Strong winds and waves also produce offshore flow down submarine canyons.

along the coast toward the canyons and, as in the case of rip currents, flow seaward over the canyons. The surf-beat waves (waves caused by the beat of incident waves of different frequency) and the shelf seiche are very long waves, whose interaction with the shelf and canyon produces current pulses capable of suspending sediment. When these pulses are of sufficient velocity to suspend sediment and to maintain a suspended load of sediment, autosuspension results and the pulsating current is converted into a self-maintaining turbidity current that travels down the canyon into the adjacent deep waters of the ocean (*9*).

Longshore Transport of Sand

Wherever there are waves and an adequate supply of sand (or coarser material), beaches form. The back-and-forth motion of waves in shallow water produces stresses on the bottom that place sand in motion. The interaction of the wave stresses with the bottom also induces a net boundary current flowing in the direction of wave travel. Over a gently shoaling bed, the back-and-forth motion of sand in the presence of the boundary current produces a net transport of sand in the direction of wave travel. Thus, waves traveling toward the shore tend to contain sand against the shore, and eventually to produce a profile that is in equilibrium with the energy dissipation of the waves (*10, 11*).

When waves approach the coast at an angle, they cause sand to be transported along the shore. For the special case of longshore transport of sand in the surf zone, the breaking wave supplies the power needed to place the sand in motion, as well as the longshore current that carries the sand load. The longshore transport of sand I_l expressed in immersed weight (for example, newtons per second) is directly proportional to the onshore flux of longshore-directed energy $\boldsymbol{P}_l$ (for example, watts per unit length of shoreline) such that:

$$I_l = KP \qquad (2)$$

where K is a dimensionless constant with a value of 0.77 (*12*). Thus it is apparent that the longshore transport of sediment along a sandy coast can be estimated if the budget of wave energy (that is, the wave climate) is known.

Unfortunately, there are few localities for which the wave climate is quantitatively known, so that the volume of littoral transport along oceanic coasts is traditionally estimated from the rate of growth of spits (Fig. 9) (*13*), or from the observed rates of erosion or accretion, most often in the vicinity of coastal structures such as groins or jetties. In general, beaches build sea-

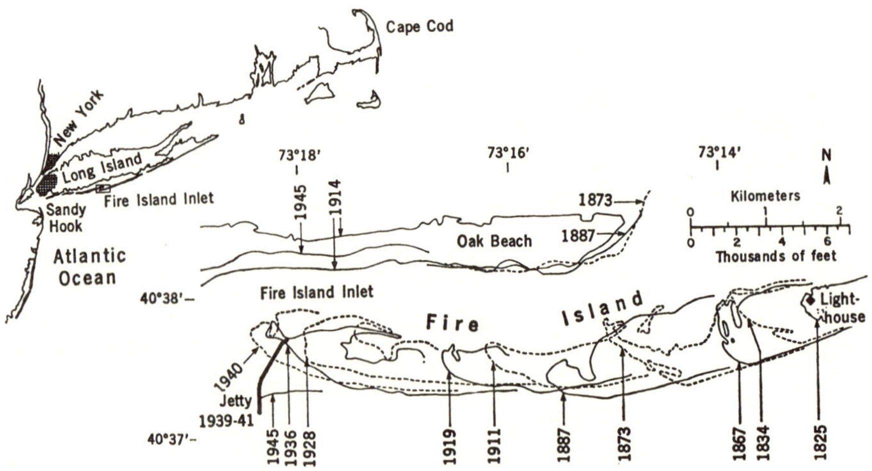

Fig. 9. Longshore sand transport has caused an extension of the Fire Island spit at an average rate of 100 m per year since 1825. [After S. Gofseyeff (*13*)] [Courtesy of J. W. Johnson, School of Engineering, University of California, Berkeley]

ward up-current from obstructions and are eroded on the lee side of the current where the supply of sand is diminished. Such observations indicate that the rate of transport of sand varies from almost zero to several million cubic meters per year, with average values of 150,000 to 1,600,000 m^3 $year^{-1}$. Along the shores of smaller bodies of water, such as the Great Lakes, the littoral transport rate can be expected to range from about 1,000 to 150,000 m^3 $year^{-1}$ (*14*). In general, these are conservative estimates, since the volume of material moved commonly exceeds that indicated either by deposition or by erosion.

Natural Balance of Energy and Sediment

The energy which drives processes in the nearshore zone comes from the sea and atmosphere. Waves and currents from the open ocean propagate toward coasts where they are concentrated and eventually dissipated. At the shoreline the breakers, storm surges, tides, and secular changes in sea level attain their greatest height.

The wind systems are the direct links between the atmosphere and the ocean, and, through momentum exchange at the ocean's surface, transfer kinetic energy at the rate of about 10^{11} kilowatts (1 kw $= 10^{10}$ erg sec^{-1}) (*15*). Winds plus ocean and earth tides are directly or indirectly responsible for most of the energy dissipated in the nearshore zones of the oceans. The most common form of nearshore energy is that associated with the wind-generated surface waves.

A wave 3 m high transmits energy at the rate of 100 kw per meter of its crest line. Toward 1 km of coastline it transmits 100,000 kw. The power of such waves is equivalent to a solid line of automobiles, each 270 horsepower, advancing side by side at full throttle (*16*). On an oceanwide basis, wind-generated waves probably have an average height of about 1 m, and they transmit power at a rate of about 10 kw per meter of wave crest. If these waves were to break continuously along the 440,000 km of the world's shoreline, they would dissipate energy at the rate of 4.4×10^9 kw. This number is probably too high for a worldwide average dissipation of energy because all shorelines are not exposed to the open ocean. A reasonable estimate of the average rate of energy dissipation of surface waves in nearshore waters is about one-half to two-thirds of this amount, say, 2.5×10^9 kw, which is equivalent to the energy production of 10,000 large power plants (Table 1).

Tides produce motion even in the deepest oceans. However, this motion is usually only a few centimeters per second so that tides dissipate very little of their energy on the deep sea bottom. The principal tidal dissipation results from the flow of strong tidal currents in shallow areas, such as the Bering Sea, the Okhotsk Sea, and the Argentine Shelf. About half of the tidal current dissipation occurs in five shallow seas having a total area of only 10 percent of the shelf area of the world. The total rate of energy dissipation by lunar and solar tides in the shallow waters of the world oceans is about 2.2×10^9 kw, or only slightly less than the rate from wave action (Table 1).

By comparison with waves and tides, other sources contribute relatively little energy to coastal waters. For example, the most spectacular waves in the sea are tsunamis, which are well known because of the loss of life and great damage to coastal structures associated with their passage. However, large tsunamis with energy contents as high as 5×10^{15} joules occur only about five times per century (*17*). The average rate of energy dissipation of a tsunami is about 10^5 kw, which is four orders of magnitude less than the total for waves and tides.

The total rate of dissipation of mechanical energy in the shallow waters of the world is about 5×10^9 kw (Table 1). Dissipation by one process or another occurs over the entire shelf (Fig. 4). The wind-generated surface waves dissipate their energy primarily nearshore, especially in the breaker zone, whereas tidal and other ocean currents and internal waves dissipate most of their energy over the outer portions of the shelf. Internal waves, edge waves, shelf seiche, and local winds may produce water motion over the continental shelf and submarine canyons. Thus, it is apparent that an understanding of processes in the nearshore environment requires a careful assessment of the amounts of energy in the various kinds and modes of wave and current motion as well as a determination of the mechanics of interaction of waves, currents, and sediments.

Table 1. Estimates of the natural rates of dissipation of mechanical energy in the shallow waters of the world (*16*) in units of 10^9 kw (1 kw = 10^{10} erg sec^{-1}).

Source	Rate
Wind-generated waves breaking against the shoreline (*39*)	2.5
Tidal currents in shallow seas (*40*)	2.2
Large-scale ocean currents in shallow seas (Guiana Current off the northeastern coast of South America, 0.13; Falkland Current over the Argentine Shelf, 0.03)	0.2
All other sources (wind stress on the beach, 0.01; internal waves, 0.01; edge waves; shelf seiche; tsunamis; rivers entering the oceans)	0.1
Total	5.0

Beaches are composed of whatever clastic material is locally in greatest abundance. The principal sources of beach and nearshore sediments are as follows: the rivers, which bring large quantities of sand directly to the coast; the unconsolidated material of the sea cliffs, which is eroded by waves; and material of biological origin, such as shells, coral fragments, and skeletons of small marine organisms. Many beaches, such as those along the east coast of the United States, are supplied by sand that has been reworked by waves and currents from ancient river and glacial material deposited during former stillstands in sea level (*18*).

Streams and rivers are by far the most important source of sand for beaches in temperate latitudes. Cliff erosion probably does not account for more than about 5 percent of the material on most beaches, except locally on trailing edge coasts such as the east coast of the United States. Wave erosion of rocky coasts is usually a slow process, even in cases where the rocks are relatively soft shales. On the other hand, erosion rates greater than 1 m year^{-1} are not uncommon in unconsolidated sea cliffs.

The contribution of sand by streams in arid countries is surprisingly high. This is so because arid weathering produces sand-size material and inhibits the growth of vegetation that would protect the land from erosion. Therefore, in an arid climate, occasional flash floods transport large volumes of sand. The maximum sediment yield occurs from drainage areas where the mean annual precipitation is about 30 centimeters (*19*).

Traditionally, geologists have estimated long-term erosion and deposition rates from the amount of material deposited during geologic time. Such estimates give erosion rates varying from about 1 to 4 cm per 10^3 years for drainage basins of moderate relief to as much as 21 to 100 cm per 10^3 years for the Himalayas (*20*). Assessment of the volume of sedimentary material on the continental United States and in its adjacent sea floors indicates that the average erosion rate of the United States during the past 600×10^6 years was 3 to 6 cm per 10^3 years (*21*).

An increasing number of measurements of river discharge have provided an independent estimate of contemporary erosion rates of the land. The average discharge of the world's rivers totals 1.1×10^6 m^3 of water per second. Measurements suggest that the average suspended plus bed load is about 480 milligrams per liter, which gives a total discharge of particulate solids of 530 metric ton sec^{-1} (Table 2) (metric units are used throughout this article except in column 1 of Table 3). This gives a contemporary average erosion rate of the land of about 6 cm per 10^3 years (*22*), which is somewhat higher than the average erosion rates estimated from deposition during geologic time. As we will show below, the increase in the erosion rate is probably associated with man's intervention.

It is of interest to compare the total fluxes of energy and sediments (particulate solids) into the coastal waters of the world. Since energy may be thought of as the ability to do work, the flux of energy associated with waves and currents gives some measure of their potential to transport sediment. For example, let us assume that the estimated total flux of wave energy (2.5×10^9 kw) is available for transporting sand along the world's coastline, and that the longshore-directed energy flux, P_l in Eq. 2, is equal to one-tenth of the incident energy flux. From Eq. 2 this gives a rate of immersed weight transport of 1.9×10^{11} newton sec^{-1}, which is equivalent to a (dry) mass transport of 3.1×10^7 ton sec^{-1} for the world's beaches, a mass flux 5.8×10^4 greater than that supplied to the coast by all of the erosion of the continents.

Table 2. Estimate of the natural run-off of fresh water and solids from the continents into the coastal waters of the world (see Fig. 4).

Source	Rate
Discharge of water into the oceans from all rivers (*41*)	1.1×10^6 m^3 sec^{-1}
Flux of dissolved solids (*42*)	125 ton sec^{-1}
Flux of particulate solids (*43*)	530 ton sec^{-1}
Average erosion rate of land (*22*)	6 cm per 10^3 years

The budget of sediment for a region is obtained by assessing the sedimentary contributions and losses to the region, and their relation to the various sediment sources and transport mechanisms. However, it is not a simple matter to determine the budget of sediment, since such a calculation requires a knowledge of the rates of erosion and deposition as well as an understanding of the capacity of the various transport agents.

Studies of the budget of sediment show that coastal areas can be divided into a series of discrete sedimentation compartments called "littoral cells." Each cell contains a complete cycle of littoral transportation and sedimentation, including sources and sinks of sediment and transport paths (*11*).

Along the coast of Southern California (Fig. 10), the principal sources of sediment for each littoral cell are the rivers, which periodically supply large quantities of sandy material to the coast. The sand is transported along the coast by waves and currents until the "river of sand" is intercepted by a submarine canyon, which diverts and channels the flow of sand into the adjacent submarine basins and depressions.

There are five littoral cells in Southern California. Each cell begins with a stretch of rocky coast where the supply of sand is limited (Fig. 10). In a downcoast direction, determined by the prevailing waves, the beaches gradually become wider and the coastline straightens where the streams supply a sufficient amount of sand. Submarine canyons terminate the littoral cell by capturing the supply of sand, thus causing the next cell to begin with a rocky coast devoid of beaches.

The concept of littoral cells and their sedimentary budget applies to all coasts, the budget differing principally in the nature of the sources and sinks for the sediment. Along coasts having large estuaries such as the east and gulf coasts and portions of the Oregon coast, the river sand is trapped in the estuaries and cannot reach the open coast (*23*). For these coasts the sediment is produced by the erosion of sea cliffs and shelf sediments deposited at a lower stand of the sea, whereas the sinks are sand deposits that tend to close and fill the estuaries (Fig. 9).

The Extent of Man's Intervention

The continental shelves are the sites of rich oil and mineral deposits, and the shallow waters covering them include much of the plant and animal life in the sea. Nearshore waters impinge on the beaches, harbors, and estuaries, where utilities, industry, recreation, and the human habitat compete

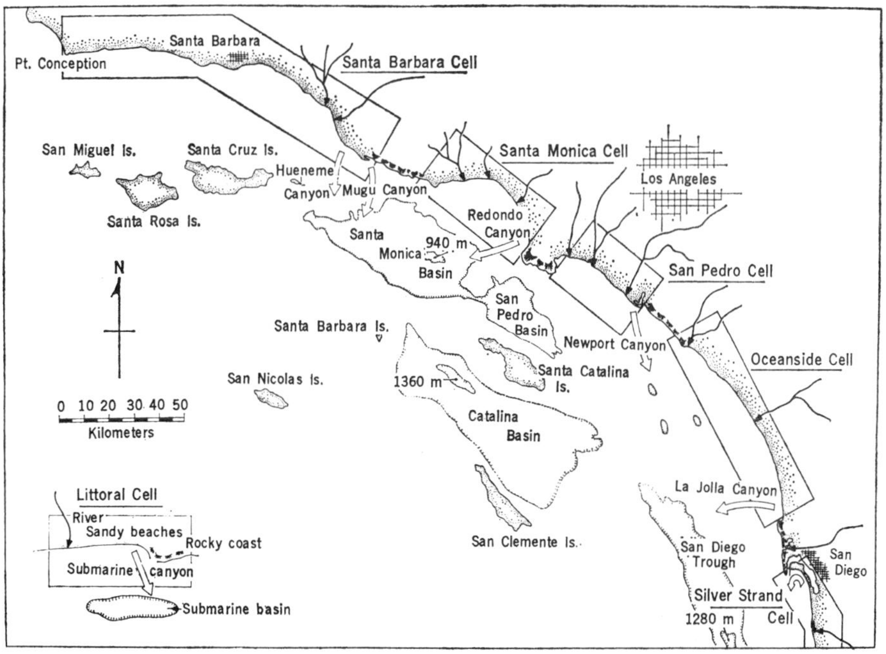

Fig. 10. Illustration of the five littoral cells along the Southern California coast. Each cell contains a complete sedimentation cycle. Most sand is brought to the coast by streams, carried along the shore by waves and currents, and lost into offshore basins by submarine canyons. [After Inman and Frautschy (*11*)]

Table 3. Potential for man's intervention in terms of the use of energy and the disposal of waste in recent years (1967–1971). The "extrapolated world value" is based on present U.S. standards of use and a population of 200×10^6 extrapolated to a present world population of 3.5×10^9. Heat units are converted to their mechanical equivalents.

Item, location, source, and amount	U.S. per capita value (metric units)	U.S. doubling time (years)	World value extrapolated from U.S. standards (metric units)
	Power, United States, 1968		
Total energy consumed, 62.5×10^{15} British thermal units (*44*)	10.4 kw	14–20	36.4×10^9 kw
Electrical energy consumed, 1.33×10^{12} kilowatt-hours (*44*)	0.8 kw	9	2.7×10^9 kw
Waste heat (coolant) from the generation of electrical energy (1.5 watts of coolant per watt distributed)	1.5 kw	8	5.3×10^9 kw
	Sewage		
Southern California, 10^7 people, 1968, 1.1×10^9 gallon day^{-1} (*45*)	416 liter day^{-1}		20×10^3 m^3 sec^{-1} (all three sewage items)
New York City, 8×10^6 people, 1970, 1.3×10^9 gallon day^{-1} (*46*)	615 liter day^{-1}		
Solids from domestic sewers into coastal waters, 500×10^3 ton $year^{-1}$ (*47*)	15 kg $year^{-1}$		
	Oil		
Spillage into world oceans (*27*)		15	2.4×10^6 ton $year^{-1}$
	Solid waste, United States		
Collected per capita in 1967, 5.2 pound day^{-1} (*44*)	2.4 kg day^{-1}	10	97 ton sec^{-1}
Dumped by barge into coastal waters, 1968, 62×10^6 ton $year^{-1}$ (*48*)	320 kg $year^{-1}$	10	36 ton sec^{-1}
	Dredge fill, United States, 1971		
Bypass, disposal, and maintenance, 3×10^8 cubic yards (*49*)	1.6 ton $year^{-1}$	4	172 ton sec^{-1}
	Mining of sand and gravel, 1970		
California, 23.5×10^6 U.S. tons (*35*)	1.1 ton $year^{-1}$		122 ton sec^{-1}
Great Britain, dredged from the North Sea, 13×10^6 British tons (*36*)	260 kg $year^{-1}$	5	29 ton sec^{-1}

for the water supply. Thus man's intervention takes three interrelated forms: (i) the impact of numbers of people, all competing in diverse and complex ways for portions of the coastal zone; (ii) the pollution and contamination of coastal waters; and (iii) the critical modification of the natural balance in the ecology of plants and animals and in the sources of sediments that constitute the world's beaches. This impact on the coastal zone is occurring not only in America but throughout the world, especially in the more technically advanced countries. Thus the problems that prevail along the coastlines of the United States are all present in various degrees along the coasts of the Black Sea, the North Sea, the Sea of Japan, the Mediterranean Sea, in fact, wherever man has access to the coast (*24*).

Population

In terms of the present world population, the amount of shoreline is even now quite limited. If everyone in the world decided to spend some time along the 440,000 km of world shoreline, each person would have less than 13 cm of shoreline. Much of the shoreline would be in the Arctic and Antarctic regions, and there would be insufficient space for each person to stand and face the sea. In the State of California alone, there are approximately 500,000 pleasure craft registered, a sufficient number, at a length of 5 m per boat, to form a solid line twice the length of the coastline of the state. In recent years there has been an increase in migration toward the coastal zone. For example, California has increased in population from 6.9×10^6 in 1940 to 20×10^6 in 1970. Thus the population of California doubles every 14 years, whereas for the United States as a whole the population doubles every 50 years.

Pollution of Coastal Waters

The coastal zone receives the bulk of man's wastes, and coastal waters are beginning to show this impact in terms of a degradation of water quality arising from the presence of numerous types of biological and chemical pollutants, thermal pollution, and acoustic pollution. Although the surf zone characteristically has high mixing rates (Fig. 6), this is not true of the shelf as a whole. Wind- and wave-induced surface currents tend to produce circulation patterns that favor the retention of particulate material near the coast (Fig. 8), whereas biological scavenging and absorption by suspended particles (both biogenous and inorganic) concentrate dissolved pollutants in coastal waters (*25*). For example, one can trace the plume of the Columbia River for nearly 400 km along the Oregon coast by measuring the ^{51}Cr in the surface waters (*26*); the spatial distribution of DDT [1,1,1-trichloro-2,2-bis(*p*-chlorophenyl)ethane] in zooplankton exhibits "hot spots" along the California coast from north of San Francisco to the Mexican border, and significant amounts for 600 km south to Punta Eugenia in Baja California, Mexico (Fig. 11).

Major oil spills of various types and their conspicuously detrimental effects on beaches, marine organisms, and birds are now an everyday fact of life (*27*). The amount of spillage, presently 2.4×10^6 tons per year (Table 3), will inevitably increase in the future with the increasing consumption of oil and the introduction of supertankers having

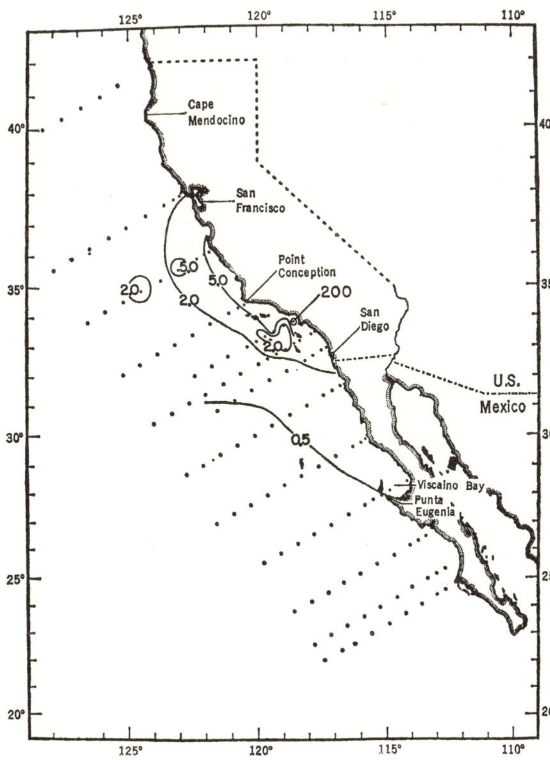

Fig. 11. Distribution of DDT and DDE [1,1-dichloro-2,2-bis(*p*-chlorophenyl)ethylene] in zooplankton (in units of 10^{-8} gram per cubic meter of surface water) along the coast of California and Baja California, Mexico, for 1969 (dots indicate sample locations). [From McClure and Barrett (*50*)] Since 99 percent of the DDT is absorbed on solid particulate material rather than in plankton, the total concentrations in seawater are approximately 100 times greater than shown here. [Courtesy of V. E. McClure and I. Barrett, National Marine Fisheries Service, La Jolla, California]

capacities of 2.5×10^5 tons and drafts in excess of 24 m. Moreover, the new big tankers have fewer bulkheads than the smaller tankers and thus have a potentiality for greater spillage when disabled.

So much plastic has been disposed of in the oceans that this material is now the most common type of flotsam on the world oceans; plastic particles in quasi-stable concentrations of 3500 pieces per square kilometer are widespread in the Sargasso Sea (*28*). Since every body of water has preferential windward shores (the prevailing wind systems cause some shores to be preferred to others), their beaches ultimately become the trash dumps for the world's flotsam. It was once romantic to find treasure at the shore—a Japanese net float, a hatch cover for a fireside bench. Now the beaches are piled with an endless supply of "disposable" plastic containers.

Coastal waters are the principal recipient of sewage and of heat from power plants, both directly by marine outfalls and indirectly from rivers. This is particularly serious in regions where the population density is high. Inland waters arrive at coastal estuaries contaminated by diluents supporting floating solids and oils, chemical discards, mining tailings, and trash. Economic pressures on upstream communities often dictate a chain of relatively inefficient sewage "treatment" plants which remove only larger chunks, sparge chlorine gas into the mixture, and relay it downstream to the next community, and the next, until finally it arrives at the sea. Utilities and industries along the way avail themselves of water for heat exchangers, condensers, and extraction processes, thus raising the mean temperature of the water. It is estimated that, by the turn of the century, power-generating plants will produce roughly enough heat to raise by 20°C the total volume of water which runs over the surface of the United States (*29*).

Power plants operating on fossil fuel require the mechanical equivalent of 1.4 watts of waste heat (coolant) for each watt of electrical power generated, whereas nuclear power plants require about 2.1 watts of coolant for each watt of electrical power. The predicted power plant requirements for the California coast for 1980 are 37,000 megawatts. According to this estimate, coolant will be used at the rate of 1.2×10^{10} calories per second, and this is equivalent to the heating of a flow of about 15,000 m^3 of seawater per second by 1°C. This flow is equivalent to one-half of the estimated average flow over the California shelf, and is 10^{-3} of the total flow of the California Current, which is one of the ocean's permanent current systems (*30*).

Southern California with approximately 10×10^6 people discharges sewage into the Pacific Ocean at the rate of 48 m^3 sec^{-1}; New York City with 8×10^6 inhabitants discharges sewage into the Hudson Estuary and ultimately into the Atlantic Ocean at the rate of 57 m^3 sec^{-1} (Table 3). The sewage effluent is rich in nitrates and phosphates which locally may contribute to red tides, fish mortality, and the contamination of edible mollusks. However, more damaging in the long term is the introduction of an entire spectrum of organic and inorganic materials (*31*). Many of these are highly toxic and become concentrated in the organs of various fish and shellfish.

A crucial fact frequently overlooked when one is considering the dispersion of pollutants in nearshore waters is that the effective rate of mixing depends both upon the mechanics of the phenomena (for example, turbulence in the breaking wave) and upon the concentration gradient of the pollutant that is being mixed (that is, the change in the concentration of the substance with distance). The concentration gradient is dependent upon the nature of the substance and its past history of dispersion in the area. The concentration gradient is particularly sensitive to the background concentration of the substance already present in the receiving waters. The presence of background concentrations decreases the effectiveness of the mixing process by recirculating material already there, and this in turn leads to an exponential buildup of concentration in the receiving waters. Thus the presence of background concentrations progressively leads to saturation and a complete breakdown in mixing even though the mechanical mixing mechanism remains undiminished. This effect is demonstrated in the comparison shown in Fig. 6, A and B, and in the coastal concentrations shown in Figs. 7 and 11.

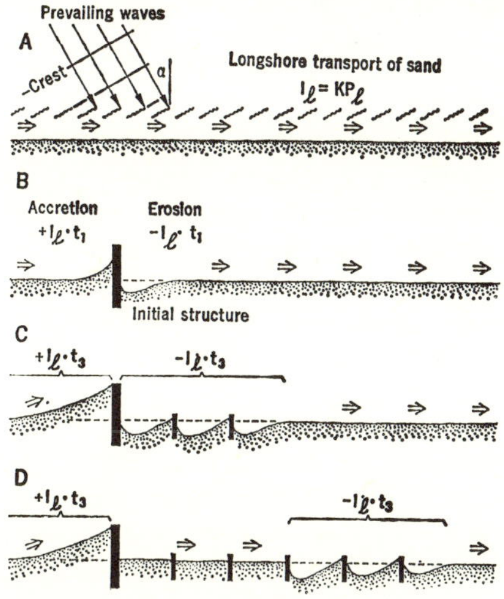

Fig. 12. Erosional chain reaction following the installation of a single coastal obstruction: (A) straight beach with prevailing waves producing a longshore transport of sand, I_l; (B) accretion and erosion showing initial obstruction; (C) downcoast erosion requiring two additional groins; and (D) continuation of the downcoast erosion requiring three more groins. Note that the first three groins are now unnecessary.

Critical Modification of Shorelines

Man's modification of the balance of sediment and his increasing depletion of the supply of sand to the beaches of the world has reached a critical state. On the one hand, man's use of the land has markedly increased the rate of erosion of the continents and increased the supply of sediment to rivers and streams. Paradoxically, the increasing construction of dams on most major streams results in the trapping and depositing of sediment behind the dams, thus shortening the life of the dam and intercepting the sediments that previously nourished the coasts and beaches. In this regard, the Aswan Dam on the Nile River is proving to be a major catastrophe.

The effect of dams on beaches is not always readily apparent. Erosion begins in the up-coast end of each sedimentation cell (Fig. 10) and progresses down the coast (Fig. 12). Most of the streams along the Southern California coast have dams and no longer supply sand to the coast. However, concurrent with the construction of dams, sand dredged from harbors and marinas has been placed on the beaches at approximately the same rate as sand supplied naturally by streams. For example, the Silver Strand littoral cell (Fig. 10, southern portion) has had no natural source of sediment since the Rodriquez Dam was completed in 1937. Silver Strand Beach has been maintained by artificially replacing 22×10^6 m^3 of sand in the period between 1941 and 1967.

The mining of beach and dune sands has further depleted this source, and the entrapment of coastal material by harbors, groins, and jetties has seriously reduced the supply of sand on beaches where it serves to protect the coast from the erosive action of waves. Moreover, sand is a valuable recreational asset that is now in very short supply. It is a natural resource which, like forests, streams, and mountains, should be protected.

It is well known that some of the activities of man, especially agriculture, construction, and deforestation, have in many places increased the rates of erosion of land by nearly an order of magnitude above the natural rate (*32*). The estimates of natural erosion listed in Table 2 are undoubtedly influenced by this factor, for man's influence on erosion can be documented to pre-Christian eras (*33*). On a global basis, it would appear that the present yield of sediment from the continents may be about twice what it was before man's introduction of large-scale agricultural and constructional activities during the past century. The increased yield of sediment causes an anomalous situation with regard to the supply of beach sand from rivers. The effective life of a dam is shortened by the rapid fill of sediment, and the very existence of beaches is imperiled by the termination of their natural source of sand (*34*).

The mining of beach sand has become an important industry in many areas. During 1970, 21.3×10^6 tons of sand and gravel were mined from beaches, river beds, and coastal sand dunes in California (*35*) (Table 3). During the same year, 112×10^6 tons were mined in Great Britain, of which 13.2×10^6 tons were mined from offshore banks, many of which provide essential protection to the coastline from wave erosion (*36*). A classic example is the erosion of the village of Hallsands which occurred in 1894 after 500,000 tons of gravel had been removed from a bank just offshore from the village (*37*).

By far, the most dramatic coastal intervention occurs when man interrupts the longshore transport of sand by the construction of jetties, breakwaters, and groins. Any artificial structure that produces a local accretion of sand by interrupting the transport of sand along a coast will cause, at least temporarily, a corresponding local erosion just down the coast from the area of accretion. Since the structure causes erosion to occur in a locality where erosion did not previously occur, or at a higher rate than occurred previously, a second structure is often needed to protect the new area of erosion. When built, this second structure in its turn causes erosion farther down the coast and a third structure is required to remedy this new situation, and so forth. Thus the construction of the first structure can set off a chain reaction that results in a requirement that the entire coastline be fronted by protective structures (Fig. 12).

The real "need" for a second or third structure may have been only temporary and may have diminished once the normal rate of longshore transport had been reestablished around the structure that caused the problem initially. However, if additional structures are built, the down-coast erosion becomes more severe with each succeeding structure, until finally a "point of no return" is reached where the need for additional protection from erosion becomes so urgent that the only choices are: (i) to continue to build protective works, (ii) to find a new source of beach sand, or (iii) possibly a combination of both.

From the foregoing it becomes apparent that the point of no return actually depends upon the longshore rate of transport of sand and the time needed to reestablish a normal transport rate around a structure. If a structure traps a large proportion of the total amount of sand transported longshore for a long period of time, the point of no return may be reached when only a single structure is built. On the other hand, it is possible to trap small amounts of sand at widely separated points along a coast provided that: (i) the total longshore transport rate is accurately known, so that the structure can be properly designed, and (ii) provisions are made to cope with the down-coast erosion until equilibrium in the longshore transport of sand is again established.

Comparison with Natural Phenomena

It is apparent that man's intervention in the coastal zone has both material and esthetic connotations. Almost all aspects of his impact on the environment are directly related to his numbers (the present population of the world is 3.5×10^9 people) and his relative affluence. An acceptable procedure for evaluating the extent and possible future consequences of man's intervention would be to assess his present intervention in the world's coastal zones and extrapolate these values to some future time, say, the year 2000, when it is estimated that the population will be 8×10^9 people. Unfortunately, the data are hopelessly inadequate except for a few of the advanced countries which constitute a small percentage of the population. Thus, an alternative, although less rigorous, procedure becomes necessary. We will assume that the present world population extrapolated to present U.S. standards of living constitutes a useful guide to man's potential for intervention in the future. Although admittedly less rigorous, this criterion is based on known and well-documented values of population and standards of living, thus providing an inherent simplicity in the assumptions and arithmetic. Global values for man's possible interventions, extrapolated to the present population from present U.S. standards, are shown in column 4 of Table 3.

It becomes readily apparent when Tables 1 and 2 are compared with Table 3 that man's intervention is indeed substantial. Man's extrapolated power consumption of 36.4×10^9 kw is over five times the total amount of natural energy dissipated in the shallow waters of the world, whereas the amount of waste heat (coolant) from electrical generators to be dissipated in the nearshore zone is slightly greater than the energy driving the dissipating mechanisms! It is more difficult to compare the effect of pollution of various types because the toxicity of the particular pollutant is not directly related to its volume or weight. However, the extrapolated volume of sewage effluent alone is 2 percent of all the discharge of the world's rivers. The extent of the distribution of DDT is dramatically illustrated in Fig. 11.

Man's solid waste disposal is much simpler to assess. Waste materials now cover significant portions of the sea floor adjacent to large metropolitan areas. Recent surveys show that waste-containing deposits cover about 210 km^2 of New York Harbor and the adjacent continental shelf (*38*). The extrapolated value for the waste disposal in coastal waters is 7 percent of the total discharge of solids from the world's rivers, whereas the extrapolated value for the bypassing of solid material by dredging is 32 percent of the world's supply of solids from land! Even more significant is the effect of the mining of sand and gravel, which would constitute 23 percent of the world's solids, when only about 10 percent of the solids supplied by rivers are in sand and gravel sizes, the remainder being silt and clay which is not suitable for beaches or building materials.

Planning Criteria

Our understanding of shore processes is still in a rudimentary stage, and it is not yet possible to describe many important phenomena and their interactions by rigorous theory. Only during the past 5 years or so have experiment and measurement progressed to the point where even general concepts can be formulated and tested. However, as this brief résumé of the physical processes shows, it is beginning to be possible to identify and to understand some of the mechanisms that are currently being studied, and to place what we do know in a planning context.

It has become evident with more intensive study that shorelines all over the world are subject to similar dynamic experiences. As a result, it will soon be possible to forecast the mixing ability of the waters over the shelf on a local regional basis. It will be possible to predict the longshore transport rate of sand, and to assay the effectiveness of various techniques for bypassing sand around coastal obstructions and harbors (Fig. 13). It is possible to estimate and pace man's intervention both in terms of utilization and waste. There is a desperate urgency associated with fitting these components into a coherent planning effort: the impact is harsh and frighteningly nonlinear, and the need is great right now!

If resolute planning is undertaken at once, the needs of society in terms of the coastal zone may be served with a directed and justified meliorism. The history of efforts to influence the utility of coasts and harbors extends far back in time, but the concern for preservation and conservation has been a recent addition. Two types of efforts have been made: historical efforts made on a palliative local basis with limited tools and technology, and more recent undertakings characterized by emerging knowledge of basic principles combined with single-remedy technology. (This type of technology may solve the single problem but may also create other problems.)

The systems concept is now a mandate; the balance of nature has its corollary in coastal undertakings. Previous self-perpetuating mistakes not only have left visible scars, they also have affected the environment in ways easily detectable to a concerned observer. Any guidelines fundamental to coastal planning necessarily involve an understanding of the natural physical processes active in the environment. It has been shown that these processes are driven by primary interrelated forces that are essentially systematic and regular in form. This observation is borne out by recent findings that the extent of the mixing of water in the surf zone can be predicted from a knowledge of the breaker height and the width of the

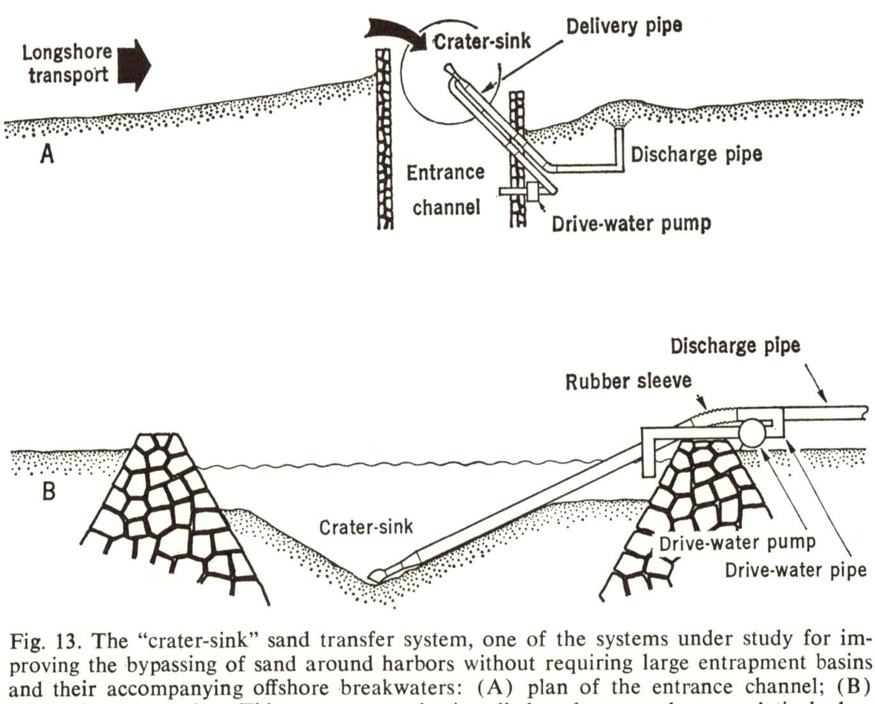

Fig. 13. The "crater-sink" sand transfer system, one of the systems under study for improving the bypassing of sand around harbors without requiring large entrapment basins and their accompanying offshore breakwaters: (A) plan of the entrance channel; (B) enlarged cross section. This system may be installed and operated at a relatively low cost. A jet pump (similar to a steam ejector) and a suction nozzle are located in the bottom of a crater-like depression in the sea floor which acts as a collecting basin for sand. The crater-sink assures that the water will be deep at the entrance. [From Inman and Harris (*51*)]

surf zone, and the longshore transport of sand is directly proportional to the longshore component of the energy flux of the incident waves. Thus a coherent approach to the formulation of unifying planning criteria seems clearly within our grasp; and it is becoming apparent that the driving forces (that is, the nearshore "climate" which consists of waves, winds, and current in nearshore waters) and the budget of sediment in a given coastal region are the principal environmental measurements that must be assessed for effective planning.

At least three distinct approaches are necessary if we are to achieve the overall goal of determining the essential qualities of unifying criteria: (i) specific studies of phenomena vital to our understanding of nearshore processes, (ii) the development of the sensors and systems for data-handling that permit effective monitoring of coastal climate and sediment transport, and (iii) the development of new planning criteria and the dissemination of information on known processes.

Coastal communities are presently in the curious position of rapidly acquiring and improving beach frontage while at the same time they lack criteria for evaluating the likelihood that the beach will still be in existence in 10 or 20 years. Certainly, the future of any coastal man-made structure placed in the path of the longshore movement in a littoral cell (Fig. 10) is questionable, and great reservations should accompany any commitment to build such a structure. Aerial photographs of Miami Beach, Florida, or Cape May, New Jersey, show the abuse to the coastline that can result when cascades of groins (Fig. 12) are erected without a master plan.

It is imperative that we develop the means to preserve the beaches and harbors that we now have and that we develop practical techniques for creating new beaches and nearshore structures that are less damaging to the environment. From an environmental standpoint there are three fundamental steps necessary for the good design of coastal structures: (i) identification of the important processes operative in an environment, (ii) understanding of their relative importance and their mutual interactions, and (iii) the correct analysis of their interaction with the contemplated design. To some extent we have considered all of these factors in earlier sections of this article, the emphasis being on the first two. There is no substitution for direct measurements in identifying and determining the relative importance of the various environmental parameters. It is also necessary that measurements be continued over a sufficient length of time so that predictions based on an earlier series of measurements can be tested for agreement with a subsequent series.

The need to develop new technologies to cope with specific problems is urgent. This type of development will likely be achieved simply because the need is specific and urgent, and there is a material objective. The stronger basic challenge lies in our ability to formulate satisfactory criteria for long-range planning and to develop an effective procedure for disseminating and implementing these criteria. It is now clear that man can no longer treat bits and pieces of a coast as individual entities; rather he must develop planning procedures that will include entire coastal zones and their adjacent ocean environments. This will require the development of planning criteria based on a far better understanding of both physical processes and environmental ecosystems than we now possess.

Human vanity does not encourage the recognition of failings, but one failing that must be recognized for survival is the fact that evolution has never prepared our species to think exponentially; this ability is a hard-won achievement of the disciplined mind. Cogent arguments abound for early and resolute planning in the coastal zone, with due allowance for the exponential hazards.

The Freeway Effect

The freeway effect (growth breeds growth) occurred in California as a result of the construction of improved, limited-access roadways. These roads were intended to relieve traffic congestion. They caused, instead, increased use of the available roadways, and people changed their living habits to suit their convenience. The freeway effect may be noted in population growth, energy production, automobile and appliance design, and agriculture. The unidirectional animus specifically threatens the coastlines, because it encourages one to believe that it is beneficial to build unlimited numbers of marinas, to erect increasing numbers of power plants, to construct housing developments adjacent to existing housing developments, or to increase the number of drilling sites, or the number of commercial and shipping facilities with insufficient evaluation of possible alternatives.

If planning efforts are ineffective and ill-informed, then people are inclined to proceed with each of the foregoing proposals as though the others should give way before it. An informed and concerned public reaction of hitherto unknown magnitude may possibly lead to solutions for the problem of slowing and redirecting the machine, without wrecking the machinery: that is the coastal challenge.

References and Notes

1. D. L. Inman and C. E. Nordstrom, *J. Geol.* **79**, 1 (1971).
2. F. P. Shepard, *Submarine Geology* (Harper & Row, New York, ed. 2, 1963), p. 257.
3. F. Ursell, *Proc. Roy. Soc. Ser. A* **214**, 79 (1952).
4. M. S. Longuet-Higgins, *Phil. Trans. Roy. Soc. London Ser. A* **245**, 535 (1953); D. L. Inman and A J. Bowen, in *Proceedings of the Eighth Conference on Coastal Engineering*, J. W. Johnson, Ed. (Council on Wave Research, University of California, Berkeley, 1963), pp. 137–150.
5. M. S. Longuet-Higgins and R. W. Stewart, *Deep-Sea Res.* **11**, 529 (1964); D. L. Inman, A. J. Bowen, V. P. Simmons, *J. Geophys. Res.* **73**, 2569 (1968).
6. A. J. Bowen and D. L. Inman, *J. Geophys. Res.* **74**, 5479 (1969); E. B. Thornton, *Proceedings of the Twelfth Coastal Engineering Conference* (American Society of Civil Engineers, New York, 1970), pp. 291–308.
7. F. P. Shepard and D. L. Inman, in *Proceedings of the First Conference on Coastal Engineering*, J. W. Johnson, Ed. (Council on Wave Research, University of California, Berkeley, 1951), pp. 50–59.
8. D. L. Inman, R. J. Tait, C. E Nordstrom, *J. Geophys. Res.* **76**, 3493 (1971).
9. D. L. Inman, *Trans. Amer. Geophys. Union* **51**, 319 (1970); R. A. Bagnold, in *The Sea, Ideas and Observations: The Earth Beneath the Sea*, M. N. Hill, Ed. (Interscience, New York, 1963), vol. 3, pp. 507–528.
10. D. L. Inman and R. A. Bagnold, in *The Sea, Ideas and Observations: The Earth Beneath the Sea*, M. N. Hill, Ed. (Interscience, New York, 1963), vol. 3, pp. 529–553.
11. D. L. Inman and J. D. Frautschy, in *Coastal Engineering* (Santa Barbara Specialty Conference) (American Society of Civil Engineers, New York, 1966), pp. 511–536.
12. Here $P = P \sin \alpha \cos \alpha$, where P is the power per unit crest length of the breaking wave and α is the breaker angle [see P. D. Komar and D. L. Inman, *J. Geophys. Res.* **75**, 5914 (1970)].
13. S. Gofseyeff, in *Proceedings of the Third Conference on Coastal Engineering*, J. W. Johnson, Ed. (Council on Wave Research, University of California, Berkeley, 1953), pp. 272–305.
14. J. W. Johnson, *Amer. Ass. Petrol. Geol. Bull.* **40**, 2211 (1956); *J. Waterways Harbors Div. Amer. Soc. Civil Eng.* **83**, 1211 (1957).
15. J. S. Malkus, in *The Sea, Ideas and Observations: Physical Oceanography*, M. N. Hill, Ed. (Interscience, New York, 1962), vol. 7, p. 98.
16. D. L. Inman, in *Oceanography: The Last Frontier*, R. C. Vetter, Ed. (Basic Books, New York, in press).
17. W. G. Van Dorn, in *Advances in Hydroscience*, V. T. Chow, Ed. (Academic Press, New York, 1965), vol. 2, pp. 1–48.
18. J. H. Hoyt and V. J. Henry, Jr., *Geol. Soc. Amer. Bull.* **82**, 59 (1971).
19. S. A. Schumm, *U.S. Geol. Surv. Prof. Pap. 454-H* (1963).
20. J. Gilluly, *Geol. Soc. Amer. Spec. Pap. 62* (1955), pp. 7–18; H. W. Menard, *J. Geol.* **69**, 154 (1961).
21. J. Gilluly, J. C. Reed, Jr., W. M. Cady, *Geol. Soc. Amer. Bull.* **81**, 353 (1970).
22. This is based on the assumption that continental rock with a density of 2.7 ton m^{-3} erodes to form 20 percent dissolved load and 80 percent solid particulate load.
23. K. O. Emery, *Amer. Ass. Petrol. Geol. Bull.* **52**, 445 (1968).
24. For example, M. I. Goldman, *Science* **170**, 37 (1970); D. W. Hood, Ed., *Impingement of Man on the Oceans* (Wiley, New York, 1971); W. Marx, *The Frail Ocean* (Ballantine, New York, 1967).
25. E. D. Goldberg, W. S. Broecker, M. G. Gross, K. K. Turekian, in *Radioactivity in the Marine Environment* (Committee on Oceanography, National Academy of Sciences, Washington, D.C., 1971), pp. 137–146.
26. C. Osterberg, N. Cutshall, J. Cronin, *Science* **150**, 1585 (1965).
27. R. Revelle, E. Wenk, B. H. Ketchum, E. R. Corino, in *Man's Impact on Terrestrial and Oceanic Ecosystems*, W. H.

Matthews, F. Smith, E. D. Goldberg, Eds. (M.I.T. Press, Cambridge, Mass., 1971), pp. 297–302; M. Blumer and J. Sass, *Science* **176**, 1120 (1972); Ocean Science Committee, *Marine Environmental Quality* (National Academy of Sciences, Washington, D.C., 1971); Massachusetts Institute of Technology, *Man's Impact on the Global Environment* (C. L. Wilson and W. H. Matthews, chairmen) (M.I.T. Press, Cambridge, Mass., 1971).

28. E. J. Carpenter and K L. Smith, Jr., *Science* **175**, 1240 (1972); E. J. Carpenter, S. J. Anderson, G. R. Harvey, H. P. Miklas, B. B. Peck, *ibid.* **178**, 749 (1972).
29. P, M. Boffey, *ibid.* **168**, 1554 (1970). These data are projected from the 1980 base established in "National Power Survey, A Report by the Federal Power Commission 1964" (Government Printing Office, Washington, D.C., 1964), p. 36; we here assume 12.3×10^{12} kilowatt-hours for the year 2000 and a water flow of 4.2×10^{11} m^3 $year^{-1}$.
30. W. S. Wooster and J. L. Reid, Jr., in *The Sea, Ideas and Observations: Composition of Seawater*, M. N. Hill, Ed. (Interscience, New York, 1963), vol. 2, pp. 253–280.
31. E. D. Goldberg and M. G. Gross, in *Man's Impact on Terrestrial and Oceanic Ecosystems*, W. H. Matthews, F. Smith, E. D. Goldberg, Eds. (M.I.T. Press, Cambridge, Mass., 1971), pp. 371–376; E. D. Goldberg, in *Fertility of the Sea*, J. D. Costlow, Jr., Ed. (Gordon & Breach, New York, 1971), pp. 143–156; see also (*25*).
32. H. W. Lull, in *Man's Impact on Terrestrial and Oceanic Ecosystems*, W. H. Matthews, F. Smith, E. D. Goldberg, Eds. (M.I.T. Press, Cambridge, Mass., 1971), pp. 240–251; R. H. Mean, *Geol. Soc. Amer. Bull.* **80**, 1265 (1969); S. Judson, *Amer. Sci.* **56**, 356 (1968).
33. S. Judson, *Science* **160**, 1444 (1968).
34. C. P. Vetter, *Iowa Univ. Studies Eng. Bull.* **34**, 5 (1952).
35. *Calif. Geol.* **25** (No. 4), 90 (April 1972).
36. A. A. Archer, in *North Sea Science*, E. D. Goldberg, Ed. (M.I.T. Press, Cambridge, Mass., in press).
37. W. A. Price, personal communication.
38. M. G. Gross, *Geol. Soc. Amer. Bull.* **83**, 3163 (1972).
39. For example, B. Kinsman [*Wind Waves* (Prentice-Hall, Englewood Cliffs, N.J., 1956), p. 154] estimates that the energy dissipated by waves hitting the shoreline is 1.9×10^9 to 2.2×10^9 kw.
40. On the assumption that the dissipation of lunar tidal energy in shallow water is 1.7×10^9 kw, and that the ratio of the flux of solar to lunar tidal energy is the same as their torques, that is, 1 : 3.4. W. H. Munk and G. J. F. MacDonald [*The Rotation of the Earth* (Cambridge Univ. Press, London, 1960), p. 201] estimate the total dissipation of lunar tidal energy as 2.7×10^9 kw, of which two-thirds is dissipated in shallow seas [G. R. Miller, thesis, University of California, San Diego (1964)].
41. Average value from K. K. Turekian, in *Impingement of Man on the Oceans*, D. W. Hood, Ed. (Wiley, New York, 1971), chap. 2, pp. 9–72, based on data of D. A. Livingstone [*U.S. Geol. Surv. Prof. Pap. 440-G* (1963), p. 64], O. A. Alekin and L. V. Brazhnikova [*Gidrokhim. Mater.* **32**, 12 (1961)], and J. N. Holeman [*Water Resour. Res.* **4**, 737 (1968)].
42. R. M. Garrels and F. T. Mackenzie, *Evolution of Sedimentary Rocks* (Norton, New York, 1971), table 5.1.
43. On the assumption that the total particulate load consists of 15 percent bed load and 85 percent suspended load, where values of the suspended load are averages from Garrels and Mackenzie (*42*), Turekian (*41*), and Holeman (*41*).
44. U.S. Department of Commerce, *Statistical Abstracts of the United States, 1970* (Government Printing Office, Washington, D.C., ed. 91, 1970).
45. Southern California Coastal Water Research Project, 1100 Glendon Avenue, Los Angeles 90024 (unpublished report, 1970).
46. M. Eisenbud, *Science* **170**, 706 (1970).
47. E. D. Goldberg, *McGraw-Hill Yearbook of Science and Technology, 1970* (McGraw-Hill, New York, 1971), pp. 65–73.
48. D. D. Smith and R. P. Brown, "Ocean Disposal of Barge-Delivered Liquid and Solid Wastes from U.S. Coastal Cities," report prepared for the Environmental Protection Agency by the Dillingham Corporation, La Jolla, California (1971); M. G. Grass, in *Man's Impact on Terrestrial and Oceanic Ecosystems*, W. H. Matthews, F. Smith, E. D. Goldberg, Eds. (M.I.T. Press, Cambridge, Mass., 1971), pp. 252–260.
49. "Disposal of Dredge Fill," report to the U.S. Congress by the Chief of Engineers, U.S. Army (unpublished report, 1971).
50. V. E. McClure and I. Barrett, in *Baseline Studies of Pollutants in the Marine Environment*, E. D. Goldberg, workshop chairman (National Science Workshop, Brookhaven National Laboratory, New York, 1972), pp. 493–497.
51. D. L. Inman and R. W. Harris, *Proceedings of the Twelfth Coastal Engineering Conference* (American Society of Civil Engineers, New York, 1970), pp. 919–933.
52. This work was sponsored by the National Oceanic and Atmospheric Administration, Office of Sea Grant, under grant 04-3-158-22 with the Institute of Marine Resources, University of California. The background knowledge of basic mechanisms described in this study evolved under long-range support by the Office of Naval Research and the Coastal Engineering Research Center, U.S. Army Corps of Engineers, with the Scripps Institution of Oceanography, University of California, La Jolla.

21

Reprinted from *Geog. Jour.* **107**(1):57–61 (1946)

COASTAL PRESERVATION AND PLANNING

J. A. STEERS

SINCE WRITING the paper on Coastal Preservation and Planning, published in the *Journal* in 1944,[1] I have been able to complete my survey of the coast and also to revisit certain important areas. The parts covered since 1944 include the Isle of Wight, the coast from Lulworth to Gravesend, and the Isles of Scilly. The surveys of parts of the Dorset, North Devon, and North Cornwall coasts, and those of Cardigan Bay, parts of South Wales, and Northumberland have been checked. Only for that section of the Dorset coast between Weymouth and Swanage have I made any change in my original assessment: the differences will be clear if the present map is compared with the folding plate accompanying the former paper. The value of subsequent visits was great, however, even if few alterations on the maps resulted from them. All my earlier surveys were made when much of the coast was accessible only with difficulty or with the help of an official pass. In 1944 and 1945 something akin to peace-time conditions prevailed, and the contrast between the Northumberland coast from Amble to Tynemouth in 1943 and the Sunday after the basic petrol ration was restored in 1945 was emphatic.

The value of a survey of the whole coastline of England and Wales is, in my view, greatly increased after taking the south-east into consideration. This near-London coast is so urbanized and artificial that if it were considered on its own merits, the result would be very misleading. To many this may be an obvious statement, but for that very reason it is worth emphasis. Those who visit the south-east coast regularly and know no other, can have little idea of what truly rural coast is like, and many who know it casually are apt to regard

[1] *Geogr. J.* 104 (1944) 7–27.

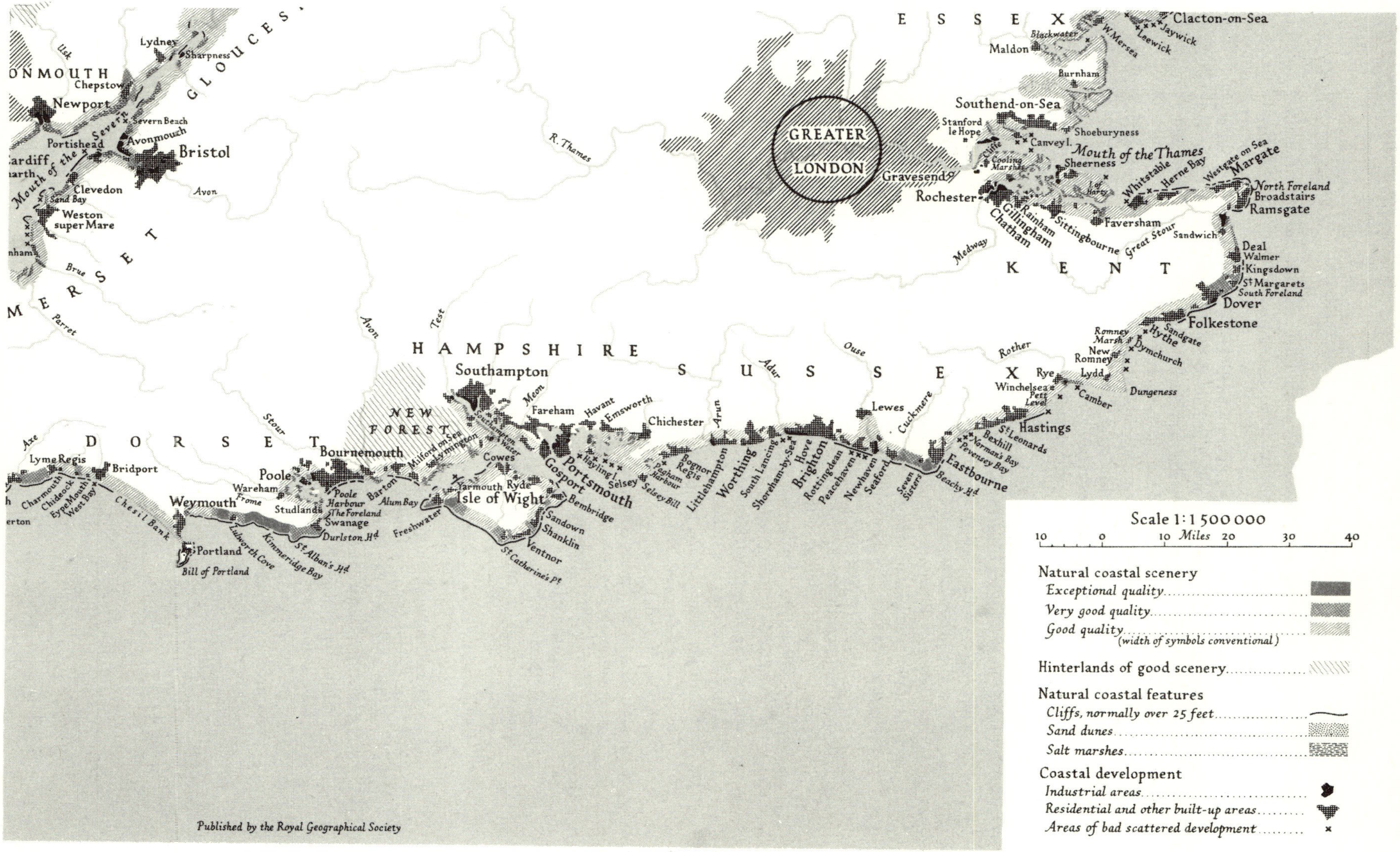

Scale 1:1 500 000
10 0 10 20 30 40 Miles
Natural coastal scenery
Exceptional quality
Very good quality
Good quality
(width of symbols conventional)
Hinterlands of good scenery
Natural coastal features
Cliffs, normally over 25 feet
Sand dunes
Salt marshes
Coastal development
Industrial areas
Residential and other built-up areas
Areas of bad scattered development
Published by the Royal Geographical Society
GREATER LONDON
E S S E X
K E N T
S U S S E X
H A M P S H I R E
D O R S E T
NEW FOREST
Mouth of the Thames
Mouth of the Severn
Isle of Wight
Clacton-on-Sea
Southend-on-Sea
Gravesend
Rochester
Chatham
Gillingham
Margate
Ramsgate
Dover
Folkestone
Hastings
Eastbourne
Brighton
Worthing
Portsmouth
Gosport
Southampton
Bournemouth
Poole
Weymouth
Portland
Bristol
Newport

it as typical of too great a proportion of the total coastline. Along the south-east coast there are still a few beautiful and unspoiled districts: Beachy Head and the Cuckmere Valley, some of the Downs between Folkestone, Dover, and Deal (although these are not completely free from the threat of industrialism based on local iron ore), the Isle of Harty, and parts of Cliffe and Cooling Marshes are examples. Pagham Harbour is another, but its full beauty is spoiled by the proximity of the shacks west of Bognor.

The evils of a lack of planning are appallingly obvious on the Sussex coast. Travel along the so-called coastal road from Portsmouth to Seaford and note the almost continuous and extremely unsightly ribbon-development, count the number of level crossings, and examine the devious ways down to the sea through building estates whose houses often extend right to the beach. Walk along the beach fringing the Sussex coastal plain and contemplate the landward views. Consider also the position of the electric light and gas works near Brighton, the beach ribbon development between Pevensey Bay and Norman's Bay and at Shoreham, notice the seaward and landward views from the main road from Brighton to Newhaven and the growth of Seaford. It is easy to point out examples of faulty and indiscriminate development, but difficult to suggest remedies. In these notes I am not concerned with such matters, but rather to urge that since the coast is a great national asset we should in future do our utmost to use it properly. It is easy to see now that if a comprehensive scheme of development of the south-eastern coast had been evolved and put into practice even after the last war, it would have been comparatively simple to have escaped the present haphazard growth. Further, it would have been possible to give as much pleasure to those owning or renting houses in the area as they enjoy now, and infinitely more to those who merely visit it for the sake of a seaside holiday. The point that matters is that in developing other parts of our coast we must bear in mind that access to the coast for all is essential, and that in order to enjoy the coast fully buildings must be properly sited and planned and must not obstruct coastal walking or appear on sky-lines. A careful study of the Sussex and Kent coasts can teach us what not to do.

The danger of ribbon-development along the coast is shown on a small scale in much of the Isle of Wight. Except for the part between Alum Bay and St. Catherine's Point and a very few short stretches elsewhere, almost the whole is built up, in striking contrast to the rural and unspoiled nature of the interior. The very varied and beautiful coastline of a small island is not improved by rows of houses.

This is not the place to discuss the immediate housing and related accommodation problems on the coasts, but I do urge again the necessity of having a national and not a local perspective. Because a particular area is very popular there is a natural tendency to build and attract more people. It is largely the position and type of the buildings which matters in such an area, and thus the central authority alone can assess adequately the national needs. There is no reason at all why all those who wish to visit the seaside should not be encouraged to do so, but a comprehensive scheme for hostels, camps, and ordinary houses is wanted in all coastal areas. The solution of limited space in part depends on staggered holidays, not only for the workers but also for schools. It means that big new camps or even new towns are necessary.

Phot. British Council

The Seven Sisters cliffs from the mouth of the river Cuckmere, Sussex

Phot. Aerofilms

Peacehaven

Phot. British Council

Sand dunes, West Beach, Littlehampton

Pevensey Bay

For example, some of the best sands on the south coast are near Camber, where a large holiday town grew up before the war. Since 1939 it has suffered considerably from bombs and in other ways; but it is surely to the national advantage to have a big and properly planned holiday town at Camber that will cater adequately for large numbers, rather than to allow the small but painful eyesores that spoil Winchelsea and Pett Level.

There is another and serious threat to our coast deriving from war-time uses. During the war many coastal areas had to be taken over by the Government Services for some specific purpose, but it would be little short of a national disaster if they were not handed back. The tank range which was

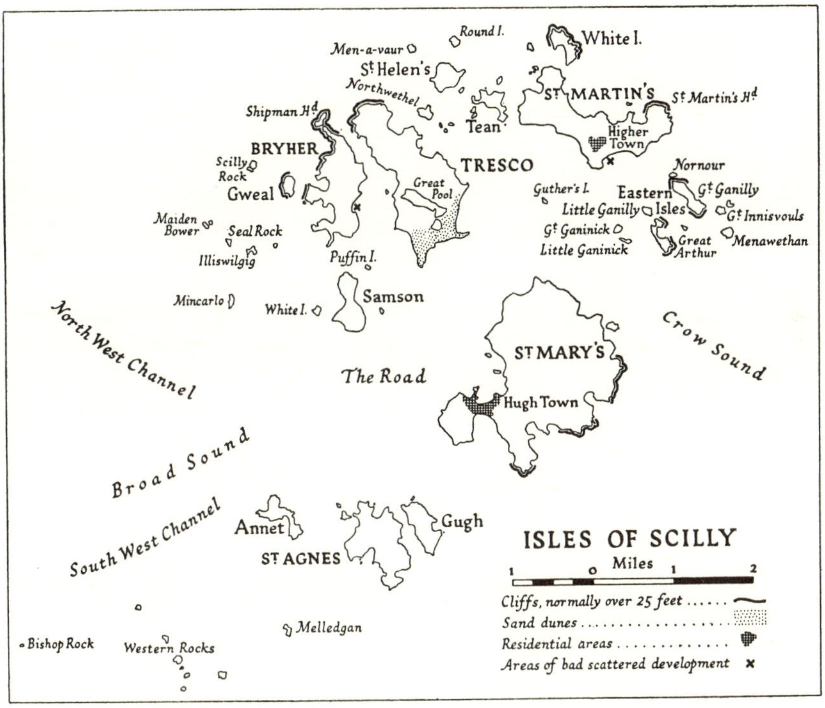

built behind Lulworth in Dorset has expanded far to the east and put out of bounds to the public for the time being some of the finest and most interesting parts of the Dorset coast between Kimmeridge and Lulworth. The beautiful limestone cliffs and upland between Stackpole and Linney Head in the Tenby Peninsula likewise are now as a rule unapproachable. The fine dunes on the Cumberland coast near Drigg, Pendine and Laugharne Burrows in Carmarthenshire, the South Haven Peninsula near Studland in Dorset, part of Ross Links in Northumberland, and Scolt Head Island in Norfolk are all, regrettably, prohibited areas. It is clear of course that space is limited in these islands, but there are plenty of open areas with less claims for preservation than those mentioned above. Once again it is the national view that is all important. It can be argued that a place like Scolt Head Island is not popular in the sense that other parts of the Norfolk coast attract visitors. But Scolt and other places are essential open areas visited in the course of a year by a surprising number of people who want unspoiled sea coast. Moreover many of these areas are of great scientific interest. Unless the coast is visualized as a whole it is only too easy to say that such and such a particular area does not matter: if the total of these areas be reckoned and proper allowance made for the recreational and scientific value, a very different assessment of value emerges.

Many people dislike the word planning and fear that in practice it may disguise some form of prohibition. This view should be dispelled. As far as the coast is concerned, if indiscriminate building were to continue as in pre-war days most of it would be ruined. Having seen the whole coast of England and Wales I cannot emphasize too strongly that if we as a nation wish to preserve one of our finest heritages for the good of the people as a whole, we must act now and act vigorously on a national scale.

The Isles of Scilly

The map gives in black and white a general assessment of the Isles of Scilly. For the most part their great natural beauty is unspoiled, and all the coastal scenery can be classed as of exceptional quality, though here and there an unsightly building shows how easily it could be spoiled. The conventions used on the map are similar to those on the general map of England and Wales, but modified a little to accord with local conditions and difference of scale.

22

Reprinted from pp. 239–258 of *Coastal Geomorphology*, D. R. Coates, ed., State University of New York, Binghamton, New York, 1973, 404 pp.

COMPARISON OF ECOLOGICAL AND GEOMORPHIC INTERACTIONS BETWEEN ALTERED AND UNALTERED BARRIER ISLAND SYSTEMS IN NORTH CAROLINA

Paul J. Godfrey and Melinda M. Godfrey

INTRODUCTION

The Outer Banks of North Carolina, a chain of narrow, sandy islands with a fringe of salt marshes and lagoons behind them, are one of the world's best examples of a continuous offshore barrier system (Fig. 1). First seen by Eruopeans during the explorations of Verrazano in 1524, the Outer Banks have long been used and abused by people. Today they are faced with increasing demands from recreation and development programs which threaten these islands more seriously then have all the years of use since white men first settled the region. Because of shortsighted development schemes combined with the belief among many coastal managers that stabilization of dunes and beaches is essential to preserve the shoreline, undesirable changes are being made in ecological and geomorphic processes which create and maintain these islands.

While such development pressure has been deplored by conservationists, there has been little research on the effects of the stabilization that necessarily accompanies development on the barrier island system. Fortunately, certain sections of the Outer Banks are still available in a natural condition to provide comparisons between developed and undeveloped barrier islands. Our research on the ecological and geomorphic changes of a natural barrier system (Cape Lookout National Seashore) has paralleled the geomorphic research of Dr. Robert Dolan (Chapter 12) on a developed beach (Cape Hatteras National Seashore), and we have come to the same general conclusions regarding the present condition of the Outer Banks independently and by different routes. These studies were initiated by the National Park Service to avoid the mistakes made at Cape Hatteras as new seashores became established, particularly Cape Lookout, and to plan development programs for barrier islands more in keeping with natural processes. Our goal is to provide guidelines for the recreational use of National Seashores that will fit into these processes instead of conflicting with them.

Department of Botany/National Park Scrvicc
University of Massachusetts, Amherst, Mass. 01002

CONTRASTS BETWEEN CAPE HATTERAS AND CAPE LOOKOUT

The northern section of the Outer Banks from Nags Head to Ocracoke Inlet comprises Cape Hatteras National Seashore, established in 1953, and the islands south of Ocracoke to Beaufort Inlet are authorized to become Cape Lookout National Seashore (Fig. 1). All these islands once looked much the same, judging from historical records, but since the 1930's the northern section has been progressively stabilized and developed. Much of the early stabilization efforts were made in the belief that the generally low islands with migrating dunes were in a deteriorating condition due to past misuse, although little evidence supports that view, except in localized places (Dunbar, 1958). Programs were instituted to plant beach grass, erect fences, and stabilize as much moving sand as possible with the general blessing of the scientific community, up to the present time.

The southern part of the Outer Banks, however, has largely been left alone except for grazing, which in places was apparently quite heavy. Livestock were removed from the Banks in the late 1950's with the exception of Shackleford Banks, where today herds of horses, cattle, goats, and sheep still hark back to the day when all of the Banks were used for grazing. While the signs of grazing pressure on Shackleford are quite evident, there does not seem to be a wholesale general deterioration of the island due to the livestock as some have maintained. What changes have occurred in the past century, such as migrating dunes covering forests, we believe are primarily due to hurricanes and natural recession. Grassland cover has increased on Shackleford since the early part of this century despite heavy grazing pressure.

Today the two sets of islands present a strikingly different appearance. The Cape Hatteras sections show a narrow and irregular berm, and continuous, artificial dune lines that are eroding in numerous places on every island (Fig. 2). Some of this erosion is serious enough to have prompted massive and expensive control projects, usually in front of some threatened facility. Nearly $½ million was spent during the early part of 1972 for beach nourishment at Buxton, only to have it all disappear 10 months later (Fig. 3). A $2 million project has been approved and is now in the planning stages to restore this same section of the beach. In spite of endless expensive erosion control efforts (between 1957 and 1968 the Park Service spent an average of $420,000 per yr on dunes, with additional sums of $1,200,000 to repair storm damage), buildings and roads close to the barrier dune are constantly vulnerable to destruction by the sea. Despite the increasing threat of eroding dunes and storm damage, buildings are now under construction just behind the barrier dune in those interior areas not owned by the Park Service. Unfortunately, the Service is committed by law to maintain the dune in front of these inholdings at public expense, Nearly all of Cape Hatteras National Seashore is an artificial system with a high, continuous barrier dune protecting the highway and towns and halting the natural process of barrier island recession, but not the erosion. Instead of "saving" the barrier islands, these programs seem to be threatening their very existence as natural

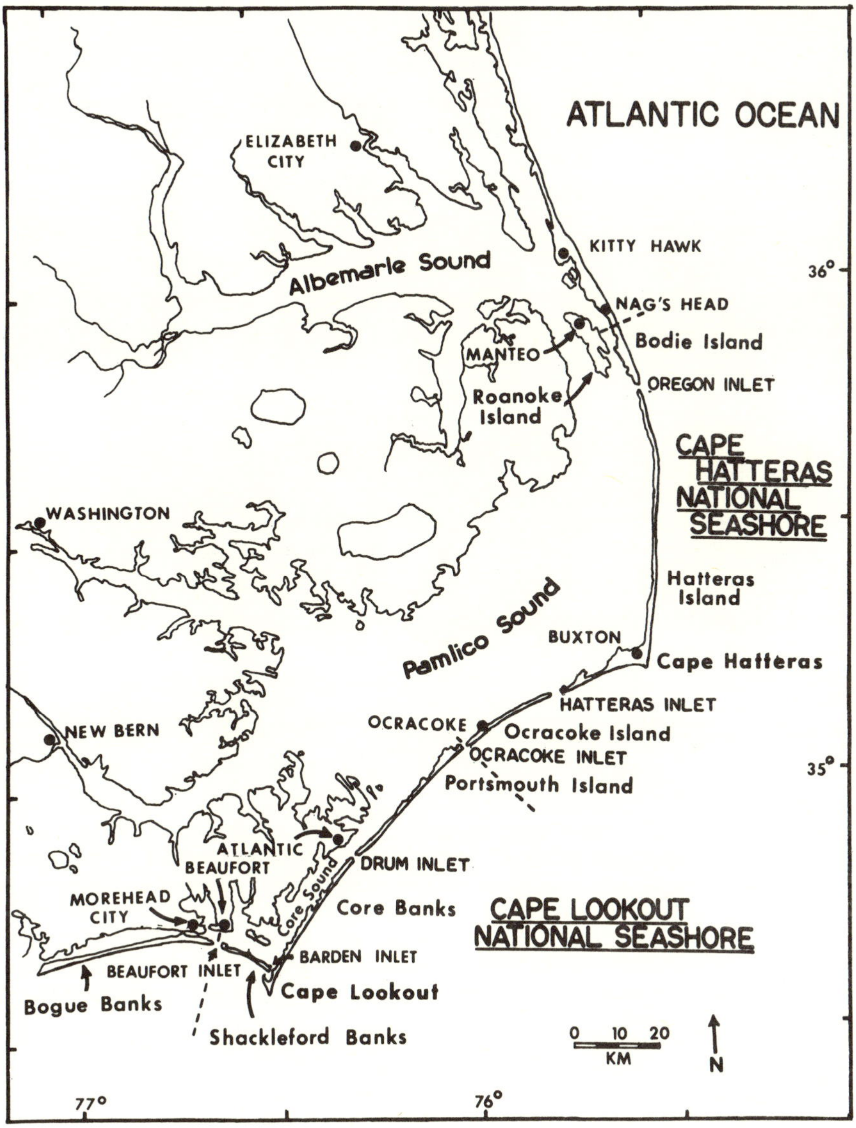

Figure 1. The Outer Banks of North Carolina, showing the locations of the Cape Hatteras and Cape Lookout National Seashores.

Top

Bottom

Figure 2. Eastern part of Ocracoke Island, looking south toward Ocracoke Village; a typical stabilized barrier island. Main features are the narrow beach, single, man-made, continuous dune; grassland vegetation directly behind the dune (in dense bands where fertilized) with advancing shrub lands; and the road. Old overwash terraces still visible behind the road. Beyond are salt marshes. Large pit in salt marsh is borrow area for sand to build secondary dike along the road. Atlantic Ocean on left, Pamlico Sound on right.

Figure 3. Buxton and the Cape Hatteras Lighthouse during a winter storm 17 November 1971. Dune line has been nearly destroyed in front of motels. Overwashes behind dune are plainly visible. Plastic sand bags to protect dune are on beach. Groins built opposite the lighthouse created the bulge in the beach line. By 17 November 1972, all of this beach nourishment material was gone and the beach is now essentially the same as this photograph taken a year previously. View is south towards Cape Hatteras. (Photo courtesy of Cape Hatteras National Seashore.)

features.

On the Cape Lookout islands, however, the berm is consistently wide, and the dunes are in a zone, rather than line, which is low and broken by frequent overwash passes (Fig. 4) and erosion occurs only in areas which are undergoing natural geomorphic change. The few permanent buildings are well back from the ocean (Fig. 5) and no money has been spent on erosion control in recent years. During major storms, ocean water floods across these islands without doing any real damage. Based on our research we believe these islands represent the natural condition of low barrier islands undergoing recession, and that they show a type of dynanic stability that has allowed their survival through time.

Sea Level and Storms

Both sets of islands are exposed to approximately the same environmental conditions. Of paramount importance are rising sea level and oceanic storms. Sea level has been rising about 30 cm per century during the past 5,000 years (Milliman and Emery, 1968), with a more dramatic rise of about 8 cm during the past decade (Hicks, 1972). Evidence for submergence of the coast and subsequent recession is present all along the Outer Banks. Low lying forests shown on maps of Shackleford Banks in 1850 are today salt marshes. Stumps and peat from an outcrop at the low tide mark on Shakleford were dated by the U.S. Geological Survey (Sample W-2307) at less than 200 yr B.P.; such outcrops are frequently found along the ocean side of the banks where they are exposed at low tide. Along with rising sea level, the regular oceanic storms are a major factor in this environment. The Outer Banks are in a particularly hazardous storm path (Mather, et al 1967) and front on a high energy sea; hurricanes frequently recurve up the East Coast and pass over or by the Banks. At least 150 hurricanes have affected the Outer Banks between 1585 and 1967 and there is no reason to expect these storms will suddenly stop (Carney and Hardy, 1967). Yearly, winter "northeasters" moving along this coast are often even more destructive than hurricanes. The storms create surges that invariably wash over the lower parts of the islands, where they meet little resistance on natural sections. However, the surges are prevented from overwashing the developed islands and instead erode the dunes. The natural response of a barrier island to storms and sea level rise is to retreat (Shepard, 1963; Hoyt, 1967), with wind and overwash moving the sand back. Such a pattern of natural retreat can still be seen on Core Banks, but no longer at Cape Hatteras. Nevertheless, the sea still tries to push these islands back and severe erosion results.

THE NATURAL SYSTEM – CORE BANKS

Core Banks (Fig. 4) is an example of an overwash barrier combined with relic inlet deposits similar to the barriers of New Jersey described recently by Kraft (1971). Storms maintain the wide, bare berm as waves dissipate their energy on the back shore and pass across the island. The berm is the "elbow room" for the

Figure 4. Central Core Banks south of Drum Inlet; a typical unstabilized barrier island. Features are a wide berm; low, open dune zone; grasslands of varying density on dunes and overwash terraces; scattered shrubs and thickets; salt marshes that formed on tidal deltas and overwash deposits. Ocracoke and other parts of Cape Hatteras National Seashore were similar before development. Atlantic Ocean on the left, Core Sound on the right.

Figure 5. Core Banks and Cape Lookout Lighthouse, looking south towards Cape Point. In contrast with Cape Hatteras, there is a wide berm and dune zone, such as Cape Hatteras had in the 1930's. The island is covered by water during storms, but no damage results.

storm surge. Whatever dunes or vegetation build up on the active berm during quiet periods are usually destroyed by the surge. Those dunes that survive are back far enough to escape the main wave force. As the overwash occurs, sand is carried from the beach and berm crest onto and across the berm to the interior of the island, where it is deposited as a veneer of sand in the form of a terrace or fan. Winds then blow the fine sand from the deposit into dunes, leaving a coarse shell pavement on the surface. Such layers of shell and coarse sand occur regularly in stratigraphic profiles and represent surfaces of the island in years past. As sand is moved into the interior of the island by overwash, and on dunes by wind, the height of the land increases and tends to keep pace with the rising sea. The buildup of the interior of Core Banks from overwash alone was measured relative to permanent beach marks set up in 1960 by the Corps of Engineers at 8.4 cm in ten years (Godfrey, 1970). The thickness of the overwash veneer and the distance it is carried is a function of the storm surge. Thin layers are added nearly every year, with major buildups during severe storms.

Based on historical aerial photographs and corings of the island, we have been able to construct a cross-sectional diagram of Core Banks showing surfaces approximatley as they were at previous times (Fig. 6). We know that the underlying marsh in the center of the island was exposed during the 1940's and was then buried by overwash. Oceanic shells indicate this material had to have come from the beach. During the late 1950's and early 1960's a series of storms pushed more sand across the island and created the present terraces. Natural vegetation and new dunes have developed on these terraces during the past decade (Fig. 7). Thus, much of the island consists of a series of quite recent overwash terraces. New dunes form on the terraces only to be buried or eroded by overwash in their turn. The sand, however, is carried farther back into the island during an overwash instead of being lost to the littoral currents, and so it is conserved.

Grassland is the typical vegetation of these barrier islands, with shrub savannas and thickets where protection by dunes or where overwash has not occurred recently. With long term protection, woodlands can develop. The presence of this grassland makes overwash a constructive rather than an erosive force. On some islands, such as Portsmouth, where there are no grasslands perhaps as a result of excessive grazing in the past plus severe storms, overwash is lowering the land surface rather than building it up. *Uniola paniculata,* (sea oats) is the primary dune builder and it will grow on the berm wherever its seeds are buried. As it grows, sand accumulates around the plant, and both the plant and dune grow vertically. The development of dunes by beach grasses such as *Uniola* has been well documented by ecologists. Other species have this ability as well; one very common grass, *Spartina patens* (salt meadow cordgrass) builds dunes as well as growing on flats and in salt marshes. It is a very versatile and resistant species that plays a major role in the bio-geomorphology of the barrier island. It grows rather sparsely in the passes between dunes, and on the dunes themselves, but becomes quite dense on the oldest, least recently overwashed terraces. Figure 8

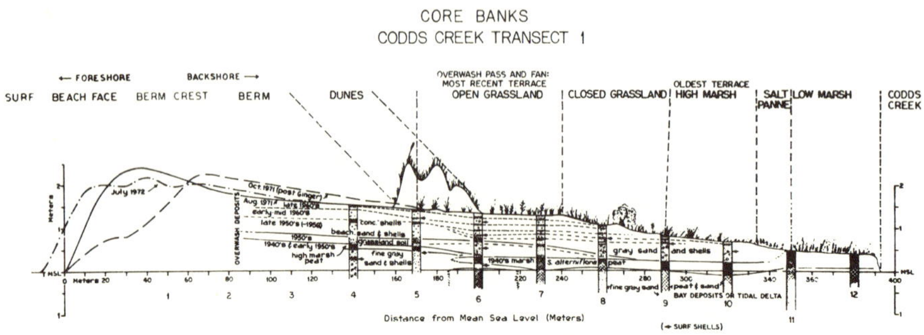

Figure 6. Stratigraphic diagram of Core Banks at Codds Creek showing locations of piston cores from which overwash strata and marsh deposits were identified. Vertical scale greatly exaggerated relative to horizontal scale. Changes of the beach profile due to Hurricane Ginger are shown; note overwash build-up and progradation of beach a year later.

Figure 7. Dune zone on Core Banks at the location of the transect shown in Figures 6 and 8. Note the open nature of the dunes, their rounded shape, and the relatively sparse vegetation of sea oats *(Uniola paniculata)* and cordgrass *(Spartina patens)*.

shows the general distribution of dominant species along a transect across Core Banks. It is quite clear that the vegetation is directly related to the overwash terraces, their elevation, and the frequency of overwashing those areas close to the berm and between dunes being overwashed most frequently.

Spartina patens is remarkably well adapted to an overwash environment. It is resistant to salt spray and salt water flooding. The tough wiry leaves and culms dissipate wave energy and catch sand. It is often completely buried, but will come up through new material by rhizomatous growth and begin a new grassland. Following an overwash burial it seems to grow more vigorously than before. Several artificial burials were made on *Spartina* grassland by covering the grass with clean beach sand in 1 m square boxes to depths of 10 and 20 cm. After two growing seasons, the growth of grasses in the boxes was the same as the controls outside. A major natural burial and recovery were observed following Hurricane Ginger in September 1971. The storm crossed the Outer Banks near Drum Inlet and storm surges deposited up to 50 cm of overwash sand on a *Spartina* grassland. A year later, it was almost impossible at first inspection, to tell that an overwash had occurred; the regrowth of the grassland was nearly complete. The overwash sediment was still there and the grasses grew up through the new layer. It is important to note, however, that for all their important roles in building dunes, dissipating wave energy, and trapping and colonizing overwash

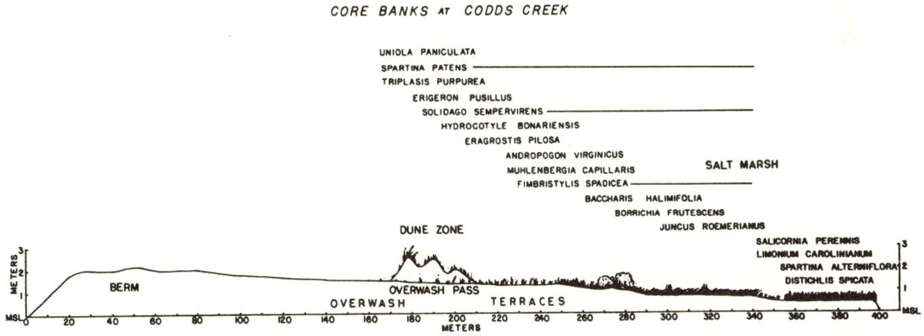

Figure 8. Diagrammatic vegetation sequence and elevation profile along Codds Creek transect. The position of a plant name indicates general zone of occurrence of that species with lines indicating more extensive distribution. Note that vegetation occurs in the overwash pass as well as on dunes and rear terraces.

material, *Spartina* and *Uniola* rarely completely stabilize the sand in the active zone, unless the area is well protected from oceanic forces.

Sand dunes play just as important a role as does overwash in barrier island dynamics. They keep overwash from being so frequent that the vegetation in the interior of the island has no time to recover between events. When a storm big enough to create overwash does occur, the dunes act as a first line of resistance. The waves are forced to dissipate some of their energy passing between the dunes, so that when they reach the inner part of the island they have slowed down and can deposit their sand load rather than cause erosion. Natural dunes can make the difference between overwash being a constructive or a destructive event. Another function of dunes is to serve as a sand reservoir; a really severe storm will flatten the first line of dunes on the berm and spread out their sand over the middle of the island. Low, broken dunes are most desirable in these respects, but even a continuous, artificial dune is better than none in some ways. For example, some parts of northern Core Banks, Portsmouth, and Ocracoke Islands became very low and bare apparently due to past abuse. Overwash was so frequent and erosive that no plants could take hold. Once on northern Core Banks and Ocracoke artificial dunes were built (Woodhouse and Hanes, 1967). However, succession proceeded to the point where a healthy vegetation cover now exists that is normally found on natural overwash terraces along the southern section of Core Banks behind the dune zone (Fig. 8); clearly, this low dune line improved the ecological conditions and the islands should now be able to withstand an overwash event without damage to the vegetation at this stage in succession. Indeed, this is now the case on northern Core Banks where the experimental dune is no longer maintained and is being pushed back in places onto the new grassland. Overwash is an example of a force which is constructive, but only if tempered, and the ability of the dunes and grasslands to temper it is one of the check and balance systems that keep the barrier island stable.

Overwash provides new substrate for salt marsh development. This happens when overwash completely crosses an island and sand spreads out as a fan into the sound. The new material is an ideal surface for colonization by marsh grass. An overwash of this type occurred on Core Banks 10 yr ago, and the new sand now supports a luxuriant stand of *Spartina alterniflora* with a standing crop three times as great as that of the neighboring, older marshes (Godfrey and Godfrey, in press).

It is also natural for a retreating barrier to be broken by temporary inlets. Fisher (1962) found geomorphic evidence, as well as historical accounts, for numerous inlets along the Outer Banks; 14 on Core Banks alone between Ocracoke Inlet and Cape Lookout. When a temporary inlet is open, usually caused by passage of a major storm and remaining open for as long as a century, or only a few decades, sand moves into the opening and spreads out to form tidal deltas. As the inlet closes, and the shoals become well developed, ecological succession begins and procedes at a rapid rate once the inlet is completely closed. The deeper flats are invaded by underwater beds of *Zostera* and *Ruppia* shoals exposed part of the time become *Spartina alterniflora* salt marsh, and the highest areas acquire a *Spartina patens* grassland in which small dunes begin to appear. These dunes may grow high enough to support shrubs and even woodlands in time. Figure 9 shows the changes at Drum Inlet which opened in 1933 and closed in 1971 and illustrates the general pattern of succession on tidal deltas. An inlet a few km south of Drum Inlet, Cedar Inlet, which was open in the last century, now has a typical salt marsh system on the former shoals and barrier island vegetation where the inlet used to be. Thus inlet dynamics combined with vegetation can greatly widen a barrier island in a relatively short time.

THE ALTERED SYSTEM – CAPE HATTERAS

On the islands of Cape Hatteras National Seashore, these natural processes have been greatly altered by the heavy hand of man. Much of this alteration – that is, dune building – in years past was done in good faith with the belief that such action was necessary to prevent the islands from being destroyed and to correct past abuses, with perhaps some justification in places. However, this approach has led to the attempts at total stabilization, the desire to hold everything in place, and to prevent the sea from washing over the islands. Such a program has come about to protect roads and buildings and with this development it has been necessary to try to hold all the sand in place and prevent dune movement, overwash, and the opening of inlets. A continuous artificial dune line has been built the length of the Seashore, planted with American Beach grass (*Ammophila breviligulata*) which must be fertilized and sprayed with biocides to combat insects and disease so that it produces a dense stand stopping all blowing sand. The beach grass is an exotic on the Outer Banks south of Nags Head while the native grass, sea oats, does not require the elaborate programs aimed at culturing the beach grass dune. As the beach grass dies out, sea oats take its place by natural propagation. The man-made dune is effectively a dike which stops all ordinary overwash. The beach has become very narrow and irregular, except at naturally accreting points such as the southwestern end of Ocracoke Island, and the south side of Cape Hatteras. Therefore storm waves impinge directly on most of the dune, since they cannot roll over the island, and rebound from the dune faces, eroding beach and dune sand into the littoral drift and out to deep water (Dolan, 1972). Grass cover cannot be stabilized or stand against this kind of erosion. The grasses are adapted to burial but not to undermining, and a dune that cannot retreat can only erode. Occasionally overwash occurs but it is hardly considered constructive (Fig. 10). There is usually an immediate response to plug the gap in the dunes and rebuild the dike.

The managed dune is strikingly different from natural dune zones. The use of fences, planting machines, and occasionally heavy equipment has produced a single, straight, steeply sloped structure compared to the rounded, irregular and scattered natural dunes. The heavy use of fertilizers produces a much denser

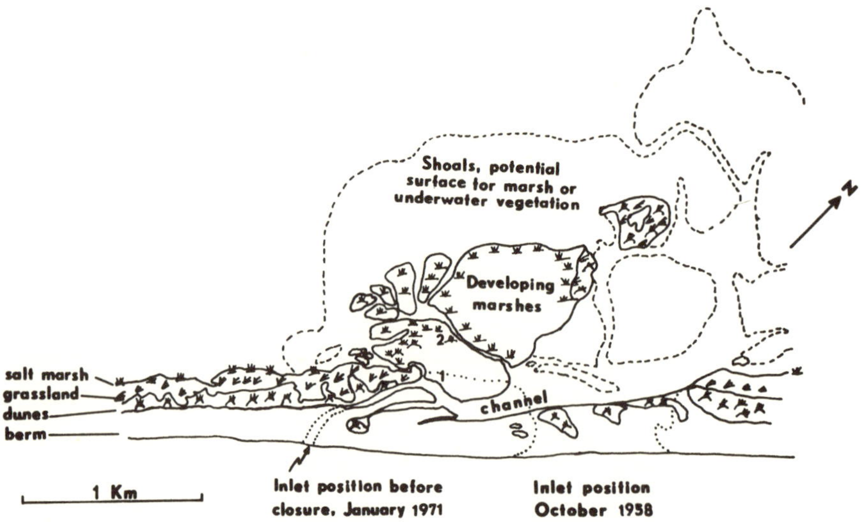

Figure 9. Map of old Drum Inlet showing inlet location in 1958, and in January 1971 just before closure. Shoals generally above low tide are being invaded by salt marsh cordgrass *(Spartina Alterniflora)*. Dunes are building on the spit as well as on some of the old shoal area. (From Godfrey and Godfrey, in press.)

Figure 10. Overwash occasionally occurs along Cape Hatteras National Seashore despite dune construction; Avon during a winter storm on 28 November 1971. Here flooding of property and damage to buildings, not on pilings, is considered most undesirable. (Photo courtesy of Cape Hatteras National Seashore.)

vegetation on the dune which effectively traps all moving sand (Fig.11). This probably accounts for the steep backslope. In most areas this dune is up to 6 m in height, the calculated height that is supposed to prevent most storm surges from going across the dune. The effect of the barrier dune on the vegetation behind it has been to accelerate ecological succession in the direction of species less tolerant of flooding and overwash burial. Shrubs and other plants which would ordinarily be expected in the most sheltered parts of the island are now coming up very close to the sea, along with normal dune plants. A storm severe enough to breach the dune would reverse this trend, eliminating all but the most tolerant species, such as *Spartina patens*.

Figure 12 is a cross-section of Hatteras Island at Sandy Bay near the area shown in Figure 11, illustrating the nature of the high man-made dune, with its dense vegetation and steep backslope, and the development of terrace-type vegetation in what was once the active dune and berm zone. Compare this profile with Figure 8. It is obvious that dune building on the berm has been a success here, but now there is concern over what will happen should the dune break. The road was built on what was once apparently part of the active beach zone; it is now below the level of the upper tide mark. The pre-development profile was probably much like Core Banks since similarities can still be seen in the basic profile, excluding the managed dune. The old overwash terrace corresponds to the major terrace on Core Banks and both probably formed during the same storm period. The low dunes to the right of the road are in apparently the normal position compared to Core Banks and probably formed before the main dune was built. Extrapolating the profile from these low dunes to the berm, suggests this was the active berm zone. The loss of the berm zone and natural overwash, because of the road and dune, has led to major difficulties in this section. Here the island is eroding from both sides – sound and ocean – since no sediments are moving back to widen the island and extend the protecting salt marsh fringe. An artificial widening process is being planned for the narrowest places by using fill and marsh propagation. (Godfrey, 1970b; Woodhouse, Seneca, and Brown, 1972).

Should a storm break through and overtop this dune line, sand from the dune will be pushed back over the road and onto the present vegetation. We do not know how long it would take for this system to recover from such a major burial; we do know that regular burial is easily survived by natural Outer Banks grassland. In the earlier stages of grassland succession, such as could be seen on the southwestern end of Ocracoke Island following the dune building programs there, recolonization should be fairly rapid since overwash tolerant species dominate in that stage. However, as time goes on and overwash is prevented, vegetation will grow which will be less likely to survive a major overwash.

To allow the normal opening and closing of inlets has of course been considered impractical on a developed barrier island. Inlets necessary for navigation are kept open by dredging, such as Oregon (which has been steadily migrating to the south), Hatteras, Ocracoke, Barden, and Beaufort Inlets, while

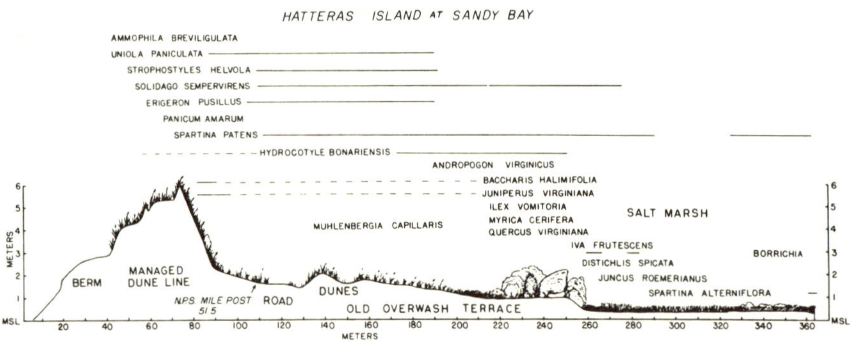

Figure 11. The man-made, managed dune line near the transect shown in Figure 12 on Hatteras Island at Sandy Bay. The dune is high, continuous, unbroken, eroding on the seaward side in numerous places, steeply sloped on the sound side, and covered with dense dune vegetation.

Figure 12. Elevation profile and vegetation sequence across Hatteras Island at Sandy Bay. Note the shape of the dune and the narrow beach compared to Core Banks. Dune vegetation is typical of the region except for the American beach grass *(Ammophila breviligulata)*. While the species distribution is similar to that of Core Banks, zones are much closer to the sea and much denser on the dune, a result of fertilization. While not listed here, typical salt marsh species such as those on Core Banks are present at Sandy Bay. Of particular note is the shrub zone developing along the backside and advancing seaward.

others are closed usually at great expense and effort, for example the inlet that opened just north of Buxton in 1962. An experiment in inlet creation was made on Core Banks by the Corps of Engineers following the closure of Drum Inlet. A new inlet was made opposite Atlantic where Core Banks was very narrow, and the spoils were placed behind the island in a more or less natural pattern. This spoil was then planted with marsh grasses in an attempt to stabilize the sand and create a new marsh thus widening the island at that point. A typical tidal delta has formed within the year that the inlet has existed and only time will tell if it will be navigable, which was the main reason for its creation.

The overall result of dune management at Cape Hatteras National Seashore is that the islands are eroding from the front but not retreating as a unit. Overwash and inlet closure are the two main processes which widen the islands, and also the two most important sources of new substrate for salt marsh development. Since the younger salt marshes are the most productive, the trophic condition of the sounds cannot help but suffer if present practices are continued indefinitely. Another function of overwash is to build up the interior of the islands; without it, the islands become not only narrower and narrower, but ever lower with respect to rising sea level. Management emphasis has been on replacing beach sand, trying to move sand from the back of the island to the front, while natural forces tend to move it from front to back. Recent projects have moved sand back up the beach from a downdrift accreting area so the sand could make another trip in the littoral drift.

The results of interfering with normal island dynamics can be seen in the contrast between the effects of Hurricane Ginger on the Cape Hatteras Islands and Core Banks, mentioned above (Dolan and Godfrey, in press). At Cape Hatteras and all along the developed beaches there was serious erosion, even though the storm passed over Core Banks. In contrast to Cape Hatteras, it was very difficult to see any permanent change on Core Banks and Cape Lookout (Figs. 13a, 13b). These islands were completely under water except for higher dunes, yet there was no damage. Instead, there was overwash buildup along the whole barrier island. The berm crest was pushed back, and the beach flattened; new sand was added to the backshore, and the level of the interior was raised; new fine sands were available for dune growth; and the grasslands recovered as usual from burial. Soon after the storm the beach was building seaward again and in 10 months had built out beyond its pre-Ginger position (Fig. 6). Meanwhile, at Cape Hatteras, the severely damaged dune scarp remained.

COMPARATIVE SEQUENCES ON NATURAL VERSUS ALTERED BARRIER ISLANDS

One of the main goals of our studies on National Park Service managed islands is to provide alternative management models, or options, that can be considered by the Service as it develops or reexamines master plans for such areas. We are

Figure 13. Core Banks at Codds Creek, showing effects of Hurricane Ginger September 1971. Figure 13a before storm and Figure 13b soon after. From the air little change could be seen, while at Cape Hatteras extensive dune erosion was evident. Ground surveys on Core Banks showed changes involving overwash deposition. Although the island was completely flooded except for higher dunes, it was very difficult to tell that a storm had passed over this area, other than by drift lines on the dunes. Dotted line indicates the location of profiles shown in Figures 6 and 8.

now in a position to compare natural coastal systems and those modified by man, and point out certain features that typify both. These comparisons can be used by resource managers to predict the type of situation they will have on a barrier island should certain fundamental decisions be made. Since geomorphic and ecological patterns vary from region to region and even within a single island, each system has to be considered individually.

Figure 14 compares a wholly natural barrier island system, such as Cape Lookout National Seashore, with a developed and stabilized system such as Cape Hatteras National Seashore. The patterns shown here in cartoon form can be readily seen on these seashores respectively, and illustrate the two basic types of change we are now witnessing – one controlled solely by natural forces, the other by man's interference with these forces. Two options are offered to the seashore manager – what happens if the natural system is allowed to operate, and what happens from stabilization efforts. Most of the information on beach processes at Cape Hatteras has been derived from Dolan (1972).

CONCLUSION

When a barrier system is not altered, its vegetation and geomorphic processes are closely interdependent. The grassland is adapted to prevailing conditions; its zonation depends upon elevation, distance from the sea, frequency of overwash, and salt exposure. This ecosystem is well able to withstand continuous natural stress, and can be expected to maintain itself in its present form for the foreseeable future.

An artificial system, such as Cape Hatteras, presents a disturbing contrast. Thanks to the barrier dune, both vegetation and developments are now very close to the ocean. If the dune were to be breached now, most of the plants right befind it belong to species which could survive overwash and flooding. If present successional trends continue, however, shrubs and small trees will take over the land just behind the dunes. These species are not able to handle overwash, and if a break in the dune occurs there will be a major ecological perturbation instead of the gradual change that we see on Core Banks. The ocean system can be expected ultimately to reassert its control of these islands, with inevitable human dislocations, unless prohibitively expensive engineering projects are undertaken. The ecological future of the area is not bright unless natural processes are allowed to restore themselves, at least partially; to deliberately allow this to happen would be politically difficult. Several alternatives are possible in managing the Seashores. Present undeveloped seashores should be preserved in their natural condition, such as Cape Lookout. On the managed seashores where no permanent structures are present, retreating beaches and dunes can be allowed to follow the natural course of events, but in a controlled fashion. Dune growth can be encouraged, perhaps as dune zones instead of lines, but it must be recognized that the dunes will retreat with the migrating island by the means of overwash and wind. It is necessary to get away from ideas that all dunes must stay in place, and, that storm flooding will destroy the island. A

A
NATURAL
B
STABILIZED
1
2
3
4
5
6
7
8
PJG

controlled overwash passing through dune fields will build the island's interior and in the process beaches will widen. Islands can be widened artificially by techniques which mimic the natural patterns of retreat. It may become possible to predict weak areas that might open into inlets and plan ahead to bridge such areas before the event occurs. The important point is that dunes, overwash, wind, grasses, inlets, and storms are all part of a very flexible, changing environment and successful management must likewise become flexible. The lesson which emerges, however, is clear; man, not the sea, is the worst enemy of barrier islands today.

Figure 14. Comparison of altered and unaltered barrier systems. In 14-1A and 1B, the barrier systems are alike. 2A shows no change in the natural system. 2B shows the stabilization of an artificial dune to protect a road. 3A shows the affect of a storm overwashing the natural barrier. 3B shows erosion of the artificial dune and foreshore and nearshore zones by storm waves, enhanced by positive feedback. 4A the overwash terrace is raised and recolonized, while 4B shows badly eroded stabilized dunes and rebuilding berm. 5A shows waves from the sound overwashing the natural barrier. 5B shows the sound storm tide eroding the land side of the dune and damaging property and vegetation, supposedly protected by it. 6A shows the natural barrier undamaged by the storm, while 6B shows further human interference, rebuilding the dune artificially. 7A a very violent storm causes overwashing, but no permanent damage to the natural barrier. 7B the same violent storm breaches the artificial dune, with disastrous results to human installations. 8A shows new salt marsh growing on overwash sediment and stability of the natural barrier. 8B artificial beach nourishment, at great cost, attempts to repair the damage on the narrowed seaward slope of the artificial barrier and erosion of the marsh further narrows the barrier from the land side. Soundside dredging for borrow material has fortunately been discontinued in many areas, because sediments are not as sandy and easily wash away. Accreting spits, inlets, or other sources are now considered more suitable for beach nourishment.

REFERENCES

Carney, C.B. and A.V. Hardy. 1967. North Carolina Hurricanes: Weather Bureau, ESSA, U.S. Dept. of Commerce, Washington, D.C., 40 p.

Dolan, R. 1972. Barrier dune systems along the Outer Banks of North Carolina: A reappraisal: Science, v.176, p.286-288

Dolan, R. and P.J. Godfrey. In press. Effects of Hurricane Ginger on the Outer Banks of N.C.: Geol. Soc. of Am. Bull.

Dunbar, G.S. 1958. Historical geography of the North Carolina Outer Banks: La. State Univ. Studies, Coastal Series No. 3, La. State Univ. Press, Baton Rouge, 234 p.

Fisher, J.J. 1962. Geomorphic expression of former inlets along the Outer Banks of North Carolina: M.A. Thesis, Univ. of N.C., Chapel Hill, 120 p.

Godfrey, P.J. 1970a. Oceanic overwash and its ecological implications on the Outer Banks of North Carolina: National Park Service, U.S. Dept. of Interior, Washington, D.C.

Godfrey, P.J. 1970b. The use of *Spartina alterniflora* to control erosion and build new salt marsh: National Park Service, U.S. Dept. of Interior, Washington, D.C. 63 p.

Godfrey, P.J. and M.M. Godfrey. In press. The role of overwash and inlet dynamics in the formation of slat marshes on North Carolina barrier islands: Physiology of Halophytes Symp. Academic Press, New York.

Hicks, S. 1972. As the oceans rise: Shore and Beach 40(1): p. 20-23.

Hoyt, J.H. 1967. Barrier island formation: Geol. Soc. of Amer. Bull., v. 78, p. 1125-1136.

Kraft, J.C. 1971. Sedimentary facies patterns and geologic history of a Holocene marine transgression: Geol. Soc. of Amer. Bull., v. 82, p. 2132-2158.

Mather, J.R., R.T. Field, and G.A. Yoshioka. 1967. Storm damage hazard along the east coast of the United States: Jour. Applied Meteor, v. 6(1). p. 20-30.

Milliman, J.D. and K.O. Emery. 1968. Sea levels during the past 35,000 years: Science, v. 162, p. 1121-1123.

Shepard, F.P. 1963. *Submarine Geology:* 2nd ed., Harper and Row, N.Y., 557 p.

Woodhouse, W.W. Jr., and R.E. Hanes. 1967. Dune stabilization with vegetation on the Outer Banks of North Carolina: U.S. Army Coastal Engr. Res. Center Tech. Memo. No. 22.

Woodhouse, W.W. Jr., E. Seneca and S. Broome. 1972. Marsh building with dredge spoil in North Carolina: Bull. 445, Agr. Exp. Stat., N.C. State Univ., Raleigh, 28 p.

23

Reprinted from *Photogramm. Eng.* **39**(5):477–488 (1973)

Remote Sensing of Tidal Marsh

Robert J. Reimold,* John L. Gallagher, and Donald E. Thompson

These techniques and production measurements are of value to those planning industrial and recreational development of the coastal zone by providing information on the spatial distribution of species and primary production in the marsh.

Introduction

ONE OF THE objectives of contemporary ecology is to describe the flow of energy and materials within various ecosystems. In many estuaries, primary production in the tidal marshes is the major source of fixed carbon. This carbon enters the estuarine food web primarily as detritus washed from the marsh surface by the ebbing tide (Odum and de la Cruz, 1967). Estimates of tidal marsh primary production have been made by Smalley, 1959; Morgan, 1961; Odum, 1961; Teal, 1962; Wass and Wright, 1969; Udell *et al.*, 1969; Stroud and Cooper, 1969; and Williams and Murdoch, 1972. A review of these data indicates that primary production varies between 150 and 3300 grams of dry matter per square meter per year. Because of the difficulties of working in the soft soils of the marshes, the limitations imposed by tidal fluxes and inaccessibility, most of the production figures are based on studies of relatively small areas and few samples. Local zonational variation in marsh production combined with small sample sizes has resulted in production estimates which may not be representative of the larger region. Reliable primary production values on large areas are necessary components in developing an understanding of estuarine energetics and nutrient cycling.

ABSTRACT: *Remote sensing is considered to provide quantitative data on primary production in tidal marsh ecosystems. Emphasis focuses on: (1) differentiation of vegetative types (species), and (2) assessment of primary production between each vegetative type (species) and within the dominant species,* Spartina alterniflora. *Remote sensing methodology utilized single-image photographs with Aerochrome Infrared 2443 at various altitudes. Thermal imagery was used to examine various bands of reflected and emitted radiation. Image parameters and plant characteristics were quantified by visual interpretation and color densitometry. Ground-truth data was used to calibrate remote sensing imagery. Ground-truth measurement included plant biomass by standard harvest methodology, stem density, plant height, size class distribution, and plant pigment concentration. The agronomic production on a 46-hectare area of* Spartina *marsh was 571 g m^{-2} year^{-1}, and 698 g m^{-2} year-1 on a 16-hectare site of* Juncus, Distichlis *and* Spartina. *The average production for the 1142-hectare Duplin Estuary marsh is 591 g m^{-2} year^{-1}.*

Remote sensing techniques, coupled with reliable ground-truth data, provide a rational approach for assessment of primary production on large land areas. The technical and physical aspects of imaging with photographic sensors was reviewed by Heller (1970). Among the advantages of photographic sensors he enumerates are: (1) high resolution capabilities, (2) good spatial orientation with precision cameras, and (3) the relative ease of identifying vegetation and making measurements. Anson (1966) com-

* Study supported in part by National Science Foundation Grant GA 29446 to the University of Georgia. Messers Reimold and Gallagher are with the University of Georgia Marine Institute, Sapelo Island, Ga. 31327. Mr. Thompson is with MAPCOtech Inc., Daytona Beach, Fla. 32015.

pared panchromatic, color and color-infrared films for effectiveness in mapping drainage, vegetation and soils types. With the exception of soils mapping, color and color-infrared were considered to be the most effective films.

Olson (1964) was able to recognize general types of marsh vegetation from panchromatic or color aerial photographs. Stroud and Cooper (1969) used Kodak Ektachrome Infrared Aero film, type 8443, to differentiate vegetative types in marshes of North Carolina. Randomly selected portions of the marsh photographs were interpreted and the acreage of each community type in the whole marsh was based on these selected areas. Color and color-infrared transparencies were used by Eyre (1971) to determine the decrease in wetlands (in Florida) due to drainage for agricultural and industrial use. Egan and Hair (1971) and Anderson and Wobber (1972) used color and color-infrared film to delimit the boundaries of wetlands in Maryland and New Jersey respectively.

Among the disadvantages of photographic sensors are the need for clear weather and lack of sensitivity (of these sensors) to portions of the electromagnetic spectrum beyond the 0.90 nm (nanometers) range (see Figure 1). Gates (1970) emphasized the importance

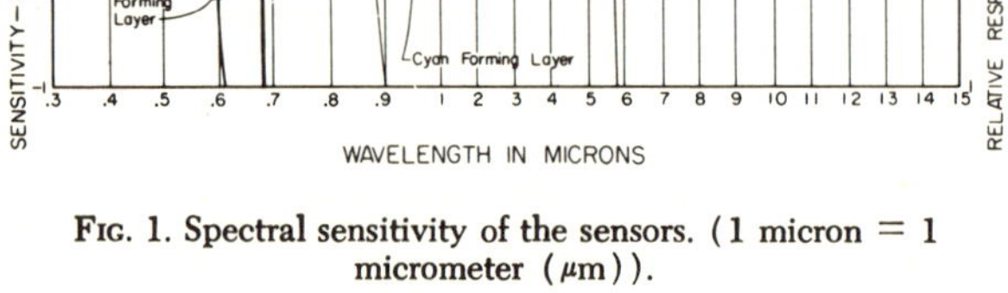

Fig. 1. Spectral sensitivity of the sensors. (1 micron = 1 micrometer (μm)).

of leaf geometry, morphology, physiological state and chemistry, as well as soil and climate, on the spectral quality and intensity of reflected and emitted energy. Reflectance and emissivity patterns of various plants can be useful in selecting bands within the electromagnetic spectrum where differences are maximized. Colwell and Olson (1966) have separated ten vegetation and terrain types in the Mud Lake Bog area of Michigan using a series of six spectral band tone signatures on concert.

Methods

Three remote sensing flights with concurrent ground truth acquisitions were conducted. Figure 2 shows the location of Sapelo Island along the coast of Georgia and indicates the location of the Duplin Estuary marsh study sites along the west side of the island.

Stereo coverage of the two study sites on east-west flight lines was acquired at scales of 1:2,500 and 1:5,000 using Kodak Aerochrome Infrared type 2443 film and a Zeiss RMK A 15/23 camera. The Duplin River estuary-marsh complex (hereafter termed the *Duplin*) was photographed with stereo coverage on north-south flight lines at scales of 1:5,000 and 1:10,000 (Plate 1). During the third mission, the *Duplin* was also photographed at scales of 1:20,000 and 1:40,00. The first photographs were taken near solar zenith and low tide. Because of solar reflection from the wet soils, the second and third flights were timed to coincide with an approximately 50° sun angle at low tide. The first photographic mission was conducted 22 June 1971; the second mission, 18 August 1971; and the third mission, 14 September 1971. These dates were chosen to be equally spaced during the theoretical maximum primary productivity period of the marsh plants.

The other remote sensing technique used was interpretation of thermal imagery with a Bendix Thermal Mapper Model LN-3. Flight lines for the first, second and third missions included an east-west flight line over the two study sites and a north-south flight line over

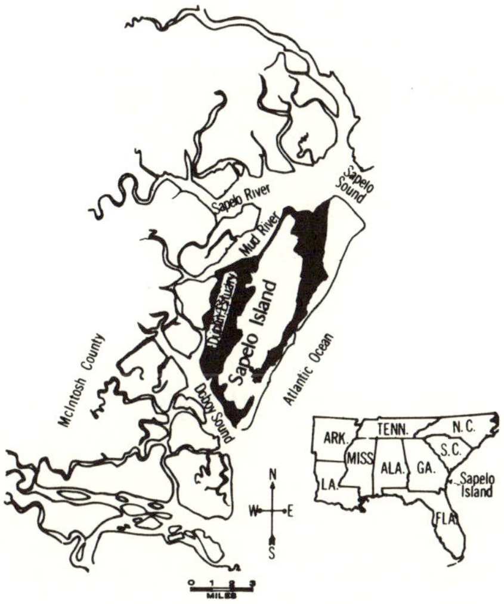

FIG. 2. Location of the Duplin Study Area on the west side of Sapelo Island, Georgia.

the entire *Duplin*. All flight lines were flown during the daylight hours. The first mission using thermal imagery was flown with a mercury-cadmium-telluride detector crystal in the thermal mapper filtered to receive the 8.0- to 12.5-micrometer (μm) wavelength band. In addition, the second mission included band widths of 2.0 to 3.5, 2.0 to 13.0 and 3.7 to 5.5 μm. On the third mission, an indium-antimonide detector, filtered to detect the 2.0 to 3.5 μm wavelength band, was used in addition to the detector-filter combinations used during the second mission. A summary of the remote sensing data acquired is shown in Table 1.

The north study site is located approximately four miles north of the mouth and on the east side of the Duplin Estuary. It was chosen because it contains a large variety of habitats. *Spartina alterniflora* grows in dense tall stands along the streams and in thin stands in other areas; large sections are composed of pure stands of *Juncus roemerianus;* and mixed stands of *Spartina alterniflora, Distichlis spicata* and *Salicornia* sp. form a band along the bare sands bordering the woods. In contrast, the south study site, located approximately two miles north and on the east side of the mouth of the Duplin Estuary, consists of a pure stand of *Spartina alterniflora.*

Prior to the photography and ground truth collection, the study sites were divided into 250 meter square grids and each point marked with a one meter, white circular plywood disk. These disks formed a collar around the posts and were free to float with the rising and falling water. They provided identi-

TABLE 1. SUMMARY OF REMOTE SENSING DATA ACQUIRED

	Data	*Mission* 1	2	3
Color	*Scale*			*
Infrared	1: 2,500	x	x	x
Photography	1: 5,000	x	x	x
	1:10,000	x	x	x
	1:20,000			x
	1:40,000			x
Thermal	*Mercury-Cadmium-*			
Imagery	*telluride*			†
	2.0 – 3.5 μm		x	x
	3.5 – 5.5 μm		x	x
	8.0 – 12.5 μm	x	x	x
	2.0 – 13.0 μm		x	x
	Indium antimonide			
	2.0 – 3.5 μm			x
Date		22 June 1971	18 August 1971	14 September* 7 October † 1971
Sun angle		90°	45–50°	45–50°
Tide stage		Low	Low	Low

fiable points on both the photographs and the ground.

Ground-truth data for the first mission were obtained by harvesting all living and dead material at the soil line in a 1.0-m^2 plot at 25-meter intervals along a line extending east and west through the middle of both the north and the south study sites. Thirty square meters were cut along the transect in the south study site and 20 square meters along the north. Each species in each plot was sorted into living and dead plants and into one-half meter height classes.

Data from the 50 square meters were organized into groups depending on the photographic color of the area where they were cut. Studies of the initial photographs indicated that perhaps six production categories could be seen.

Sampling sites for the second mission were selected from the vegetative patterns portrayed on the first mission photography. Variability within species and areal coverage were used as the bases for determining the number of replicates to be taken. Five samples were collected in the *Juncus* and *Distichlis* areas and 20 in the *Spartina* where four categories were established. Square white markers, similar to those used for the grid system, were placed on the plots following harvest and the photographs subsequently taken. Samples of plants from each of these areas for pigment analysis were returned to the lab and frozen. Pigments were extracted in acetone and chlorophyll *A* determined by the method of Odum *et al.* (1958).

For the third mission the sampling stations were once again selected by predicting areas of similar production based on the six color categories noted on previous photographs. Areas for sampling were selected to fall within the limits of each of the six colors rather than trying to select all of exactly the same hue.

Results and Discussion

The ground truth from the third mission is shown in Table 2, Dry weight of living matter varied from 1665 g m^{-2} for stream bank *Spartina* to 246 g m^{-2} for the *Distichlis* stand. These values represent the maximum standing crop and were used to compute the study site and watershed values. Chlorophyll *A* was assessed only in the leaves of this collection because the leaves appear to be the major contributors to the spectral reflectance or spectral response of the plants (Table 2). Within *Spartina*, chlorophyll *A* differences are clear at the extremes (red and blue), but between the intermediate groups (light red and blue-red) variations within each group preclude statistical separation. The spectral reflectance of *Juncus* and tall and short *Spartina* as measured in the laboratory with a Beckman DK-2A spectrophotometer are shown in Figure 3 (courtesy Dr. Harold Gausman, Rio Grande Soil and Water Research Center, Weslaco, Texas). The characteristic color of *Juncus* compared to *Spartina* on the color-infrared film was used to determine the proportion of these two major angiosperms in the *Duplin*. *Juncus* covers 67

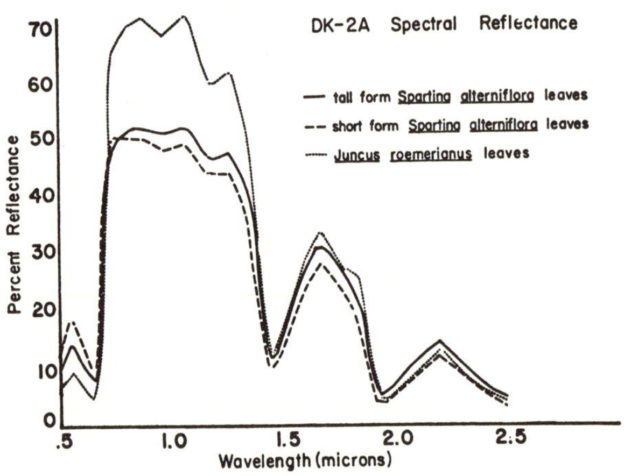

Fig. 3. Reflectance spectra for leaves of *Juncus roemerianus* and two forms of *Spartina alterniflora*. (1 micron = 1 micrometer (μm)).

TABE 2. GROUND-TRUTH DATA FROM THE THIRD MISSION (14 SEPTEMBER 1971) ON THE DUPLIN ESTUARY MARSH

	PRODUCTION CLASSIFICATION											
	Juncus		*Distichlis stand*		*Spartina*							
					Red		*Light red*		*Blue-red*		*Blue*	
Data	$Y^{n=5}$	*CV*	$Y^{n=5}$	*CV*	$Y^{n=5}$	*CV*	$Y^{n=5}$	*CV*	$Y^{n=5}$	*CV*	$Y^{n=5}$	*CV*
DENSITY												
Stems m^{-2}	786	20.5			98	19.2	167	25.7	142	26.2	206	38.9
Height Class (Percent abundance)												
0.01 – 0.50 m	15.5				28.2		35.7		51.0		78.8	
0.51 – 1.00 m	42.0				16.0		64.3		47.1		17.1	
1.01 – 1.50 m	42.5				23.5				1.9		2.9	
> 1.51 m					32.4						1.3	
LIVING MATTER												
Dry Weight (g m^{-2})	912.7	28.3	246.0	11.4	1664.9	41.3	629.5	41.9	499.1	24.7	331.3	34.2
Dry Weight (Percent)	47.5		42.8		27.6		35.7		34.5		35.7	
DEAD MATTER												
Dry Weight (g m^{-2})	624.6	30.3	357.3	30.8	301.1	71.2	169.4	25.0	225.8	35.2	57.1	46.2
Dry Weight (Percent)	63.2		62.0		32.3		40.6		32.5		34.9	
CHLOROPHYLL "A"												
(mg g^{-1} of fresh weight of leaf)	0.730		1.690		1.216		0.714		0.636		0.571	

** *Y*, mean; *CV*, coefficient of variation; *n*, sample number.

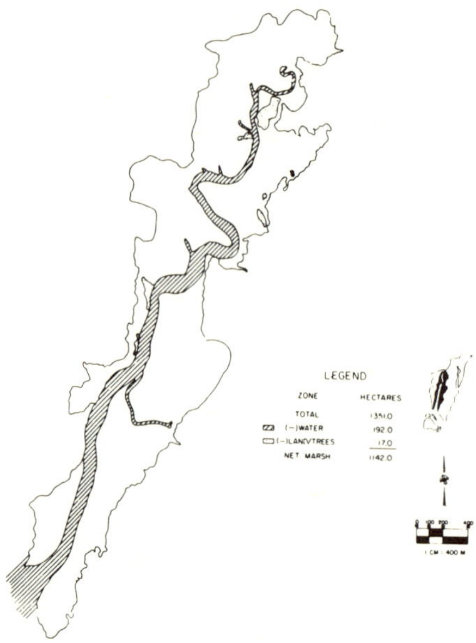

FIG. 4. Visual interpretation of marsh area in Duplin Watershed from color-infrared photographs, scale 1:40,000, 14 Sept. 1971.

hectares and *Spartina* covers 1075 hectares. The *Duplin* consists of 192 hectares of water and 1142 hectares of marsh as outlined in Figure 4.

Three distinct colors corresponding to *Juncus, Spartina* and a mixed stand composed primarily of *Distichlis* and *Salicornia* can be seen in the marsh areas of the photographs. Within the *Spartina* areas four fairly distinct colors are discernible: (1) bright red, tall robust plants; (2) light red, medium height robust plants; (3) blue-red, medium height-medium density plants; and (4) blue, low density short or medium height plants. (A typical area depicting these variations appears on the cover of this journal.)

The net reflectance from the various marsh areas depends on a number of factors including plant density, geometry, morphology, and the reflectance of the soil. The color produced if color-infrared film is exposed (with the proper filter) is due to the combined effects of all incoming energy with wavelengths between 0.5 and 0.9 μm. The colors produced (although their basis of formation is not completely documented) are useful in distinguishing various categories of marsh vegetation.

Because healthy, verdurous plants generally have a high reflectance in the near-infrared (see Figure 3), exposure of the color-infrared film layers should result in a bright red or magenta color. Differences in chlorophyll and carotenoid concentrations caused by factors such as mineral limitations, disease, genetic differences, physical edaphic condi-

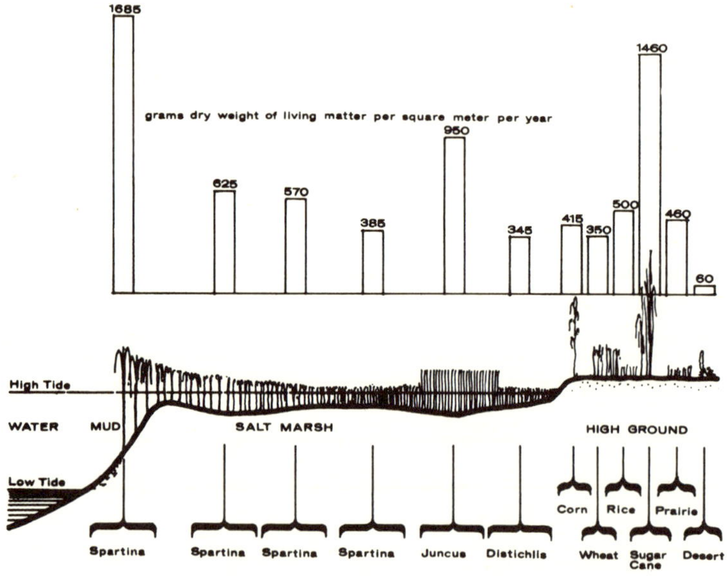

tions and salt stress dominate the reflectance patterns at the lower end of the film's spectral sensitivity. In the short-infrared range, the reflectance differences of the plants are associated with differences in moisture in the mesophyll and differences in leaf anatomy (Gates, 1970). Based on the relationship where dry weight divided by wet weight multiplied by 100 equals the percent dry weight, the lower percent dry weight figures (as summarized from Table 2 and depicted in Figure 5) for the stream bank *Spartina* (red on color-infrared film) as opposed to that in the high marsh (blue) may indicate differences in the proportion of various cell types, cell size and intercellular spaces. Stream bank *Spartina* averaged 27.6 percent dry weight whereas the high marsh was 35.7 percent (significantly different at the 0.001 level).

The areas of each color class in the north and in the south study sites were interpreted on a Kern PG-2 stereo plotter and their areal extent determined with H. Dell Foster quantifying accessories. The maps produced are shown in Figures 6 and 7. Using the September living plant biomass figures as an indicator of agronomic production, the mean values for the south study site were 571 g m^{-2} year^{-1} and for the north 698 g m^{-2} year^{-1}. The average production for the 1142 hectare Duplin Estuary marsh is 591 g m^{-2} year^{-1}.

All of the film used for the three missions flown was purchased at one time and only the amount needed for a single mission was unfrozen and removed from the roll at a given time. Because of variations that may occur from one batch of film to the next and the cost of film, it is desirable to cover as much area as possible with a single 9×9-inch photograph. In the third mission, photographs were taken at 1:20,000 and 1:40,000 scale. A comparison of the interpretation of the north and south study sites was made on enlargements of the 1:40,000 scale photographs. Figures 8 and 9 show these 1:40,000 scale interpretations. The comparisons of the interpretations at the two scales is shown in Table 3. Total species areas are distinguished with equal ease on either scale photograph. Within the *Spartina* there is a tendency to interpret a larger portion of the area as blue-red on the 1:40,000 scale photograph than on the 1:5,000.

Figure 10 shows images of the north study site formed using the four bands with the mercury-cadmium-telluride detector plus the one with indium-antimonide dectector. Ther-

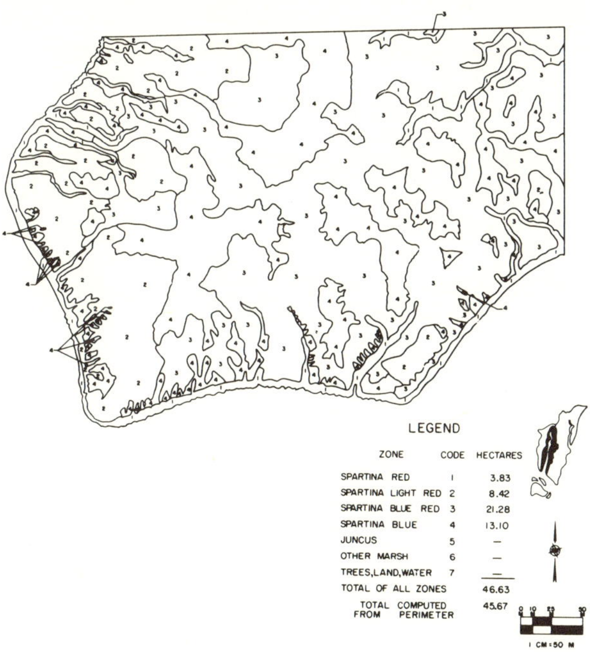

FIG. 6. Marsh grass zones in South Study Area defined by visual interpretation of color infrared photographs, scale 1:5,000, 14 Sept. 1971.

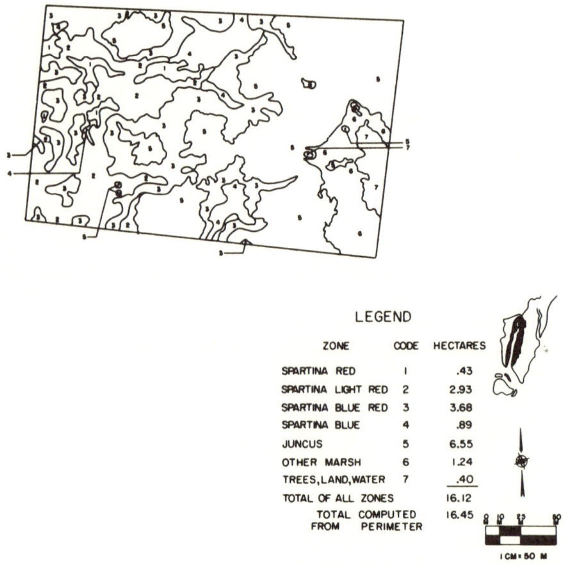

FIG. 7. Marsh grass zones in North Study Area defined by visual interpretation of color-infrared photographs, scale 1:5,000, 14 Sept. 1971.

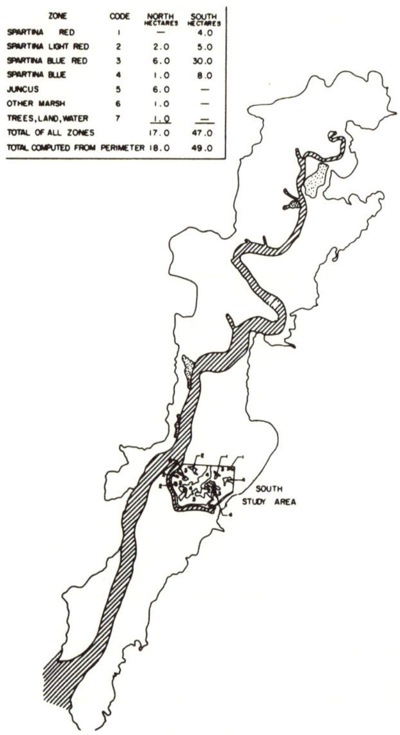

FIG. 8. Marsh grass zones in Duplin South Study Area defined by visual interpretation of color-infrared photographs, scale 1:40,000, 14 Sept. 1971.

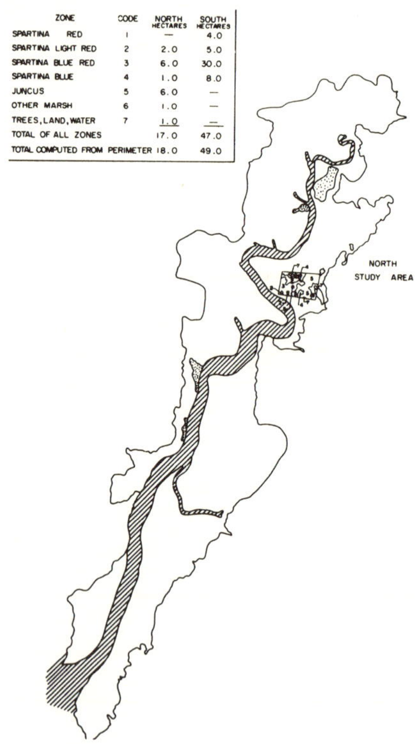

FIG. 9. Marsh grass zones in Duplin North Study Area defined by visual interpretation of color-infrared photographs, scale 1:40,000, 14 Sept. 1971.

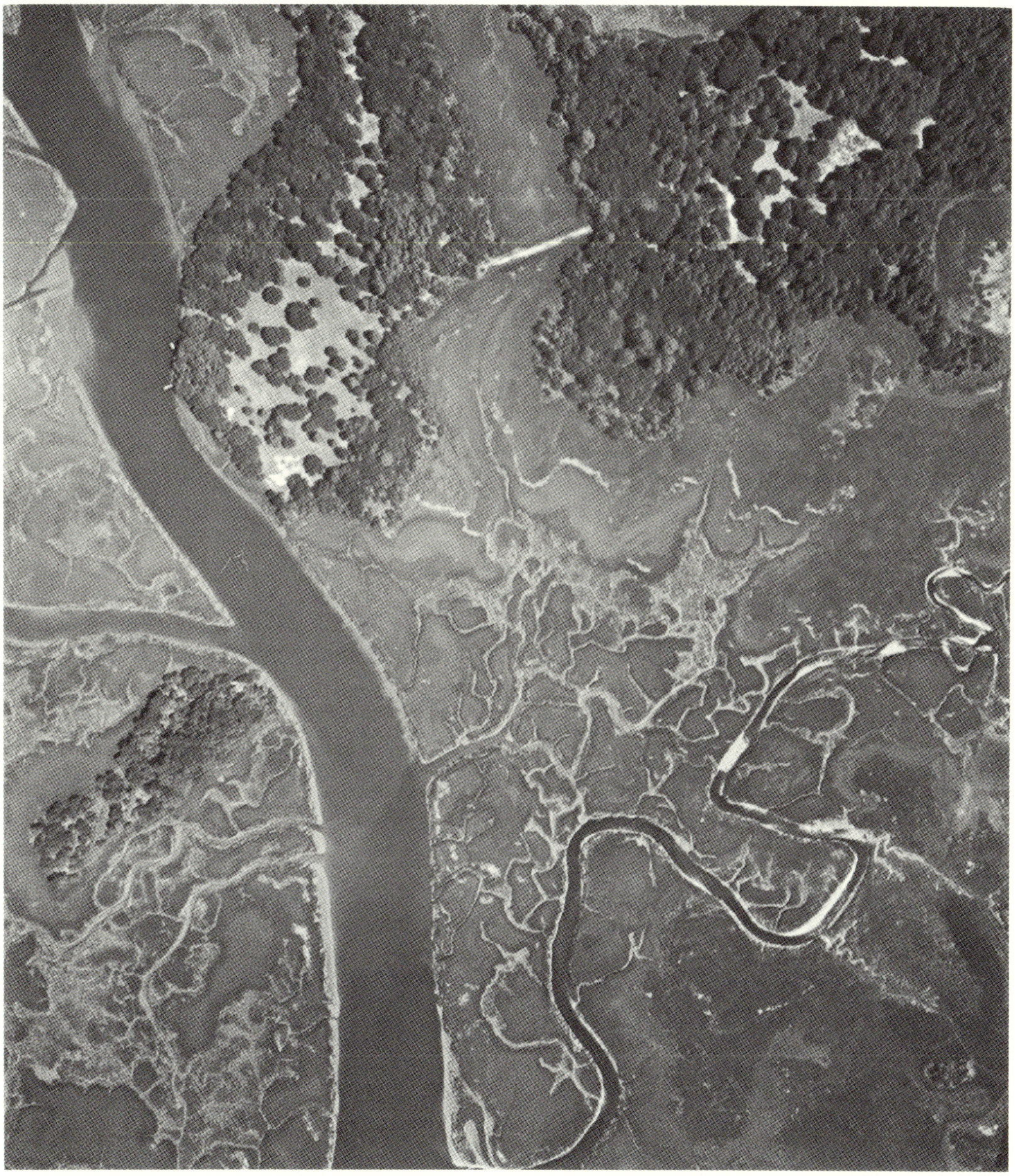

Plate 1. A characteristic color-infrared photograph of an area north of the Duplin North Study Area. The original scale was 1:5,000; here it is approx. 1:6,400. 14 Sept. 1971.

TABLE 3. COMPARISON OF VISUAL INTERPRETATION OF THE 1:5,000- AND 1:40,000-SCALE COLOR-INFRARED PHOTOGRAPHS OF THE TWO STUDY SITES

Production Categories	*Study Sites*			
	North		*South*	
	1:5,000	1:40,000	1:5,000	1:40,000
	Hectares			
Juncus	6.5	6.0	–	–
Distichlis stand	1.2	1.0	–	–
Spartina				
red	0.4	0	3.8	4.0
light red	2.9	2.0	8.4	5.0
blue-red	3.7	6.0	21.3	30.0
blue	0.9	1.0	13.1	8.0

mal imagery with the 2.0- to 3.5-μm bands and either the mercury-cadmium-telluride or indium-antimonide detector gives clear distinction of *Juncus* and *Spartina* vegetative zones.

A possible use of the thermal imagery, not anticipated at the outset of the study but discovered while working with the Datacolor System Color Densitometer, is the capability to discriminate the tidal flooding pattern of the marsh by color enhancement of the thermal imagery taken at low water. Preliminary examination of successive black-and-white infrared photographs taken over a flooding tidal period verify this new method.

SIGNIFICANCE OF RESEARCH

It is evidently possible to increase the accuracy of primary production estimates of large marsh areas by employing remote-sensing techniques. The interpretations of physiographic features seen in the photographic and nonphotographic images, the creation of vegetation maps and the measurement of primary production over large areas of the tidal marsh are already proving to be useful to ecologists and other scientists (U.S. Army Corps of Engineers, Georgia State Highway Department, and other state and federal agencies). These remote-sensing techniques and production measurements are also of value to those planning industrial and recreational development of the coastal zone by providing information on the spatial distribution of species and primary production in the marsh.

Primary production estimates for large areas obtained by coupling ground-truth

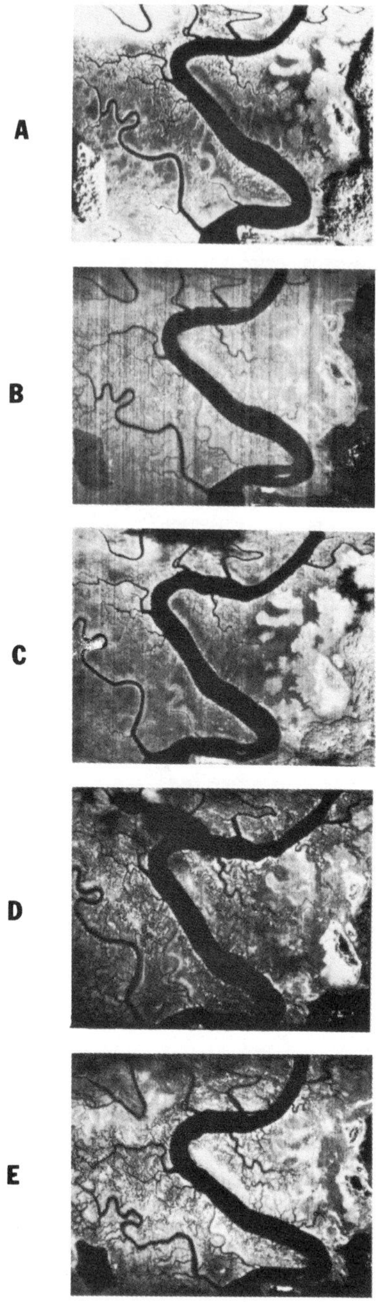

FIG. 10. Imagery of the North Study Area acquired with a Bendix LN-3 thermal mapper. *A*—Indium-antimonide detector, 2.0-3.5 μm. *B*—Mercury-cadmium-telluride detector, 3.7-5.5 μm. *C*—Mercury-cadmium-telluride detector, 2.0-3.5 μm. *D*—Mercury-cadmium-telluride detector, 2.0-13.0 μm. *E*—Mercury-cadmium-telluride detector, 8.0-12.5 μm.

measurements and photographic and non-photographic images provide essential basic information for mathematical modeling of the salt marsh ecosystem. Information on species spatial relations and primary production category distribution has proven useful for selecting sampling sites for other studies. As other ecosystem processes are associated with species and production distribution, the information obtained in this study is additionally useful in evaluating the contribution of particular areas to numerous aspects of the biological activity of the ecosystem.

Literature Cited

Anderson, R. R. and F. J. Wobber, 1972. "Wetlands Mapping in New Jersey." p. 530 to 536. *In* Kosco, W. J., Chairman. *Proceedings* of the 38th annual meeting, American Society of Photogrammetry, Falls Church, Virginia. 636 p.

Anson, A. 1966. "Color photo comparisons." *Photogrammetric Engineering.* 32:286–297.

Colwell, R. N. and D. L. Olson. 1966. "Thermal infrared imagery and its use in vegetation analysis by remote aerial reconnaissance." p. 77 to 91. *Selected Papers on Remote Sensing of Environment,* American Society of Photogrammetry, Falls Church, Virginia. 291 p.

Egan, W. C. and M. E. Hair. 1971. "Automated delineation of wetlands in photographic remote sensing." p. 199 to 200 *Summaries Seventh International Symposium of Remote Sensing of Environment.* University of Michigan, Ann Arbor, Michigan. 218 p.

Eyre, L. A. 1971. "High-altitude color photos." *Photogrammetric Engineering.* 37:1149–1153.

Gates, D. W. 1970. "Physical and physiological properties of plants." p. 224 to 252. *Remote Sensing with Special Reference to Agriculture and Forestry.* National Academy of Sciences, Washington, D.C. 424 p.

Heller, R. C. 1971. "Imaging with photographic sensors." p. 35 to 72. *Remote Sensing with Special Reference to Agriculture and Forestry.* National Academy of Sciences. Washington, D.C. 424 p.

Morgan, Marcia H. 1961. *Annual angiosperm production on a salt marsh.* M.Sc. Thesis. Univ. of Delaware. 34 p.

Odum, E. P. 1961. *The role of tidal marshes in estuarine production.* N.Y. State Conserv. 35:12–15.

Odum, E. P. and A. A. de la Cruz. 1967. "Particulate organic detritus in a Georgia salt marsh-estuarine ecosystem." p. 383 to 388. *In* Lauff, G. H. (ed.) *Estuaries,* American Association for the Advancement of Science. Washington, D.C. 757 p.

Odum, H. T., W. McConnell and W. Abbott. 1958. "The chlorophyll "A" of communities." *Publ. of the Inst. of Mar. Sci.* 5:66–96.

Olson, D. P. 1964. "The Use of Aerial Photographs in Studies of Marsh Vegetation." *Marine Agr. Expt. Sta. Bull.* 13. 62 p.

Stroud, Linda M. and A. W. Cooper. 1969. "Color-infrared Aerial Photographic Interpretation and Net Productivity of a Regularly Flooded North Carolina Salt Marsh." Water Resources Research Institute of the Univ. of N.C. *Report No.* 14. 86 p.

Smalley, A. E. 1959. "The role of two invertebrate populations, *Littorina irrorata* and *Orchelimum fidicinium,* in the energy flow of a salt marsh ecosystem." Ph.D. Thesis. Univ. of Georgia. 126 p.

Teal, J. M. 1962. "Energy flow in the salt marsh ecosystem of Georgia." *Ecology* 43:614–624.

Udell, A. F., *et al.* 1969. "Productivity and nutrient values of plants growing in salt marshes of the town of Hempstead, Long Island." *Bull. Torrey Bot. Club* 96:42–51.

Wass, M. S. and T. D. Wright. 1969. "Coastal Wetlands of Virginia. Interim Report to the Governor and General Assembly." Va. Inst. of Mar. Sci. *Spec. Rept. in Applied Mar. Sci. and Ocean Eng.* No. 10. 154 p.

Williams, R. B. and Marianne B. Murdoch. 1972. "Compartmental analysis of the production of *Juncus roemerianus* in a North Carolina salt marsh." *Chesapeake Sci.* 13:69–79.

24

Reprinted from *Remote Sensing in Oceanography Symp., 1973, Proc.,* Pt. 2, pp. 819–839

SHORELINE GEOLOGY: DEFINITION BY REMOTE SENSING

Roy E. Hunt
Partner
Joseph S. Ward and Associates
P. O. Box 91
Caldwell, New Jersey 07006

BIOGRAPHICAL SKETCH

Roy E. Hunt, P.E. attended Upsala College (1952 B.S., Geology), Newark College of Engineering (undergraduate studies in civil engineering) and Columbia University (1956 M.S., Soil Mechanics and Foundation Engineering). He has 21 years of experience, including positions with Greer and McClelland Esso Research and Engineering Company; and Woodward, Clyde, Sherard and Associates. He joined Joseph S. Ward and Associates in 1965; and is licensed as a Professional Engineer in New York and New Jersey. He is a Fellow of ASCE; and Member of AEG and AICE.

ABSTRACT

In recent years, the application of interpretation of remote sensing imagery to define geologic conditions along shorelines, has become common practice for many engineering studies. From the interpretations, geologic maps are prepared illustrating offshore and onshore conditions.

Two recent applications from projects in Bridgetown, Barbados, W.I., and the Cape Fear Riyer Basin, North Carolina are presented. The Barbados Project includes a sewer system, treatment plant and ocean outfall in an environment of coral reef formations. The project in North Carolina was an investigation of dredge spoil disposal sites and extended for a 34 mile stretch of the Cape Fear River from Wilmington to the sea.

INTRODUCTION

Engineering projects always require information on geologic conditions, and evaluations of the influence of the conditions on planning, design and construction. During preliminary planning for projects, often several alternate locations, routes, or solutions must be evaluated from the aspects of construction cost estimates and other economic factors, as well as environmental factors.

The interpretation of remote sensing imagery provides a means of rapidly and thoroughly obtaining information on geologic conditions. Its application is important on almost any project but it is particularly valuable during preliminary studies of large areas when time and budget limitations preclude the undertaking of extensive exploration programs.

The surficial extent of the various geologic formations or differing geologic conditions, are delineated and mapped from the imagery study. Additional important information is often obtained on distribution and stratigraphy from the literature, but some confirmation is necessary with data obtained by explorations, usually test borings. Test boring programs can be minimized by limiting explorations to areas typical of each surficial condition delineated from the imagery study. Thereby it is possible to extend the test boring data for similar deposits throughout the entire study area.

Two project examples follow: the first is for an ocean outfall to be constructed in shallow offshore waters with coral reef formations where test boring is extremely difficult and costly; the second is for a study of dredge spoil disposal along 34 miles of river channel that enters the ocean.

BARBADOS W.I. - OCEAN OUTFALL

Regional Setting and Geology

Barbados is a small coral-capped non-volcanic island in the Atlantic Ocean, located 200 miles northeast of Trinidad and 100 miles east of the Lesser Antilles. About 85% of the surface area of the pear shaped island (23 miles in length and 15 miles in width) is Pleistocene coral rock and associated sedimentary rocks. The remaining area, consists of older Tertiary sedimentary rocks.

The geology is characterized by two physiographic provinces: The gentle sloping, terraced surface of hard coral rock rising from the western and southern shores; and the steep sided hills and gullies in poorly indurated Tertiary sediments in the northeast. Sand beaches are present about most of the island, except for areas where the sea has cut cliffs in the coral. Living coral reefs, for the most part fringe reefs, are present on the western, or leeward, coast. Extensive banks of coral rubble off the southern coast and the dead coral reefs along the eastern and northern coasts support only small patches of coral growth.

Study Purpose

The capitol and population center, Bridgetown, is located on the southwest coast. Significant population growth combined with increased tourism in recent years has caused a substantial increase in the disposal of liquid waste into the ground through septic tanks. Recognizing a potential public health hazard, the Government of Barbados is planning a water pollution control system, including sewers to serve the greater Bridgetown area, a treatment plant and an ocean outfall. The study areas are on Figure 1.

The ocean outfall is planned to be a 24-inch pressure pipe, ranging up to 4000 feet in length, and placed in trenches up to 10 feet in depth. Three alternate alignments were studied, all located in the general area of the new deep water port, just north of the mouth of the Constitution River. As shown on Figure 2 two of the alignments are located in the shallow offshore waters of the fringe reef area, and one would pass through the reclaimed land of the port facility.

The design engineers required an evaluation of the influence of sea floor conditions on the design and construction of the outfall. Specifically, they were concerned with excavation methods and costs and with pipe bedding requirements.

Data Collection

To obtain information on coral reef development and morphology, the literature was searched and studied. Most of the underwater information in the literature was obtained by investigations using scuba gear. Aerial photography in two scales, 1:6000 and 1:10,000, were made available by the Government and stereoscopic study and interpretation of coastline conditions were made. Figure 3 is a copy of an aerial photograph (photo scale 1:6000) covering the outfall study area. Clearly shown are the shoreline, a portion of the deep water port, and the shallow water sea floor conditions. Evident are the inner fan-shaped fringe reef, deposits of sand, and an inner rubble reef. With the stereo viewer waves can be seen breaking over the fringe reef zone. The dark hole in the reef in the center of the photo represents a dredged borrow area for the port construction; water there is 50 feet in depth.

The 1:10,000 scale photos provided greater offshore coverage and enabled mapping of the outer reef. Water depth estimates based on photo interpretation compared favorably with the limited test boring data available. The estimates were made by comparing depths apparent beneath boats with the known heights of structures on land. Depths were zoned in the ranges 0 to 5 feet, 5 to 10 feet, 10 to 20 feet, and greater than 20 feet.

Sea Floor Conditions

The sea floor conditions as interpreted from the aerial

photographs are generalized on Figure 2. The interpretations were correlated with information from the logs of test borings made for the deep water harbor expansion. Four general zones characterize the near-shore conditions along the proposed outfall alignments: shallow water covers the fringe coral reef zone; coral heads and sands in about 14 feet of water; rubble coral reef and sands in 30 to 40 feet of water; and the deep water zone.

The fringe reef zone extends from the proposed plant site in the shape of a fan, to a distance varying from 500 to 1200 feet offshore. At mean low tide it is covered by 2 to 4 feet of water. Most of the zone is termed "Reef Flat Zone;"[1] "an almost horizontal reef-rock surface is partly covered by coral debris, sand, sediment-trapping filamentous green algae and encrustations of coralline algae." Along the seaward margin of the fringe reef is the "Reef Crest Zone, which is characterized by a series of spurs and grooves extending seaward and approximately perpendicular to the coastline. The spurs have a relief of 12 to 20 feet with respect to the bottom of the grooves." These spurs are evident on the aerial photos; the grooves are filled with coral rubble and sand.

As mentioned earlier, the large hole in the coral off the mouth of the Indian River, evident on the aerial photos, is the result of dredging for borrow for the new port facilities. Water is 50 feet deep in this area and the seabed cliff is vertical.

Outfall Alignment - The sea floor conditions along the routes of Alternates A and B are described in light of the findings from the literature review and air photo study.

As shown on Figure 2 Alternate A leaves the Fringe Reef Zone about 600 feet offshore and enters a zone where water is about 8 to 14 feet deep. This zone appears to be characterized by numerous coral heads, forming a very irregular surface. The openings in the coral growth range greatly in size and can be expected to be filled with coral debris and sand. In Borehole 6, located near the seaward edge of the zone and near the alignment of Alternate A, loose to compact silty coral sand with occasional small coral fragments was encountered from the seabed to a depth of 25 feet below seabed; dense coral sand with fragments of brittle coral were reported to depths of 65 feet below seabed. At low tide, water is 14 feet deep.

---

[1]Macintyre, Ian G. " Preliminary Mapping of the Insular Shelf of the West Coast of Barbados, W.I." Caribbean Journal of Science, Vol. 8 No. 1-2, March-June 1968.

In the third zone entered into by alternate routes A and B, water appears to range from about 36 to 42 feet in depth. Borehole No. 7, drilled in 42 feet of water, penetrated 10 feet into hard porous coral with intersititial coral sand. This zone appears to be the zone referred to by Macintyre as the "Dead coral rubble band," an area "composed of large fragments of coral debris under which a coral-rock base can be recognized in many areas." He suggests that in depths less than 30 to 40 feet, the insular shelf is a gently sloping (1-3°) wave cut platform. This would indicate that the coral formation could be expected to be massive and hard as it is inland.

Beyond the limits of the rubble reef, the sea-floor slopes at a much increased rate, in the order of 10 to 30°.

Engineering Evaluations

From a review of the existing data it was suggested to the design engineers that the preparation of preliminary designs and construction cost estimates for the two offshore alternates should be based on the following:

a. The coral rock in the fringe reef zone will probably require blasting for excavation. As water is very shallow in the fringe reef area, the construction of a causeway with a rubble coral fill will be necessary to provide access for contruction equipment. Wave and current action will require periodic maintenance of the causeway. The alignment of Alternate B should be relocated to avoid the 50 foot deep hole north of Borehole 8. Approximately 700 feet of causeway will be required for route A, and 1400 feet of causeway for route B, relocated to the west to avoid the deep hole.

b. The intermediate reef zone, where water is about 14 feet deep probably contains hard coral heads of various sizes with interstices filled with sand and coral debris. Excavation will probably require a dredge with a powerful cutting head to break up the coral stands and heads.

c. The outer reef zone, where water is 36 to 42 feet deep, probably is characterized by hard, shallow, massive coral. The report prepared for the port facilities harbor expansion noted that the large dredge used for the 1957-1959 operation may have experienced difficulty "in chewing off the harder coral at the outer margin of the reef (near Borehole 7)." The dredge had horse power ratings of 750 for the cutter and 2000 for the suction pump. If this is the case, excavation of the hard coral by blasting may be required for some areas of the outer reef zone.

d. Pipe bedding will require careful consideration because of the possibility of large voids among the coral stalks and heads. Migration of fine grained sediments from current flow is a distinct possibility. Bedding should consist of a minimum of two feet of coral debris in the coarse gravel to cobble sizes (1/2 to 8 inches), with care taken to fill large voids revealed by excavation. This will require trench wall and bottom inspection and mapping by divers.

CAPE FEAR RIVER BASIN, NORTH CAROLINA - DREDGED SPOIL DISPOSAL STUDY

Regional Setting and Geology

The study area encompassed a 4 mile width of the Cape Fear River Basin, extending from Cape Fear on the Atlantic Ocean to Wilmington, North Carolina, 35 miles to the north. (Figure 4 is a copy of an ERTS-A satellite image of the area and Figure 5 is the geology map of the area.)[2] The physiography is typical coastal plain; flat tidal marshlands and sandy uplands which seldom rise over an elevation of 25 feet (MSL). Many geomorphic features characteristic of a submergent coastal plain are evident: bars and spits; barrier beaches; sand dunes; tidal marsh; and fresh water swamp and marsh. The Castle Hayne Limestone, a white, sandy shell limestone varying to gray dense silicified limestone underlies the coastal plain soils at Wilmington at depths in the order of 20 to 40 feet. It outcrops in the town of Castle Hayne.

History of Dredging Operations

The section of the Cape Fear River covered in the study varies in depth from 25 feet at the northern end at Wilmington to 40 feet at the southern end near Cape Fear. The land use over the length of the channel studied varies from urban and industrial in the north around Wilmington; recreational and residential to undeveloped marshland along the east shore; and residential to military usage (Sunny Point) to urban (Southport) along the west shore.

Dredging of the Cape Fear River Channel began in 1829 and has continued into the present. During this period the channel depth was increased from 7 to 38 feet and is presently maintained at that depth. This requires yearly dredging in localized areas of shoaling. The annual amount of dredged spoil is about one million cubic yards.

Historically the dredging operation was performed to open or

---

[2]From "Geologic Map of North Carolina," 1958

maintain a deeper channel. The dredged material was discharged back into the river (termed "overboarding") as close as possible to the line of dredge but far enough away to prevent it from sloughing back into the channel. As the depth of the channel increased and with the continual maintenance program, it became necessary to discharge the spoil material farther away from the channel. Since the spoil material consists mostly of organic soil and fine sands, the pumped material stands on a very flat slope and tends to spread laterally over a very wide area. As dredging and disposal in the river continued large areas of shallow water were created all along the river channel.

Where existing islands were available the spoil material was discharged onto the islands within low dikes, in the order of 5 feet in height. The lateral extent of the dikes was determined by how far a bulldozer could go and still remain out of the water. Eagle Island was filled in this manner with the dikes created by the rock removed from the channel. The dikes at Eagle Island vary from about 20 feet high along the Cape Fear River to 9 feet along the Brunswick River.

The situation has been reached wherein additional disposal will fill the shallows, existing island spoil areas are reaching capacity, and the soil material is creating a maintenance problem by sloughing back into the channel.

In 1972 an upland disposal site was designed and constructed on the Military Ocean Terminal at Sunny Point. This site consists of a dike up to 40 feet high enclosing an area of approximately 400 acres. Dredged material from the Military Ocean Terminal's channels, docking and turning areas will be discharged here. It is expected that the maintenance of the channels will be reduced since the pumped spoil will not refill the dredged areas.

Study Purpose

The purpose of the study was to determine and evaluate geologic conditions within the project area, review the past and present dredging and disposal methods and assist in the selection, from an engineering point of view, of sites to be used for disposal over the next 50 years.

In addition, the results of the study will be used by other consultants to evaluate the following alternatives:

a. Abandonment of the dredging operation.

b. Reducing the depth and/or length of channel.

c. Continuing the existing program.

d. Moving the channel.

e. Increasing the depth and/or length of channel.

Data Collection

Geologic conditions were determined by review of literature including the logs of previously performed borings supplied by the Corps of Engineers, examination of aerial photographs, site reconnaissance, test borings, probes and laboratory soil testing.

The initial task was to interpret stereo pairs of black and white aerial photographs (scale 1:24,000) and to prepare the "Surficial Soil Map" of the study area shown in Figures 6 and 7. Using the map a program of test borings was planned to obtain information on stratigraphy and to recover samples of the subsoils for laboratory testing. Because of the extensive information provided on the "Surficial Soil Map," it was possible to obtain adequate information on subsurface conditions over the 140 square mile study area with only 17 test borings and 25 probes.

Geologic Conditions

The Coastal Plain of the Atlantic Coast is a result of emergence of the sediments of the ocean floor. During emergence major rivers such as the Cape Fear, and their many small tributaries, eroded valleys in the uplifted land. Recent rises in sea level, however, have drowned the valleys and the tributaries. During the recent period of gradual submergence, the tributary streams have become filled with alluvium and tidal marshes have formed on the small deltas being created at their mouths in the Cape Fear River. Dredging operations over the past 144 years have also filled in the river channel. An example of an alluvial-filled, drowned tributary valley is given on Figure 8, an aerial photograph. Figure 9 is an aerial photograph showing a drowned valley filled with tidal marsh, and spoil islands in the Cape Fear River.

Ocean waves have thrown up barrier beaches and deposited spits and bars along the shoreline. Sediments from the Cape Fear River have filled in behind the barrier beach to create a quiet, shallow water environment which permits the formation of tidal marsh deposits. Spit, bar, barrier beach, beach ridge and tidal marsh deposits are illustrated on Figure 10, an aerial photograph of Cape Fear.

Soil formations were differentiated on the Surficial Soil Map as follows:

Beach Deposits - Recent deposits of sands along the shoreline deposited in barrier beaches, bars and spits. Vegetation cover is sparse.

Dune Deposits - Recent wind blown deposits of loose sands; vegetation is thin brush and grass cover.

Coastal Plain Soils - Pleistocene deposits of varied colored silt and sand mixtures, containing gravel and shells in the deeper portions. Occasionally interbedded with varied colored clay, the Pleistocene soils are loose to medium compact in the upper 20 feet, gradually becoming compact with depth.

Alluvial Deposits - Recent deposits of loose organic silty fine sands to organic silts along drowned fresh-water stream channels. Thickness usually ranges from 5 to 10 feet, although in the larger tributary streams they were found to extend as deep as 32 feet (Jackeye Creek).

Marine Tidal Marsh - Soft organic soils deposited in a saltwater environment overlying alluvial organic silts and sands. Thicknesses along the shoreline are usually in the order of 5 to 10 feet, although very soft deposits in the river itself were found to extend to depths of 50 feet (at Snow Marsh), and they can be expected to be quite thick behind the barrier beach.

Old Spoil and Marine Tidal Marsh - Undifferentiated areas that are either natural or spoil deposits of soft organic soils and loose sand.

Swamp - Soft organic soil in a freshwater environment as determined from the aerial photographs by a marked difference in vegetation. Thickness was found to vary between 5 and 11 feet.

Not shown on the map are River Channel Deposits - These are recent deposits along the entire river channel which were found in the test borings to be primarily very soft to soft "dark gray organic silts, with fine sand, shells" to depths ranging from 28 feet near Cambell Island to 45 feet near Snow Marsh. The upper portions of the strata are coarser grained upstream near Wilmington, and also near the mouth of the Cape Fear River. Underlying the organic soils in the upper reaches of the channel near Wilmington are the Castle Haynes Limestone; the lower reaches are underlain by compact silty coarse to fine sands.

Engineering Evaluation

From the geotechnical aspects the primary factors influencing spoil disposal are the availability of materials for dike construction and the suitability of the surficial soils to support dikes. The economic factors to be evaluated are the cost of hauling fill for dike construction, the amount of fill material required for dike construction, the cost of site preparation required such as excavation of weak organic soils, and the cost of pumping the spoil materials (as a function of distance).

Dike construction over organic soils must be evaluated from the aspects of settlement and stability. If the dike height is raised too rapidly, the strength of the organic soils will be exceeded and a failure of the dike will occur. Long term settlement must be determined in order to provide adequate free board. The stability of the dike under the loads imposed by the spoil material must be taken into consideration when the cross section of the dike is designed. These problems are eliminated if the thinness of the organic deposits justify excavation.

The various soil types differentiated in the study were evaluated from the foregoing considerations. The areas identified as beach deposits and sand dunes, are primarily sandy materials and can provide suitable support to dikes. The construction of dikes with these poorly graded, highly pervious materials will require a lining or flat slopes and a graded filter to prevent excessive seepage and piping through the dike. The coastal plain deposits will provide adequate support for dikes and are suitable for use as dike materials.

The areas delineated as alluvial deposits, swamp, and the onshore and inshore marine tidal marsh deposits represent conditions where the weak surficial soils are relatively shallow (with a few localized exceptions). It is probably more economical to strip these soils prior to constructing dikes. They are not suitable as dike material.

The areas delineated as old spill and marine tidal marsh (offshore locations) represent areas that can be expected to be underlain by deep deposits of weak organic soils. Construction of dikes in these areas will require careful procedures to preclude failures both during and after construction.

The following conclusions were reached regarding the dredging operations:

The present dredging operation can continue using a hydraulic dredge in the northern section of the channel and a hopper dredge where the channel extends into the Atlantic Ocean. The disposal methods of overboarding, and disposal on existing spoil islands, should be discontinued as they will result in increased maintenance and eventual filling to capacity of the spoil islands. Creation of new spoil islands will be expensive. Creation of new land by diking and upland undiked disposal is possible but its feasibility is a function of ecological aspects, land usage and value. Upland diked disposal is a relatively expensive solution but one that has already been done at Sunny Point.

Location of optimum disposal areas should take into consideration the conditions given on the Surficial Soil Map. Creation of new land, upland undiked and diked disposal can be located anywhere on existing dry land. Some soil areas are more desirable than others, depending upon surficial soil conditions.

The final selection of disposal sites should be made from an overall view of the impact on the natural environment and the needs, wants and services of the community around the Cape Fear River.

CLOSURE

Substantial information on the geologic environment of coastal regions can be obtained from interpretation of remote sensing imagery. The information must be complimented with data from the literature and some subsurface explorations. It should be apparent, that studies of coastal areas, using conventional test boring programs and geophysical methods without the use of imagery, can be extremely costly and time consuming; and, that such studies will probably result in less complete data on the geologic environment.

ASP PAPER - SHORELINE GEOLOGY

FIGURES

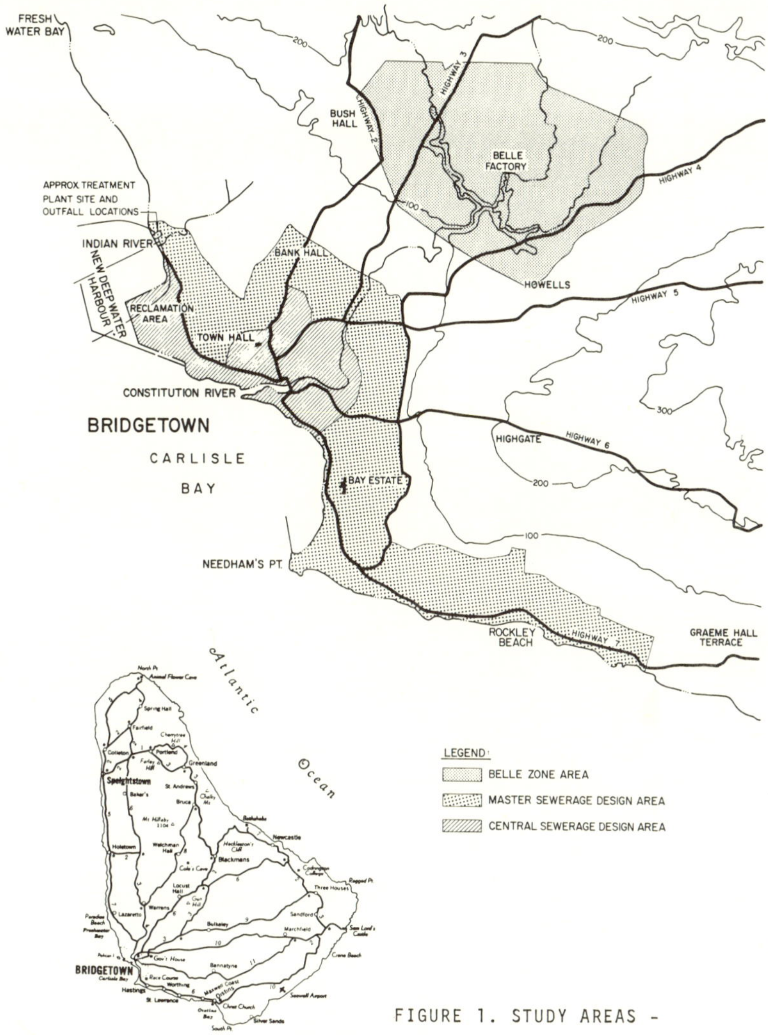

FIGURE 1. STUDY AREAS - BARBADOS WATER POLLUTION CONTROL SYSTEM

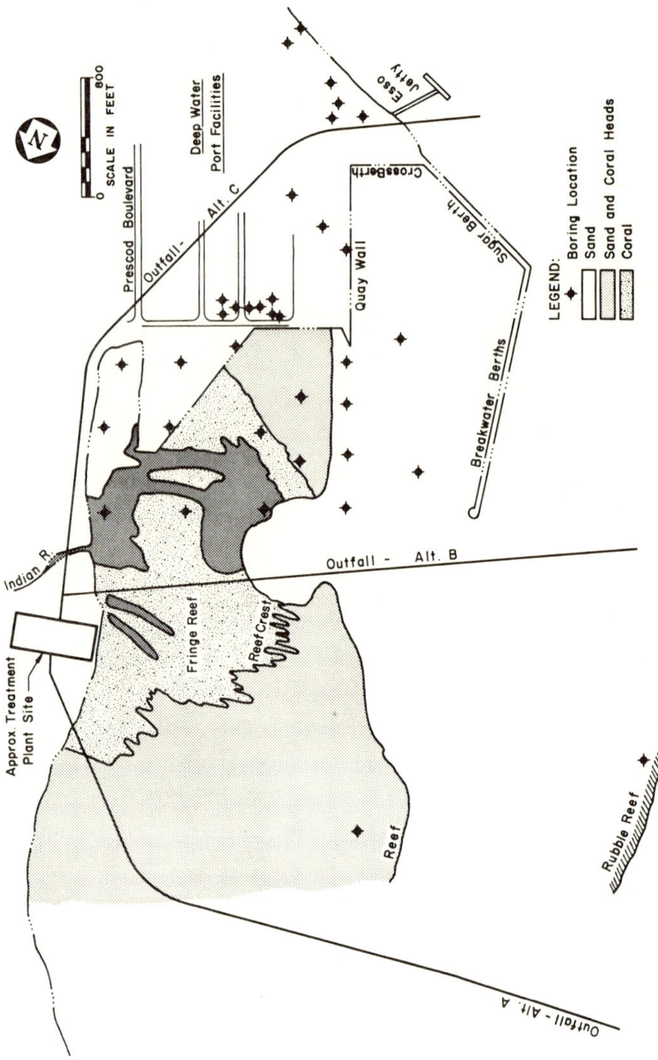

FIGURE 2. MAP OF SEA FLOOR CONDITIONS - OCEAN OUTFALL, BARBADOS

FIGURE 3. AERIAL PHOTOGRAPH - SEA BOTTOM CONDITIONS AT OCEAN OUTFALL LOCATIONS

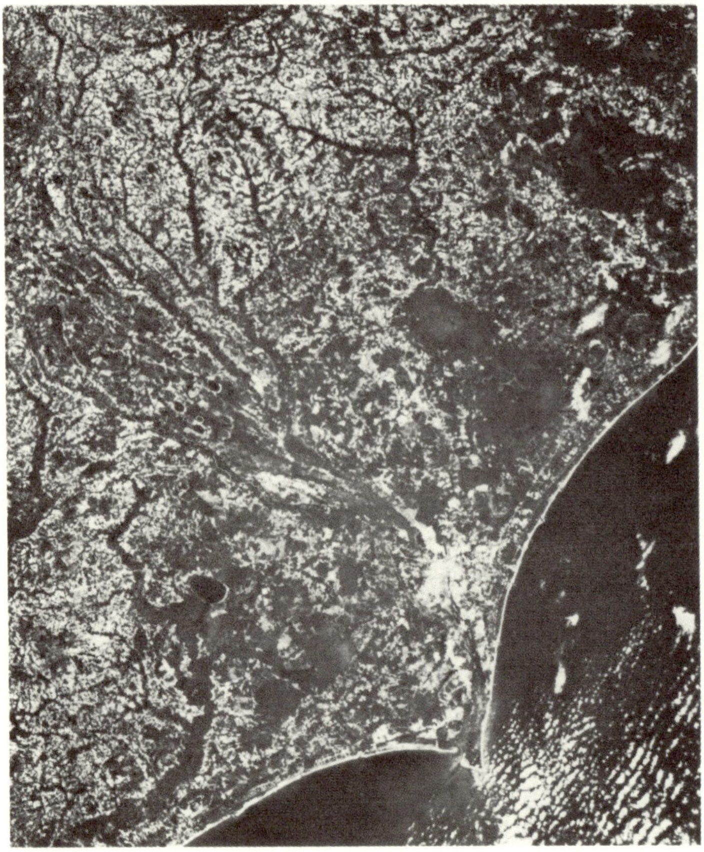

FIGURE 4. ERTS-A IMAGE - CAPE FEAR, NORTH CAROLINA

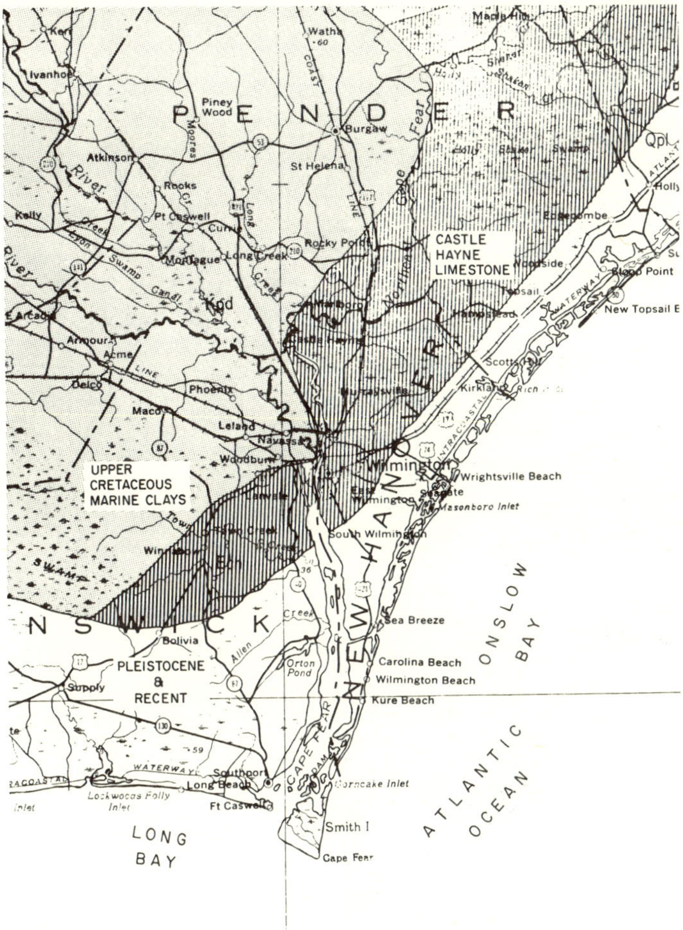

FIGURE 5. GEOLOGY MAP - CAPE FEAR, NORTH CAROLINA

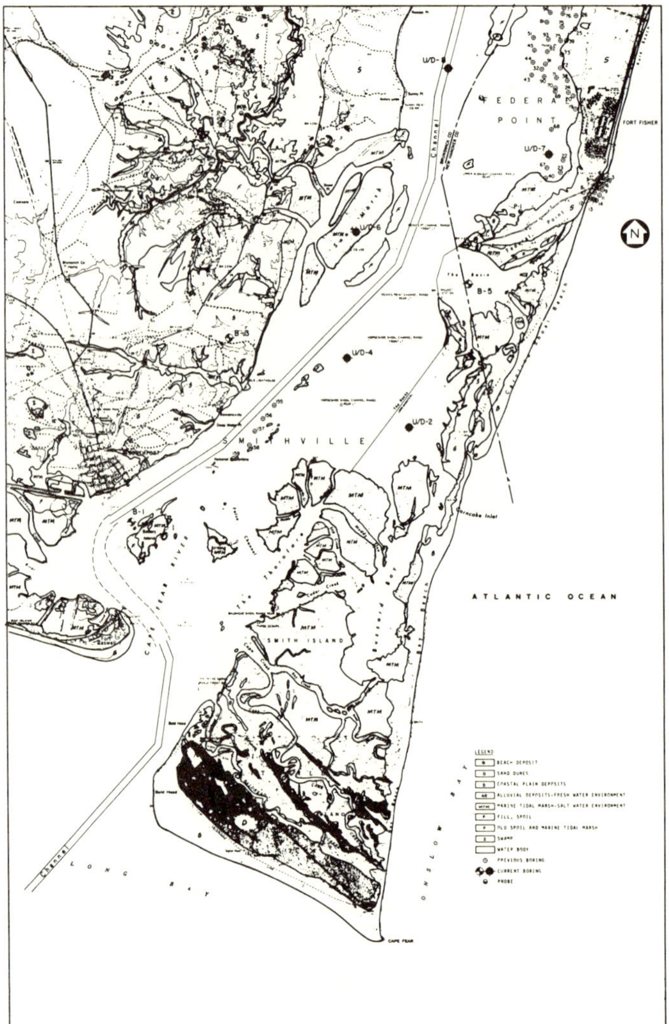

FIGURE 6. SURFICIAL SOIL MAP - CAPE FEAR RIVER BASIN

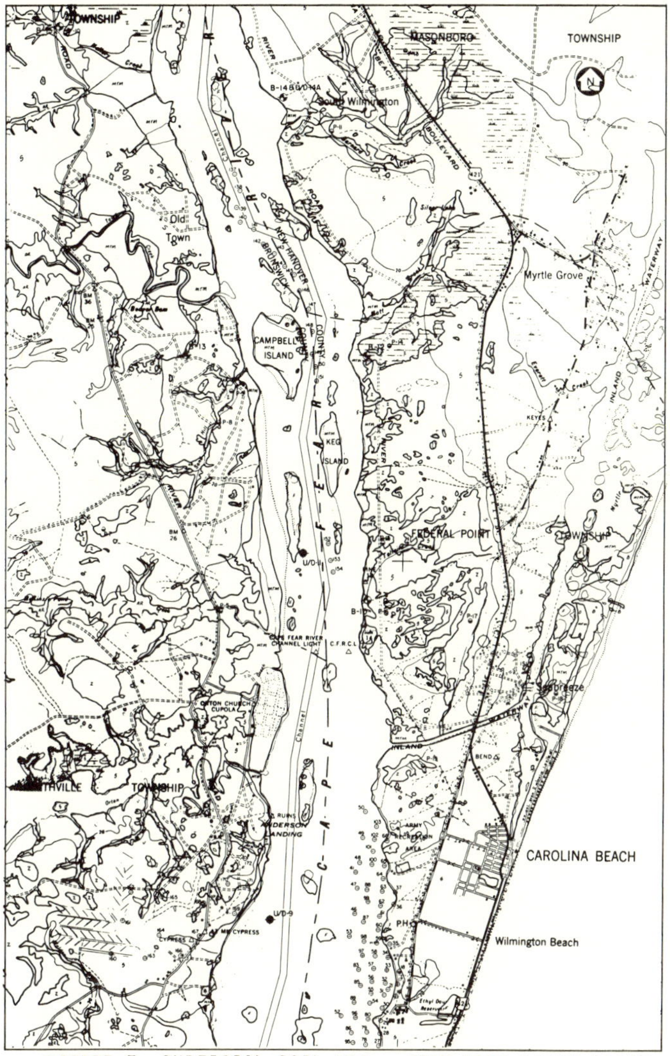

FIGURE 7. SURFICIAL SOIL MAP - CAPE FEAR RIVER BASIN

FIGURE 8. AERIAL PHOTOGRAPH - CAPE FEAR RIVER BASIN, ILLUSTRATING ALLUVIAL FILLED TRIBUTARY VALLEY

FIGURE 9. AERIAL PHOTOGRAPH - CAPE FEAR RIVER BASIN, ILLUSTRATING TIDAL MARSH AND CHANNEL SPOIL AREAS

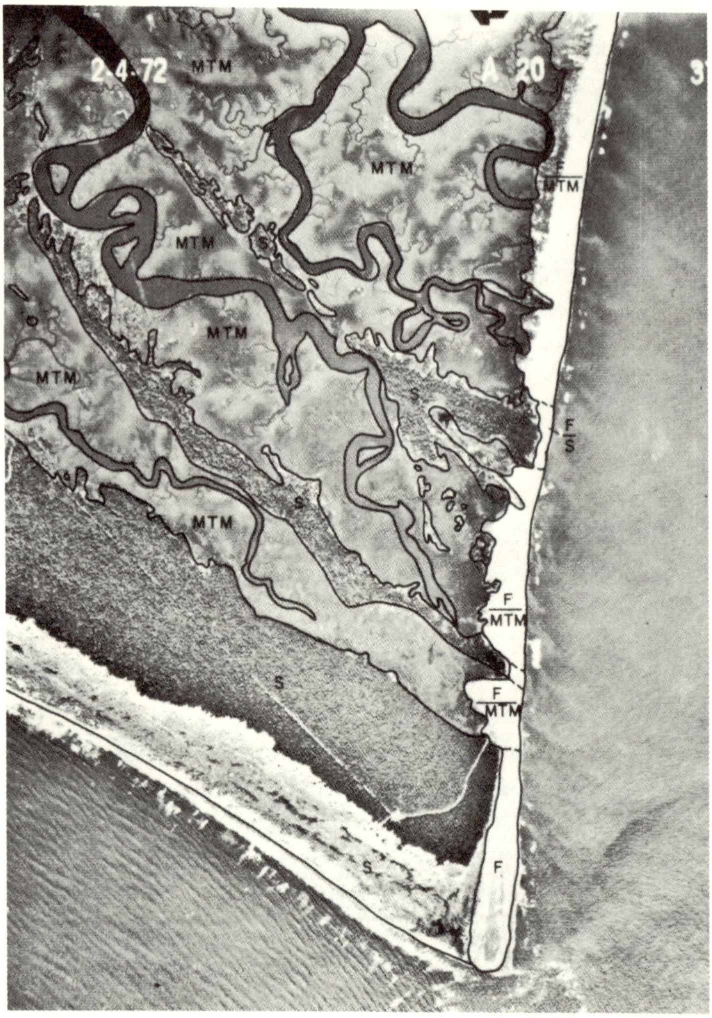

FIGURE 10. AERIAL PHOTOGRAPH - CAPE FEAR RIVER BASIN, ILLUSTRATING THE SPIT, BARRIER BEACH, AND OLD BEACH RIDGES AT CAPE FEAR

APPENDIX

SOURCES OF AIR PHOTOGRAPHY

Conventional Air Photography

United States

Information regarding the coverage and agency holding negatives for conventional air photography in the United States used to be published by the U.S. Geological Survey, Map Information Office, in a brochure entitled "Status of Aerial Photography in the U.S." However, due to the increased volume and variety of air photographic activities, the brochure has become impractical and its publication has been discontinued. Individual state "Status Maps," depicting imagery held by the Geological Survey, Forest Service, Soil Conservation Service, and the Agricultural Stabilization and Conservation Service are now available upon request from:

National Cartographic Information Center
U.S. Geological Survey
Mail Stop 507
National Center
Reston, Virginia 22092

The center can also provide more specific information on the availability of conventional air photography or space imagery for a specific geographic point or area of interest. Information on low and high altitude aircraft photography and satellite imagery over the United States and territories, and satellite imagery over foreign areas can also be obtained from the above address or from:

EROS Data Center
U.S. Geological Survey
Sioux Falls, South Dakota 57198

while information on low and/or high altitude aircraft photographic coverage over foreign areas may be obtained from:

Defense Intelligence Agency
DS-4A (FOIA)
Washington, D.C. 20301

An inquiry to the Geological Survey should include the geographic coordinates of the area of interest, or the geographic name of a specific locality and a radius of about ten miles. A computer printout will be mailed to the inquirer indicating the type of coverage (regular air, NASA aircraft, Apollo/Gemini, ERTS/LANDSAT, SKYLAB, etc.), quality of photography (inferior to excellent), cloud cover, date of exposure, scale of imagery, altitude, forward overlap, format size, and other additional information.

Governmental agencies in the United States, in addition to the U.S. Geological Survey, from which photographic coverage of coastal areas is available include:

Coastal Mapping Division, C3415
National Ocean Survey, NOAA
Rockville, Maryland 30852

A brochure entitled "National Ocean Survey Reproductions of Aerial Photographs" gives information for purchasing as well as type, size, scale, and price of U.S. photography. "Aerial photography index sheets" are also available from the same agency. These sheets are in the form of maps of the coastal area (by state) with the flight lines plotted on them and the centers of the photographs represented by dots. The marginal information gives the date and scale of photography. These indexes are most helpful in choosing photography and in determining the extent of land and water coverage on each photograph.

It is important to mention here that photography of the former U.S. Coast and Geodetic Survey is available from this same agency. This includes panchromatic, color, black-and-white infrared and some false-color infrared.

Some of the oldest coastal photography in the United States was taken by the Agricultural Stabilization and Conservation Service (ASCS). Air photography of eastern states is available from:

ASCS-USDA
Aerial Photography Division
Eastern Laboratory
45 South French Broad Ave.
Asheville, North Carolina 28802

while photography of western state is available from:

ASCS-USDA
Aerial Photography Division
Western Laboratory
2505 Parley's Way
Salt Lake City, Utah 84109

In addition to the above mentioned federal agencies, some states have their own air photographic services. Most of these services are located in the state's department of highways and inquiries related to available photography should be addressed to the same. Private firms also do considerable photography and mapping work for different concerns as well as in foreign countries. For a list of these firms consult "Photogrammetric Engineering and Remote Sensing," which is published monthly by the American Society of Photogrammetry.

Australia

Division of National Mapping
P.O. Box 548
7–9 Morrissett Street
Queanbeyan, N.S.W. 2620
(near Canberra)
Australia

Canada

Information on air photographic coverage in Canada is available from:

National Air Photo Library
Room 180, Surveys and Mapping Bldg.
615 Booth Street
Ottawa 4, Canada

The Department of Photogrammetry, Laval University, also has the names and addresses of all central government agencies and private firms in the entire world from which air photography can be purchased. Their World Data Bank consists of approximately

100,000 entries. The address is:

Department of Photogrammetry
Laval University
Quebec, P.Q.
G1K 7P4
Canada

England

Director of Military Survey
Ministry of Defense
Elmwood Ave.
Seltham, Middlesex
England

Finland

National Board of Survey
(Maanmittaushallitus)
Nirkkokatu 3
00170 Helsinki 17
Finland

or

The Topographic Service
(Topografikunta)
PL 167
Linnakuja 17
00161 Helsinki 16
Finland

France

Institut Géographique National
2, avenue Pasteur
94160 - Saint-Mandé, Paris
France

India

Survey of India
P.O. Box 37
17 EC Road
Hathibarkala
Dehra Dun 248001, Uttar Pradesh
India

Italy

Instituto Geographico Militare
Firenze
Italy

Japan

Geographical Survey Institute
2–24–13 Higashiyama Meguro
Toky 150, Tokyo, Honshu
Japan

Mexico

Director Departmento Topographico Militar
Secreteria de la Desensea National
Lomade Fotelo, D.F. 10
Mexico

The Netherlands

Topographische Dienst
Westvest 9
Delft
The Netherlands

South Africa

Directorate General of Surveys
Private Bag
Rhodes Avenue
Mowbray 7705
Cape Town, Cape Province
South Africa

West Germany

Institut für Angewandte Geodäsie
Richard-Strauss-Allee II
6000 Frankfurt am Main 70, Hessen
West Germany

Weather Satellite Imagery

The National Climatic Center
NOAA
Federal Building
Asheville, North Carolina 28801

Apollo and Gemini Photography

Information is available from EROS Data Center (address indicated below) as well as from:

Technology Applications Center
University of New Mexico
Albuquerque, New Mexico 87106

ERTS/LANDSAT and SKYLAB

U.S. Geological Survey
EROS Data Center
Sioux Falls, South Dakota 57198

Follow procedures listed under conventional photography in the United States from the same agency. An index, "Index to LANDSAT Coverage, Aug. 1972–July 1974," summarizing nominal scene, worldwide coverage of LANDSAT-1 (formerly ERTS-1) images obtained by NASA is also available from this agency.

Side-Looking Radar Imagery (SLR)

Goodyear Aerospace Corp.
Arizona Division, Dept. 408-A
Building 13-2S, C. A. Anderson
Litchfield Park, Arizona 85340

Goodyear is the depository of SLR imagery acquired by the U.S. Air Force on test flights across the United States. As of November, 1974, Goodyear had ten million square miles of SLR coverage.

Gemini V photograph (1965) of the coastal areas and the Namib Desert of Southwest Africa. (NASA photo)

REFERENCES

Abecasis, C. K. 1958. Littoral Drift Problems in Portugal with Special Reference to the Behavior of Inlets on Sandy Beaches. *Coastal Eng. Conf., 6th, Proc.,* pp. 406–438.

American Society of Photogrammetry. 1960. *Manual of Photographic Interpretation*. Falls Church, Va.

American Society of Photogrammetry. 1966. *Manual of Photogrammetry.* 3rd ed., 2 vols., Falls Church, Va.

American Society of Photogrammetry. 1968. *Manual of Color Aerial Photography*. Falls Church, Va.

American Society of Photogrammetry. 1972. *Proceedings of a Symposium on Coastal Mapping*. Falls Church, Va.

American Society of Photogrammetry. 1975. *Manual of Remote Sensing*. 2 vols., Falls Church, Va.

Anderson, R. R. 1969. The Use of Color Aerial Photography in Marshland and Estuarine Studies. In *New Horizons in Color Aerial Photography*, Seminar Proceedings, Am. Soc. of Photogramm., pp. 281–287.

Athearn, W. D., and Ronne, F. C. 1963. Shoreline Changes at Cape Hatteras: An Aerial Photographic Study of a 17-Year Period. *Research Rev. Off. Naval Research* **16**(6): 17–24.

Avery, T. E. 1968. *Interpretation of Aerial Photographs*. 2nd ed., Burgess Pub. Co., Minneapolis, Minn.

Babington-Smith, C. 1957. *Air Spy: The Story of Photo Intelligence in World War II.* Harper & Row Publishers, N.Y.

Bajorunas, L., and Duane, D. B. 1967. Shifting Offshore Bars and Harbor Shoaling. *Jour. Geophys. Research*, **72**:6195–6205.

Barwis, J. H. 1976. *Annotated Bibliography on the Geologic, Hydraulic, and Engineering Aspects of Tidal Inlets*. U.S. Army Corps of Engineers GIII Rpt. No. 4.

Bascom, W. N. 1954. The Control of Stream Outlets by Wave Refraction. *Jour. Geology,* **62**(6): 600–605.

Belcher, D. J., Amouzegar, J., Hodge, R. J., Ladenheim, H. C., and Lueder, D. R. 1951. *A Photo-Analysis Key for the Determination of Ground Conditions*. 3 vols., Tech. Rpt. no. 3 for Off. Naval Research, Cornell Univ., Ithaca, N.Y.

Berg, D. W. 1969. Systematic Collection of Beach Data. *Coastal Eng. Conf., 11th, Proc.,* pp. 273–297.

Berg, R. C., and Collinson, C. 1976. Bluff Erosion, Recession Rates, and Volumetric Losses on the Lake Michigan Shore in Illinois. *Illinois Geol. Survey Environmental Geology Notes no. 76* Urbana, Ill.

Bradley, E., and Armstrong, J. A. 1972. Description and Analysis of Coastal Zone and Shoreland Management Programs in the United States. *Michigan Univ. Sea Grant Pub. no. MICHU–SG–72–204,* Ann Arbor, Mich.

Brahtz, J. P., ed. 1972. *Coastal Zone Management: Multiple Use with Conservation.* J. Wiley & Sons, N.Y.

Brown, L. F., Jr., and Fisher, W. L. 1974. Environmental Geologic Atlas of the Texas Coastal Zone. In *Approaches to Environmental Geology*, E. G. Wermund, ed., Univ. Texas Bureau of Economic Geol., Austin, Texas, pp. 25–31.

Brown, L. F., Jr., Morton, R. A., McGowen, J. H., Kreitler, C. W., and Fisher, W. L. 1974. *Natural Hazards of the Texas Coastal Zone*. Univ. Texas Bureau Economic Geol., Austin, Texas.

Brown, W. L., Polcyn, F. C., Sellman, A. N., and Steward, S. R. 1971. Water Depth Measurement by Wave Refraction and Multispectral Techniques. *Infrared and Optics Lab. Rept. 31650–31–T*, Willow Run Lab., Inst. of Sci. and Tech., Univ. of Michigan, Ann Arbor, Mich.

Bruun, P. 1955. Coastal Development and Coastal Protection. *Florida Univ. Eng. Prog. Bull*. Series 76 **9** (11).

Bruun, P. 1962. Sea Level Rise as a Cause of Shore Erosion. *Amer. Soc. Civil Engineers Proc., Jour. Waterways and Harbors Div.* **88**:117–130.

Bruun, P. 1962. Engineering Aspects of Sediment Transport. *Geol. Soc. America Rev. Eng. Geology*. **1**:39–103.

Burton, I., Kates, R. W., and Snead, R. E. 1969. *The Human Ecology of Coastal Flood Hazard in Megalopolis.* Dept. of Geography Research Papers, Univ. of Chicago, Chicago, Ill.

Cameron, H. L. 1952. Water Current Measurement by Parallax Methods. *Photogramm. Eng.* **18**(1):99–104.

Cameron, H. L. 1965. The Shifting Sands of Sable Island. *Geog. Rev.* **55**(4): 463–476.

Cannon, P. J. 1974. The Application of Radar Imagery to Environmental Geologic Mapping. In *Approaches to Environmental Geology*, E. G. Wermund, ed., Univ. Texas, Bureau of Economic Geol., Austin, Tex., pp. 229–235.

Carey, A. E. and Oliver, F. W. 1918. *Tidal Lands, A Study of Shore Problems*. Blackie and Son, London.

Cassidy W. F. 1961. The Trend Toward Dual Purpose Inlet Control. *Shore and Beach*, **29**(4):22–26.

Chieruzzi, R., and Baker, R. F. 1958. A Study of Lake Erie Bluff Recession. *Ohio State Univ. Eng. Expt. Sta. Bull.* **27**(6):1–100.

Clark, J. 1974. *Coastal Ecosystems: Ecological Considerations of Manage-*

ment of the Coastal Zone. The Conservation Foundation, Washington, D.C.

Colcord, J. E. 1972. *Remote Sensing and Photo Interpretation.* National Tech. Inf. Service, U.S. Dept. of Commerce, PB–214 864.

Coleman, C. G. 1948. Photographic Interpretation of Coasts and Beaches. *Photogramm. Eng.* **14**(4):463–472.

Colton, J. B., Jr., Knapp, F. D., and Burns, B. R. 1974. Plastic Particles in Surface Waters of the Northern Atlantic. *Science* **185**:491–497.

Colwell, R. N. 1961. Some Practical Applications of Multiband Spectral Reconnaissance. *Am. Scientist* **49**(1):9–36.

Council on Environmental Quality. 1970. *Ocean Dumping—A National Policy*. U.S. Government Printing Office, Washington, D.C.

Cronin, L. E. 1967. The Role of Man in Estuarine Processes. In *Estuaries*, G. H. Lauff, ed., *Am. Assoc. Adv. Sci. Pub. no. 83*, 667–689.

Davidsson, J. 1963. *Littoral Processes and Morphology on Scanian Flat—Coasts*. Meddelanden Fran Lunds Universitets Geografiska Institution Avhandlingar XLII, Gleerupska Univ., Bokhandeln, Lund.

Davis, J. H. 1956. Influences of Man upon Coast Lines. In *Man's Role in Changing the Face of the Earth*, W. L. Thomas, ed., Univ. of Chicago Press, pp. 504–521.

DeBoer, G. 1967. Cycles of Change at Spurn Head, Yorkshire, England. *Shore and Beach* **35**(2):13–20.

Demathieu, P. G., and Verger, F. H. 1974. The Utilization of ERTS–1 Data for the Study of the French Atlantic Littoral. *NASA/Goddard Space Flight Center ERTS–1 Symp., 3rd, Proc.,* v. 1, Sec. B, pp. 1447–1450.

Deuel, L. 1969. *Flights into Yesterday: The Story of Aerial Archeology*. St. Martin's Press, N.Y.

Dolan, R. 1972. Barrier Dune System along the Outer Banks of North Carolina: A Reappraisal. *Science* **176**:286–288.

Dolan, R., Godfrey, P. J., and Odum, W. E. 1973. Man's Impact on the Barrier Islands of North Carolina. *Am. Scientist* **61**:152–162.

Dolan, R., and Vincent, L. 1973. Evaluation of Land Use Mapping from ERTS in the Shore Zone of CARETS. *Goddard Space Flight Center Symp. on Significant Results from ERTS–1, Proc.,* v. 1, pp. 939–948.

Dunn, G. E., and Miller, B. I. 1960. *Atlantic Hurricanes*. Louisiana State Univ. Press, Baton Rouge, LA.

El-Ashry, M. T. 1966. Photointerpretation of Shoreline Changes in Selected Areas along the Atlantic and Gulf Coasts of the United States. Unpublished Ph.D. Thesis, Dept. of Geology, University of Illinois, Urbana, Ill.

El-Ashry, M. T. 1971. Causes of Recent Increased Erosion along United States Shorelines. *Geol. Soc. America Bull.* **82**:2033–2038.

El-Ashry, M. T. 1973. An Example of Coastal Studies from Space and Aerial Photographs. *Jour. Geol. Education*, pp. 75–78.

El-Ashry, M. T., and Wanless, H. R. 1963. Photointerpretation of Coast-Line Changes due to Hurricanes. Abstract, *Geol. Soc. America Spec. Paper no. 76*, p. 53.

El-Ashry, M. T., and Wanless, H. R. 1965. The Birth and Early Growth of a Tidal Delta. *Jour. Geology* **73**(2):404–406.

El-Ashry, M. T., and Wanless, H. R. 1968. Photointerpretation of Shoreline

Changes between Capes Hatteras and Fear (North Carolina). *Marine Geol.* **6**:347–379.

Ellis, R. H. 1972. Coastal Management System: A Combination of Tools. In *Tools for Coastal Zone Management*, Marine Tech. Soc., Washington, D.C.

Emery, K. O. 1963. An Aerial Study of Hawaiian Wave Patterns. *Pacific Sci.* **17**:255–260.

Emplaincourt, J. L. G., and Wielchowsky, C. C. 1974. Detection of Shoreline Changes from ERTS–1 Data. *Southeastern Geographer* **14** (1):38–45.

Estes, J. E., and Senger, L. W., eds. 1974. *Remote Sensing.* Hamilton Pub. Co., Santa Barbara, Calif.

Fairbridge, R. W. 1961. Eustatic Changes in Sea Level. In *Physics and Chemistry of the Earth*, L. H. Ahrens, ed., Pergamon Press, N.Y., v. 4, pp. 99–185.

Fairbridge, R. W., and Teichert, C. 1948. The Low Isles of the Great Barrier Reef: A New Analysis. *Geog. Jour.* **111**:67–88.

Feinberg, E. B., and Mairs, R. L. 1974. Impact of ERTS–1 Images on Management of New Jersey's Coastal Zone. *NASA/Goddard Space Flight Center ERTS–1 Symp, 3rd, Proc.,* v. 1, Sec. A, pp. 497–503.

Gagliano, S. M. 1973. *Canals, Dredging, and Land Reclamation in the Louisiana Coastal Zone*. Center for Wetland Res., Louisiana State Univ. Rept. no. 14, 108 pp.

Gagliano, S. M., and Van Beek, J. L. 1975. An Approach to Multiuse Management in the Mississippi Delta System. In *Deltas, Models for Exploration*, M. L. Broussard, ed., Houston Geol. Soc., pp. 223–238.

Galvin, C. J., Jr. 1967. Longshore Current Velocity: A Review of Theory and Data. *Rev. Geophys.* **5**.

Geary, E. L. 1968. Coastal Hydrography. *Photogramm Eng.* **34** (1):44–50.

Geological Society of America. 1976. Impact of Barrier-Island Development—Geologic Problems and Practical Solutions. Report of the Committee on Environment and Public Policy.

Gierloff-Emden, H. G. 1961. Luftbild und Küstengeographie am Beispiel der deutchen NordseeKüste. *Landes Kudliche Luftbildaus-wertung Mitteleuropäischen*, v. 4.

Gierloff-Emden, H. G. 1976. *Manual of Interpretation of Orbital Remote Sensing Satellite Photography and Imagery for Coastal and Offshore Environmental Features (Including Lagoons, Estuaries, and Bays),* Münchener Geographische Abhandlungen, Müchen, vol. 20.

Gofseyeff, S. 1952. Case History of Fire Island Inlet, N.Y. Coastal Eng. Conf., 3rd, Proc., pp. 272–305.

Guilcher, A. 1958. *Coastal and Submarine Morphology*. Methuen & Co., London.

Guréva, Z. I. et al. 1968. *Geologo-geomorfologičeskoe morskich melkovodij i beregov po materialam aerofotos emki*. Akademiya Nauk S.S.S.R., Laboratoriya aerometodov, Leningrad.

Guss, P. 1972. Tidelands Management Mapping for the Coastal Plains Region. *Symp. on Coastal Mapping Proc.*, Amer. Soc. Photogramm., Falls Church, Va., pp. 243–262.

Hands, E. B. 1970. A Geomorphic Map of Lake Michigan Shoreline. *Great Lakes Research Conf., 13th, Proc.,* pp. 250–265.

Harris, W. D., and Jones, B. G. 1964. Repeat Mapping for a Record of Shore Erosion, Cape Kennedy, Florida, *Shore and Beach* **32**(2):31–34.

Hartford, F., and Tanner, W. F. 1976. Current Great Lakes Shore Damage. *Shore and Beach* **44**(1):16–19.

Hartley, R. P. 1965. Effect of Large Structures on the Ohio Shore of Lake Erie. *Shore and Beach* **33**(2):24–36.

Hays, M. O. 1967. Hurricanes as Geological Agents. *Texas Univ. Bur. Econ. Geology Rept. Inv.* **61:**1–56.

Herath, L. 1962. Shoreline Development and Protection at Negombo Beach, Ceylon: An Aerial Photographic Approach. *Transactions of the Symposium of Photo-Interpretation, International Archives of Photogrammetry*, Int. Soc. of Photogramm., Delft, The Netherlands, v. 14, pp. 453–459.

Herbich, J. B., and Hales, Z. L. 1970. Remote Sensing Techniques Used in Determining Changes in Coastlines. *Texas A&M Univ., Remote Sensing Center, Tech. Rpt. RSC–16.*

Herron, W. J., Jr. 1960. Beach Erosion Control and Small Craft Harbor Development at Point Hueneme. *Shore and Beach* **28**(4):11–15.

Holmes, R. W. 1969. The Santa Barbara Oil Spill. In *Oil on the Sea*, P. P. Hoult, ed., Plenum Press, N.Y., pp. 15–27.

Holz, R. K., ed., 1973. *The Surveillant Science: Remote Sensing of the Environment*. Houghton Mifflin Co., Boston, Mass.

Hood, D. W., ed. 1971. *Impingement of Man on the Oceans*. Wiley Interscience, N.Y.

Ikramuddin, A. S., and Laming, D. J. C. 1966. The Alma River Intertidal Delta in the Bay of Fundy at Alma. *Maritime Sediments* **2**(1):3–4.

Jamison, D. W. 1971. EROS as a Tool in Coastal Zone Management. *Marine Technology Soc., 7th Ann. Conf.*, pp. 365–371.

Johannessen, C. L. 1964. Marshes Prograding in Oregon: Aerial Photographs. *Science* **146**:1575–1578.

Johnson, D. W. 1919. *Shore Processes and Shoreline Development*. J. Wiley & Sons, N.Y.

Johnson, J. W. 1959. Littoral-Drift Problems at Shore-Line Harbors. *Am. Soc. Civil Engineers Trans.* **124:**525–546.

Johnson, J. W. 1961. Historical Photographs and the Coastal Engineer, *Shore and Beach* **29**(1):18–24.

Jones, D. A. 1969. Mapping the Coastal Zone for Boundary Demarcation. *Shore and Beach,* **37**(2):34–37.

Keene, D. F., and Pearcy, W. G. 1973. High-Altitude Photographs of the Oregon Coast. *Photogramm. Eng.* **39**(2):163–168.

Keller, M. 1963. Tidal Current Surveys by Photogrammetric Methods. *Photogramm. Eng.* **29**(5):824–832.

Kelly, M. G. 1969. Aerial Photography for the Study of Near-Shore Ocean Biology. In *New Horizons in Color Aerial Photography*, Seminar Proceedings, Am. Soc. of Photogramm., pp. 347–355.

Kenney, N. T. 1962. Our Changing Atlantic Coastline. *Natl. Geog. Mag.* **122**(6):860–877.

Ketchum, B. H., ed. 1972. *The Water's Edge: Critical Problems of the Coastal Zone*. MIT Press, Cambridge, Mass.

Kidson, C. 1963. The Growth of Sand and Shingle Spits across Estuaries. *Zeitschrift für Geomorphologie* **1**, pp. 1–22.

King, C. A. M. 1972. *Beaches and Coasts*. 2nd ed., St. Martin's Press, N.Y.
Komar, P. D. 1976. *Beach Processes and Sedimentation*. Prentice-Hall, N.J.
Komar, P. D., Lizarraga-Arciniega, J. R., and Terich, T. A. 1976. Oregon Coast Shoreline Changes due to Jetties. *Am Soc. Civil Engineers Proc., Jour. Waterways and Harbors Div.*, **102**(WWI):13–30.
Krumbein, W. C. 1950. Geological Aspects of Beach Engineering. *Application of Geology to Engineering Practice, Geol. Soc. America* (Berkey Volume), pp. 195–223.
Kuenen, Ph. J. 1954. Eustatic Changes of Sea-Level. *Geol. Mijnbouw* (NW Ser) 16e Juargang, pp. 148–155.
Lahee, F. H. 1951. Use of "Air" instead of "Aerial," *Amer. Assoc. Petroleum Geologists Bull.* **35** (12), pp. 2608.
Lee, W. T. 1922. The Face of the Earth as Seen from the Air. *Am. Geog. Soc. Spec. Pub. 4*, N.Y.
Lemke, R. W. 1967. Effects of the Earthquake of March 27, 1964 at Seward, Alaska. *U.S. Geol. Survey Prof. Paper 542–E.*
Lewis, A. J., and MacDonald, H. C. 1970. Significance of Estuarine Meanders Identified from Radar Imagery of Eastern Panama and Northwestern Colombia. *Mod. Geol.* **1**:187–196.
Longuet-Higgins, M. S. 1970. Longshore Currents Generated by Obliquely Incident Sea Waves, *I & II, Jour. Geophys. Research,* **15**: 6778–6801.
Lueder, D. R. 1959. *Aerial Photographic Interpretation—Principles and Applications*. McGraw-Hill Book Co., N.Y.
Macdonald, G. A., Shepard, F. P., and Cox, D. C. 1947. The tsunami of April 1, 1946, in the Hawaiian Islands. *Pacific Sci.* **1**(1):21–37.
Magoon, O. T., Jarman, J. W., and Berg, D. W. 1971. Use of Satellites in Coastal Engineering. *Internat. Conf. on Port and Ocean Eng. Under Arctic Conditions, 1st, Proc.*, **2**:1234–1255.
Mainguet, M. 1974. *Le Modèle Des Grés—Problèmes Généraux. Études De Photo-Interprétation*, Inst. Géog. Nat., Paris, 2 vols.
Mairs, R. L. 1970. Oceanographic Interpretation of Apollo Photographs. *Photogramm. Eng.* **36**(10):1045–1058.
Marine Technology Society. 1972. *Tools for Coastal Zone Management*. Washington, D.C.
Marmer, H. A. 1949. Sea Level Changes along the Coasts of the United States in Recent Years. *Am. Geophys. Union* Trans., **30**(2):201–204.
Maruyasu, T., and Minatoku, R. 1973. A Study on the Erosion of Niigata Beach from ERTS–A Imagery. *Goddard Space Flight Center Symp. on Significant Results from ERTS–1*, v. 1, pp. 665–672.
Matthews, E. R. 1934. *Coast Erosion and Protection*. Charles Griffin & Co., London.
McBeth, F. H. 1956. A Method of Shoreline Delineation. *Photogramm. Eng.* **22**(2):400–405.
McCurdy, P. G. 1947. *Manual of Coastal Delineation from Aerial Photographs*. U.S. Navy Hydrographic Off., Pub. no. 592, Washington, D.C.
McCurdy, P. G. 1950. Coastal Delineation from Aerial Photographs. *Photogramm. Eng.* **16**(4):550–555.

McGowen, J. H. 1974. Coastal Zone Shoreline Changes: A Function of Natural Processes and Man's Activities. In *Approaches to Environmental Geology*, E. G. Wermund. ed., Univ. Texas Bureau of Economic Geol., Austin, Texas, pp. 184–203.

Menard, H. W. 1961. Some Rates of Regional Erosion. *Jour. Geology* **69**(2):154–161.

Miller, G. H., and Berg, D. W. 1976. An ERTS–1 Study of Coastal Features on the North Carolina Coast. *U.S. Army Corps Engineers Coastal Eng. Research Center Misc. Rept. 76–2*, Fort Belvoir, Va.

Morgan, J. P., Nichols, L. G., and Wright, M. 1958. Morphological Effects of Hurricane Audrey on the Louisiana Coast. *Tech. Rpt. no. 10*, Coastal Studies Inst., Louisiana State Univ., Baton Rouge, La.

Morton, R. A. 1976. Effects of Hurricane Eloise on Beach and Coastal Structures, Florida Panhandle. *Geology,* **4**, pp. 277–280.

Mosby, J. E. G. 1939. The Horsey Flood, 1938: An Example of Storm Effect on a Low Coast. *Geog. Jour.* **93**(5):413–418.

Munk, W. H., and Traylor, M. A. 1947. Refraction of Ocean Waves: A Process Linking Underwater Topography to Beach Erosion. *Jour. Geology,* **55**:1–26.

National Academy of Sciences. 1975. *Practical Applications of Space Systems—Marine and Maritime Uses*. Supporting Paper 8, Panel on Marine and Maritime Uses, Washington, D. C.

National Aeronautic and Space Administration. 1972. *Data User's Handbook*. Goddard Space Flight Center.

Nichols, M. 1970. Coastal Processes from Space Photography. *Coastal Eng. Conf., 12th, Proc.,* pp. 641–649.

Nichols, M., and Kelly, M. 1972. The Sensing and Analysis of Coastal Waters. *Internat. Symp. on Remote Sensing of the Environment, 8th, Ann Arbor, Mich., Proc.,* pp. 969–981.

Nichols, R. L., and Marston, A. F. 1939. Shoreline Changes in Rhode Island Produced by Hurricane of Sept. 21, 1938. *Geol. Soc. America Bull.* **50**(9):1357–1370.

Niering, W. A. 1970. The Dilemma of Coastal Wetlands: Conflict of Local, National, and World Priorities. In *The Environmental Crisis*, H. W. Helfrich, ed., Yale Univ. Press, New Haven, Conn., pp. 143–156.

Noble, R. M. 1971. Shoreline Changes, Humboldt Bay, California. *Shore and Beach* **39**(4):11–18.

Owens, J. S., and Case, G. O. 1908. *Coast Erosion and Foreshore Protection*. St. Bride's Press, London.

Painter, W. T. 1973. Problems of Great Lakes Shore Erosion. In *Water for the Human Environment* V. T. Chow & Others, eds. *World Congress on Water Research, 1st, Proc.,* v. 3, pp. 40–60.

Pierce, J. W. 1970. Tidal Inlets and Washover Fans. *Jour. Geology* **78**(2):230–234.

Polcyn, F. C. 1970. Measurement of Water Depth by Multispectral Ratio Techniques. *3rd Ann. Earth Res. Prog. Rev.,* v. 3, Hydrology and Oceanography, Sec. 61, NASA, Manned Spacecraft Center, Houston, Tex., pp. 1–11.

Polcyn, F. C., Brown, W. L., and Sattinger, I. J. 1970. The Measurement of Water Depth by Remote Sensing Techniques. *Infrared and Optics*

Lab. Rept. 8973–26–F. Willow Run Lab., Inst. of Sci. and Tech., Univ. of Michigan, Ann Arbor, Mich.

Polcyn, F. C., and Lyzenga, D. R. 1974. Updating Coastal and Navigational Charts Using ERTS–1 Data. *NASA/Goddard Space Flight Center ERTS–1 Symp., 3rd, Proc.,* v. 1, Sec. B, pp. 1333–1346.

Polcyn, F. C., and Sattinger, I. J. 1969. Water Depth Determinations Using Remote Sensing Techniques. *Internat. Symp. on Remote Sensing of the Environment, 6th, Ann Arbor, Mich., Proc.,* pp. 1017–1028.

Price, W. A. 1952. Reduction of Maintenance by Proper Orientation of Ship Channels through Tidal Inlets. *Coastal Eng. Conf., 2nd, Proc.,* pp. 243–255.

Rankin, J. K. 1952. Development of the New Jersey Shore. *Coastal Eng. Conf., 3rd, Proc.,* pp. 306–317.

Reade, T. M. 1888. Tidal Action as an Agent of Geological Change. *Philosophical Magazine* **25**:338–343.

Richards, A. F. 1960. Rates of Marine Erosion of Tephra and Lava at Isla San Benedicto, Mexico. *Int. Geol. Cong., 21st Sess., Rpt. Pt. 10,* pp. 59–64.

Roy Mann Assoc., Inc. 1975. *Aesthetic Resources of the Coastal Zone*. Off. Of Coastal Zone Management, NOAA.

Rudd, R. D. 1974. *Remote Sensing: A Better View*. Duxbury Press.

Russell, R. J., and Morgan, J. P. 1952. Photo-interpretation Keys of Selected Coastal Marshland Features. *Louisiana State Univ. Coastal Studies Inst., Tech. Rept. 1.*

Saville, T., Jr., and Watts, G. M. 1969. Coastal Regime—Recent U.S. Experiences. Int. Navigation Cong., 22nd, Paris, Sec. II, Subject 4, pp. 1–23.

Scherz, J. P., and Stevens, A. R. 1970. An Introduction to Remote Sensing for Environmental Monitoring. *Univ. of Wisconsin Remote Sensing Prog., Rpt. 1.*

Scherz, J. P., Sydor, M., and Van Domelen, J. F. 1974. Aircraft and Satellite Monitoring of Water Quality in Lake Superior near Duluth. *NASA/ Goddard Space Flight Center ERTS–1 Symp., 3rd, Proc.,* v. 1, Sec. B, pp. 1619–1636.

Schwartz, M. 1965. Laboratory Study of Sea-Level Rise as a Cause of Shore Erosion. *Jour. Geology* **73**:528–534.

Schwartz, M. 1967. The Bruun Theory of Sea-Level Rise as a Cause of Shore Erosion. *Jour. Geology* **75**:76–92.

Shepard, F. P. 1950a. Photography Related to Investigation of Shore Processes. *Photogramm. Eng.* **16**(5):756–769.

Shepard, F. P. 1950b. Longshore Bars and Longshore Troughs. *U.S. Army Corps of Engineers Beach Erosion Board TM 15.*

Shepard, F. P. 1960a. Gulf Coast Barriers. In *Recent Sediments, Northwest Gulf of Mexico, A Symposium*, Am. Assoc. Petroleum Geologists, Tulsa, Okla., pp. 197–220.

Shepard, F. P. 1960b. Stabilization of Sedimentary Coastlines. *Nature* **118**(4749):467–469.

Shepard, F. P. 1963. *Submarine Geology*. 2nd ed., Harper & Row Publishers, N.Y.

Shepard, F. P., Emery, K. O., and LaFond, E. C. 1941. Rip Currents: A Process of Geological Importance. *Jour. Geology* **49**(4) pp. 337–369.

Shepard, F. P., and Grant, U.S., IV. 1947. Wave Erosion along the Southern California Coast. *Geol. Soc. America Bull.* **58** pp. 919–926.

Shepard, F. P., and Inman, D. L. 1950. Nearshore Water Circulation Related to Bottom Topography and Wave Refraction. *Amer. Geophys. Union Trans.* **31**(2):196–212.

Shepard, F. P., and Wanless, H. R. 1971. *Our Changing Coastlines*. McGraw-Hill Book Co. N.Y.

Slaughter, T. H. 1973. Seasonal Changes of Littoral Transport and Beach Width and Resulting Effect on Protective Structures. *Goddard Space Flight Center Symp. on Significant Results from ERTS-1*, v. 1, pp. 1259–1267.

Smith, H. T. U. 1954. Coastal Dunes. *Off. of Naval Research Coastal Geog. Conf.* pp. 51–56.

Smith, J. E., ed. 1968. *Torrey Canyon—Pollution and Marine Life*. Cambridge Univ. Press, N.Y.

Solecki, R. S., ed. 1960. Photo Interpretation in Archaeology. In *Manual of Photographic Interpretation*. Amer. Soc. of Photogramm., Falls Church, Va., pp. 717–733.

Sonu, C. J. 1964. Study of Shore Processes with Aid of Aerial Photogrammetry. *Photogramm. Eng.* **30**(6):932–941.

Sonu, C. J., and Van Beek, J. L. 1971. Systematic Beach Changes on the Outer Banks, North Carolina. *Jour. Geology* **79**:416–425.

Stafford, D. B. 1972. Coastal Engineering Applications of Aerial Remote Sensing. *Coastal Eng. Conf., 13th, Proc.*, pp. 2045–2064.

Stafford, D. B., Bruno, R. O., and Goldstein, H. M. 1973. An Annotated Bibliography of Aerial Remote Sensing in Coastal Engineering. *U.S. Army Corps. Engineers Coastal Eng. Research Center Misc. Paper 2-73*, Fort Belvoir, Va.

Stafford, D. B., and Langfelder, J. 1971. Air Photo Survey of Coastal Erosion. *Photogramm. Eng.* **37**(6):565–575.

Stanley, K. W. 1968. Effects of the Alaska Earthquake of March 27, 1964 on Shore Processes and Beach Morphology. *U.S. Geol. Survey Prof. Paper 543-J*.

Steers, J. A. 1944. Coastal Preservation and Planning. *Geog. Jour.* **104**(1):7–18.

Steers, J. A. 1951. *The Coastline of England and Wales*. 2nd ed., Cambridge Univ. Press, London.

Steers, J. A. 1962. Coastal Cliffs: Report of a Symposium. *Geog. Jour.* **128**(3):303–307.

Stephenson, T. A., Tandy, G., and Spender, M. 1931. The Structure and Ecology of Low Isles and Other Reefs. *Great Barrier Reef Exped. 1928–1929*, **3**(2):17–112.

Stoddart, D. R. 1962. Catastrophic Storm Effects on the British Honduras Reefs and Cays. *Nature* **196**(4854):512–515.

Stone, K. 1974. Developing Geographical Remote Sensing. In *Remote Sensing*, J. E. Estes and L. W. Senger, eds. Hamilton Pub. Co., Santa Barbara, Calif., pp. 1–13.

Strong, A. E. and DeRycke, R. J. 1973. Ocean Current Monitoring Employing a New Satellite Sensing Technique. *Science* **182**: 482–484.

Sunamura, T., and Horikawa, K. 1969. A Study of Erosion of Coastal Cliffs by Using Aerial Photographs. *Coastal Eng. Japan*, **12**(2):99–120.

Swanson, L. W. 1960. Photogrammetric Surveys for Nautical Charting—Use of Color and Infrared Photography. *Photogramm. Eng* **26**(1):137–141.

Swanson, L. W. 1968. Shoreline Mapping. In *Manual of Color Aerial Photography,* Am. Soc. of Photogramm., Falls Church, Va., pp. 410–411.

Talley, B. B. 1938. *Engineering Applications of Aerial and Terrestrial Photogrammetry*. Pitman Pub. Co., N.Y.

Tanner, W. F. 1961. Mainland Beach Changes due to Hurricane Donna. *Jour. Geophys. Research* **66**(7):2265–2266.

Taylor, J. L., and Solomon, C. H. 1968. Some Effects of Hydraulic Dredging and Coastal Development in Boca Grega Bay, Florida. *Fisheries Bull.* **67:**213–241.

Teichert, C., and Fairbridge, R. W. 1950. Photo-interpretation of Coral Reefs. *Photogramm. Eng.* **16**:744–755.

Teleki, P. G., Rabchevsky, G. A., and White, J. W. 1973. On the Nearshore Circulation of the Gulf of Carpentaria, Australia—A Study in Uses of Satellite Imagery (ERTS) in Remotely Accessible Areas. *Symp. on Remote Sensing in Oceanography, Proc., Am.* Soc. Photogramm., pp. 717–736.

Tewinkel, G. C. 1963. Water Depth from Aerial Photographs. *Photogramm. Eng.* **29**(6):1034–1042.

Tricart, J. 1967. Deltas Lagunaires, Embovchure du Rio Santa Maria SE d'Aguaduke Panama, *Photo Interpret.* **67**(2):pp. 1–7.

Tuyahov, A. J., and Holz, R. K. 1973. Remote Sensing of a Barrier Island. *Photogramm. Eng.* **39**(2):177–188.

Umbgrove, J. H. F. 1928. De Korallriffen in de Baai van Batavia, Dienst Mijnb. in Ned.-Indie Wet. Med., n. 7.

Umbgrove, J. H. F. 1929. De Korallriffen der Duizend-Eilanden (Java Zee). Dienst Mijnb. in Ned.-Indie Wet. Med., n. 12.

United Nations Economic and Social Council. 1975. *Coastal Area Management and Development-Report of the Secretary General*. 59th Session, E/5648.

U.S. Army, Coastal Engineering Research Center. 1973. *Shore Protection Manual*. Fort Belvoir, Va., 3 vols.

U.S. Congress. 1975. *Effects of Man's Activities on the Marine Environment*. Printed for the Committee on Commerce, 94th Congress, 1st Session, U.S. Government Printing Office, Washington, D.C.

Van Lopik, J. R., Pressman, A. E., and Ludlum, R. L. 1968. Mapping Pollution with Infrared. *Photogramm. Eng.* **34**(6):561–565.

Vonder Haar, S. P. and Gorsline, D. S. 1975. Flooding Frequency of Hypersaline Coastal Environments Determined by Orbital Imagery: Geologic Implications. *Science* **190**:147–149.

Wagener, H. D. 1970. Beach Erosion in the Charleston Harbor Area. *South Carolina Div. Geol. Environmental Geology Ser. 1*, Columbia, S.C.

Wagret, P. 1959. Polderlands. (cited in Gagliano, S.M., 1973).

Waller, R. M. 1966. Effects of the Earthquake of March 27, 1964, in the Homer Area, Alaska. *U.S. Geol. Survey Prof. Paper 542–D*.

Waugh, J. E. 1962. Storm Damage Survey. *Photogramm. Eng.* **28** (3):516–517.

Waugh, J. E. 1964. Photogrammetric Measurements of Tidal Currents. *Am. Soc. Civil Engineers Proc., Jour. Surveying and Mapping Div.*, no. SU–1, pp. 17–29.

Way, D. S. 1973. *Terrain Analysis*. Dowden, Hutchinson & Ross, Stroudsburg, Pa.

Weber, J. D. 1970. Photographic Monitoring of Shoreline Movement. *Shore and Beach* **38**(2):36–38.

Weigel, R. L. 1964. *Oceanographical Engineering*. Prentice-Hall, N.J.

Wheeler, W. H. 1902. *The Sea Coast; Destruction, Littoral Drift, Protection*. Longmans, Green and Co., London.

Williams, R. S., Jr., and Carter, W. D., eds. 1976. ERTS–1: A New Window on Our Planet. *U.S. Geol. Survey Prof. Paper 929.*

Williams, W. W. 1947. The Determination of Gradients on Enemy-Held Beaches. *Geog. Jour.* **109**(1):76–93.

Williams, W. W. 1960. *Coastal Changes*. Routledge and Kegan Paul, London.

Williamson, A. N., and Grabau, W. E. 1974. Sediment Concentration Mapping in Tidal Estuaries. *NASA/Goddard Space Flight Center ERTS–1 Symp., 3rd, Proc.,* v. 1, Sec. B, pp. 1347–1386.

Wobber, F. J. 1967. Space Photography: A New Analytical Tool for the Sedimentologist. *Sedimentology* **9**:265–317.

Wobber, F. J., and Anderson, R. R. 1973. Simulated ERTS Data for Coastal Management. *Photogramm. Eng.* **39**(8):593–598.

Wong, V. D. 1970. Moss Landing Harbor, California: A Case History. *Shore and Beach* **38**(4):26–39.

Woodwell, G. M., Wurster, C. F., Jr., and Isaacson, P. A. 1967. DDT Residues in an East Coast Estuary: A Case of Biological Concentration of a Persistent Pesticide. *Science* **156**:821–824.

Zeigler, J. M., Hayes, C., and Tuttle, S. 1959. Beach Changes during Storms on Outer Cape Cod, Mass. *Jour. Geology* **67**(3):318–336.

Zeigler, J. M., and Ronne, F. C. 1957. Time Lapse Photography—Aid to Studies of the Shorelines. *Naval Research Rev., n. 4, U.S. Office of Naval Res.*, Washington, D.C., pp. 1–6.

Zenkovich, V. P. 1967. *Processes of Coastal Development*. Oliver & Boyd, Edinburgh.

Zenkovich, V. P. 1973. Geomorphological Problems of Protecting the Caucasian Black Sea Coast. *Geog. Jour.* **139**(3):460–466.

AUTHOR CITATION INDEX

SUBJECT INDEX

About the Editor

MOHAMED T. EL-ASHRY is presently staff scientist with the Environmental Defense Fund in Denver (on leave as Professor and Chairman of the Department of Environmental Sciences at Wilkes College). Dr. El-Ashry completed his undergraduate work at Cairo University receiving a B.S. with honors in geology in 1959 and in 1966 received his Ph.D. from the University of Illinois. He has also held teaching and research positions at Cairo University, Pan-American-U.A.R. Oil Company, and East Stroudsburg State College.

Dr. El-Ashry has published or presented more than twenty-five papers, mostly on photointerpretation and coastal geomorphology. He is a fellow of the Geological Society of America and the American Association for the Advancement of Science, and is listed in *American Men and Women of Science*.